Logistics

McGraw-Hill SOLE Press Series
Benjamin S. Blanchard and Anthony E. Trovato, CPL, Editors

Logistics

Principles and Applications

John W. Langford, CPL, CCM

Second Edition

SOLE PRESS

New York   Chicago San Francisco Lisbon   London   Madrid
Mexico City   Milan   New Delhi   San Juan   Seoul
Singapore   Sydney   Toronto

The *McGraw·Hill* Companies

Library of Congress Cataloging-in-Publication Data

Langford, John W.
  Logistics : principles and applications / John W. Langford — 2nd ed.
    p.   cm.
  Includes index.
  ISBN-13: 978-0-07-147224-1
  ISBN-10: 0-07-147224-X
  1. Production management.   2.Business logistics.   I. Title.   II. Series: SOLE
Press series.
  TS155.L2534 2006
  658.5—dc22

                                                                    2006048522

1 2 3 4 5 6 7 8 9 0   DOC/DOC   0 1 3 2 1 0 9 8 7 6

ISBN-13: 978-0-07-147224-1
ISBN-10: 0-07-147224-X

*The sponsoring editor for this book was Larry S. Hager, the editing supervisor was
David E. Fogarty, and the production supervisor was Pamela A. Pelton. It was set in
Century Schoolbook by D & P Editorial Services, LLC. The art director for the cover
was Brian Boucher.*

Printed and bound by RR Donnelley.

This book was printed on acid-free paper.

McGraw-Hill books are available at special quantity discounts to use as premiums and
sales promotions, or for use in corporate training programs. For more information, please
write to the Director of Special Sales, McGraw-Hill Professional, Two Penn Plaza, New
York, NY 10121-2298. Or contact your local bookstore.

SHAFFER THURMAN DAY (1917–1990)

This book is respectfully dedicated to the memory and logistics genius of Shaffer T. Day. During his final tenure as Assistant for Logistics for the U.S. Air Force, the author was privileged to have this distinguished and knowledgeable gentleman as his boss and mentor. Shaffer's vast expertise—drawn from 41 years of professional service ranging from hands-on technician maintenance to the highest levels of management—inspired, encouraged, and enriched many successful careers in logistics. His legacy is a beacon to future achievements in logistics technology.

About the Author

John W. Langford is a Certified Professional Logistician and Certified Configuration Manager with international experience coupled with a distinguished background in contract management, logistics system engineering, and configuration management. He has served as an adjunct professor at Virginia Polytechnic Institute, Florida Institute of Technology, and George Mason University, where he taught logistics engineering and management courses.

Contents

Part 3   Logistics Systems Management and Operations

Foreword

Throughout the industrial, government, and academic sectors, the field of logistics is continually evolving and assuming a higher degree of importance than in the past. In both the government and commercial acquisition processes logistics is proving to be a deciding factor in the procurement of new systems and technology. Frequently, the life cycles of many systems are being extended while the life cycles of individual technologies are becoming shorter. Globalization requirements and international competition are increasing significantly from year to year, and the challenges of being able to first introduce new systems into the inventory and then maintain such throughout their respective life cycles are greater than ever before. The logistics for a given system (or mix of systems) is *life-cycle* oriented, and the implementation of program-related requirements in this area necessitates a highly *interdisciplinary* approach.

The currently changing emphasis on the broader spectrum and application of logistics requires that the practitioner in the field be not only knowledgeable of the various elements of system maintenance and support (e.g., procurement and supply support, transportation and handling, support equipment, personnel, data/information, facilities), but also familiar with the overall system life cycle, including the design and manufacturing processes and the overall customer environment. The interfaces and interactions are numerous throughout the life cycle and, in particular, design and/or management decisions made in the initial system development stage will have an impact on the activities and resource requirements in all subsequent phases of the life cycle. Thus, the logistics practitioner must not only demonstrate expertise in the various facets of logistics, but also be conversant with the many other elements of a given system's life cycle. The logistician must be technically competent; must be knowledgeable of available design and analysis tools/models and their application; and must be able to effectively communicate with other internal project personnel, suppliers, customer, contractor, and operational (user) personnel across the board.

This revised and updated second edition of John W. Langford's *Logistics: Principles and Applications* continues to represent the author's long-standing contribution to the literature, expanding the understanding of the totality of logistics. The twenty-six chapters of the book, presented in three distinct parts, provide an organized trip both through the disciplines of the logistics solution and their application in and relationship to the systems engineering process. Many excellent examples, analytical approaches, and problem exercises are included throughout.

Mr. Langford has moved beyond the announced objectives of his first edition and has given us a book that is more inclusive and descriptive of the entire field of logistics. The author is a noted and highly respected logistics expert, and has conducted many workshops and training programs for government and industry throughout the world. He has also served as a visiting lecturer at several different universities, both nationally and internationally.

This edition is the third offering in the newly established SOLE Press: it serves as an excellent guide not only for the practicing logistician, but also for those professionals outside of the field seeking additional knowledge in logistics. It is readily adaptable for practitioners at all levels of expertise, and the material presented allows for an easy understanding of the details of the profession.

BENJAMIN S. BLANCHARD, CPL, FELLOW
Virginia Polytechnic Institute & State University
Co-Editor, SOLE Press

ANTHONY E. TROVATO, CPL
Full Spectrum Logistics, Inc.
Co-Editor, SOLE Press

Preface

Logistics is a blend of art and science. From a scientific perspective, it involves applying the quantitative techniques of engineering and analysis to infuse logistic considerations into product design, development, production, and operation. From an artistic perspective, it entails the integration of human experience, intuition, and creative judgment with the scientific outputs to produce logical results.

Logistics as a professional discipline can therefore be described as *the application of engineering, operational and managerial skills to provide a product with prerequisite quality, reliability, maintainability and supportability and to sustain safe and cost-effective utilization of that product for its intended purpose throughout its projected service life.*

The subsidiary disciplines of logistics have significance for customers as well as for producers and suppliers. The marketplace acknowledges that the "educated consumer" is the best customer. An educated consumer is also an effective logistician. The potential purchaser of an automobile and a U.S. Navy program manager planning the acquisition of complex SONAR systems for the nuclear submarine fleet would be similarly concerned with the principles of logistics—irrespective of the differences in products, unit costs, and economies of scale. The technological significance of logistics applies throughout the total life cycle of a system or product. To assure optimum economy and efficiency throughout the operational life of the system, the logistician must be involved in the conception, design, engineering, production, and operational support of the system, and ultimately its termination and disposal. All logisticians must, therefore, be sensitive to and appreciate this life-cycle concept. Aside from the life-cycle perspective, it is noted that logistics embodies a multiplicity of subsidiary disciplines. Accordingly, it is incumbent upon logistics professionals who possess a high degree of expertise in a specialty such as inventory, transportation, or maintenance management to acquire working knowledge of those other disciplines, such as technical data, configuration management, reliability, maintainability, human factors engineering, critical path analysis, etc., which inevitably affect their individual areas of specialization. This book addresses logistics on the basis of the triumvirate of logistics systems engineering, acquisition and production, and logistics system management and operation. While the three groupings are logically sequential, it must be recognized that there is a great deal of confluence and many peripheral interfaces among their constituent subdisciplines.

This book is designed to be a reference and is oriented toward first- and second-year college undergraduates, technicians who have acquired skills in specific logistics specialties and aspire to an understanding of the total spectrum of logistics technology, and other professionals who seek an appreciation of the basic principles and applications of logistics.

The typical reader should be capable of understanding basic business mathematics, algebra, integral calculus, and principles of statistics which pertain to logistics engineering and management. The ability to relate to basic quantitative aspects of logistics technology reinforces and enhances comprehension of the governing logistics concepts and their applications. The purpose of this book is to impart a basic understanding of those subsidiary disciplines which make up logistics. Having established an initial threshold of knowledge, it should be possible to develop further expertise in any area of logistics technology.

To provide international perspective, the reader is advised that the *Système International* (SI) measurement (metric) units have been incorporated, where appropriate, in sections of the text along with the U.S measurement units. The purpose is to accommodate the scope of and effectively convey tutorial significance to the international readership. Equivalent SI measurement units are, in most cases, cited parenthetically by the notations of U.S. measurement units. Where necessary, the SI units are incorporated by footnotes or provided by separate sections in the text in conjunction with the discussions involving U.S. measurement units. Appendix C provides a crosswalk guide, within the context of this book, developed by the author for conversion of U.S. measurement units to equivalent SI measurement units.

JOHN W. LANGFORD
Vienna, Virginia

Acknowledgments

The author acknowledges with grateful appreciation the helpful assistance provided by the following in the development of this text:

Thomas Ronaldi, Deputy Director of the Advanced System Technology Office, Naval Sea Systems Command, Washington, D.C. Tom has additionally distinguished himself in serving as Director, along with related leadership management roles, for the Submarine Combat Systems Group; Gun Weapon Systems program; and the Surface Anti-Submarine Warfare (ASW) System. The author is privileged to have enjoyed a professionally rewarding and educational working relationship with this gentleman as well as benefiting from his unbounded expertise. The professional dedication and commitment to excellence of Mr. Ronaldi and his staff typify the inventive genius which has established the U.S. Navy's weapon system programs as world-class models for system reliability and operational effectiveness.

Faith Chang and Joon Chang, my friends and colleagues, whose unexcelled computer expertise greatly contributed to the completion of this book. Their capabilities in negotiating the complex world of software programs and graphics enabled me to incorporate pictorial and tabular illustrations, which gave meaning to the intricate technical details and algorithmic methodologies. Their support ensured the production of the manuscript and salvation of the author's credibility. The publication of this book is a tribute to their competence, commitment, patience, and tenacity of purpose..

My wife, Kitty; our daughter, Mary, who is now Mrs. Ramiro Donoso; our daughter, Meg; our son, Dirk and his lovely wife, Claire. Their supportiveness during this literary endeavor reinforces the doctrine that the family provides the infrastructure for individual achievement.

Logistics

Logistics Systems Engineering

Logistics Statistics

Overview

A grasp of fundamental statistical principles is a prerequisite to understanding logistics systems engineering. The mathematical techniques addressed in this chapter relate to the basic functions of reliability, maintainability, availability, and quality control. Within this perspective, the focus is on the quantitative techniques of exponential notation (also referred to as scientific notation), factorials, logarithms, and development of probability density functions which are used in elementary logistics engineering applications.

Exponential Notation

An exponential expression involves a base number and an exponential factor applied to that base number. The exponent is also described as the *power* to which a base number is *expanded*.

An exponent indicates to what extent a base number is multiplied by itself. The rules set forth below illustrate the most commonly used exponential functions. The student of logistics will find the exponential command function on an electronic calculator to be a valuable learning aid.

1. When x = exponent and n = *base* number, the exponential function is given as n^x. Assume that n = 10 (the base number). If x = 1, then

$$n^x = 10^1 = 10 \text{ (the number itself)}$$

or

"10" to the first power equals "10"

If $x = 2$, then

$$n^x = 10^2 = 10 \times 10 = 100$$

or

"10" squared equals "100"

If $x = 3$, then

$$n^x = 10^3 = 10 \times 10 \times 10 = 1000$$

or

"10" cubed equals "1000"

2. When $-x$ = exponent and n = base number, the exponential function is given as n^{-x}. An expression with a negative exponent is the same as the reciprocal of the same expression with a positive exponent, so that

$$n^{-x} = \frac{1}{n^x}$$

Assume that $n = 10$ (the base number). If $x = 1$, then

$$n^{-x} = 10^{-1} = \frac{1}{10} = 0.1$$

If $x = 2$, then

$$n^{-x} = 10^{-2} = \frac{1}{10^2} = \frac{1}{10 \times 10} = \frac{1}{100} = 0.01$$

If $x = 3$, then

$$n^{-x} = 10^{-3} = \frac{1}{10^3} = \frac{1}{10 \times 10 \times 10} = \frac{1}{1000} = 0.00$$

3. When $1/x$ = exponent and n = base number, the exponential function is given as $n^{1/x}$. The exponent $1/x$ also connotes the xth root of the base number n. For example, when $x = 2$, $1/x = \frac{1}{2}$, or the square root of n, also referred to as n to the $\frac{1}{2}$ power. When $x = 3$, $1/x = 1/3$ or the cube root of n, also referred to as n to the 1/3 power. Assume that $n = 10$ (the base number). If $x = 1$, then

$$n^{1/x} = 10^{1/1} = 10 \text{ (the number itself)}$$

If $x = 2$, then

$$n^{1/x} = 10^{1/2} = 10^{0.5} = 3.16227766$$

(That is, $10^{0.5}$ is the same as $\sqrt{10}$, the square root of 10.) If $x = 3$, then

$$n^{1/x} = 10^{1/3} = 10^{0.3333} = 2.15443469$$

(That is, $10^{0.3333}$ is the same as $\sqrt[3]{10}$, the cube root of 10.) If $x = 4$, then

$$n^{1/x} = 10^{1/4} = 10^{0.25} = 1.77827941$$

4. When an exponential function n^x is itself expanded by the power of another governing exponent y, the exponential function is noted as $(n^x)^y$. Also,

$$(n^x)^y = n^{(x)(y)}$$

The base-number exponent x is multiplied by the governing exponent y to derive a final exponential value xy, which is applied to the base number. Assume that $n = 10$ (the base number). If $x = 1$ and $y = 2$, then

$$(n^x)^y = (10^1)^2 = 10^{(1)(2)} = 10^2 = 10 \times 10 = 100$$

If $x = 2$ and $y = 3$, then

$$(n^x)^y = (10^2)^3 = 10^{(3)(2)} = 10^6 = 10 \times 10 \times 10 \times 10 \times 10 \times 10 = 1{,}000{,}000$$

If $x = 2$ and $y = -2$, then

$$(n^x)^y = (10^2)^{-2} = 10^{(2)(-2)} = 10^{-4} = \frac{1}{10^4}$$

$$= \frac{2}{10 \times 10 \times 10 \times 10} = \frac{1}{10{,}000} = 0.0001$$

A product of a positive number and a negative number always has a negative result.

5. When an exponential function n^x is divided by another exponential function with the same base number but a different exponent (n^y), the exponential function is given as n^x/n^y or $n^x \div n^y$, which may also be expressed as n^{x-y}. In this case, the denominator exponent y is subtracted from the numerator exponent x with the resultant exponential factor applied to the base number n. Assume that $n = 10$ (the base number). If $x = 1$ and $y = 2$, then

$$\frac{n^x}{n^y} = \frac{10^1}{10^2} = 10^{1-2} = 10^{-1} = \frac{1}{10} = 0.1$$

If $x = 3$ and $y = 2$, then

$$\frac{n^x}{n^y} = \frac{10^3}{10^2} = 10^{3-2} = 10^1 = 10$$

If $x = 2$ and $y = -3$, then

$$\frac{n^x}{n^y} = \frac{10^2}{10^{-3}} = 10^{2-(-3)} = 10^{2+3} = 10^5 = 10 \times 10 \times 10 \times 10 \times 10 = 100,000$$

Note that applying a minus sign to a negative number converts that number to a positive value (e.g., $-(-y) = +y$).

Factorials

Factorials have applicability to combinatorial analysis and the construction of probability density functions. The factorial value of a designated number is determined by multiplying that number by each of its preceding whole integers. The factorial function is expressed as $n!$, where n is the designated number.

The following examples illustrate the application of the factorial function. The student of logistics will find the factorial command function on an electronic calculator to be a valuable aid in achieving the learning objectives of this section.

Examples of factorial functions

1. Given that $n = 5$, then

$$n! = 5! = (5)\,(4)\,(3)\,(2)\,(1) = 120$$

2. Given that $n = 8$, then

$$n! = 8! = (8)\,(7)\,(6)\,(5)\,(4)\,(3)\,(2)\,(1) = 40{,}320$$

3. Given that $n = 10$, then

$$n! = 10! = (10)\,(9)\,(8)\,(7)\,(6)\,(5)\,(4)\,(3)\,(2)\,(1) = 3{,}628{,}800$$

Note that with each integral increment in the value of the designated whole number $n \geq 0$, the corresponding factorial value increases rapidly. Fig. 1.1 illustrates this characteristic. Also note that by convention, $0! = 1$.

n	0	1	2	3	4	5	6	7	8	9	10
$n!$	1	1	2	6	24	120	720	5040	40,320	362,880	3,628,800

Figure 1.1 Example of factorial amplification.

Combinatorial Analysis

Combinatorial analysis involves the *combination* function and the *permutation* function. Each of these combinatorial analysis functions is a structured relationship of factorials.

The *combination* factorial function determines in how many ways n objects can be combined r at a time. This function is expressed as

$$_nC_r = \frac{n!}{(r!)(n-r)!}$$

where n = number of objects
 r = how many objects are to be combined at one time

1. As an example, to determine how many ways objects A, B, and C can be combined two at a time, let $n = 3$ and $r = 2$. Then

$$_nC_r - \frac{n!}{r!\,(n-r)!} = \frac{3!}{(2!)(3-2)!} = \frac{3!}{(2!)(1!)} = \frac{(3)(2)(1)}{(2)(1)(1)} = \frac{6}{2} = 3$$

Thus, objects A, B, and C can be combined two at a time in three ways, i.e.,

$$AB \qquad AC \qquad BC$$

2. To combine five objects A, B, C, D, and E three at a time, let $n = 5$ and $r = 3$. Then

$$_nC_r = \frac{n!}{r!\,(n-r)!} = \frac{5!}{(3!)(5-3)!} = \frac{5!}{(3!)(2!)} = \frac{(5)(4)(3)(2)(1)}{(3)(2)(1)(2)(1)}$$

$$= \frac{120}{12} = 10$$

Thus, objects A, B, C, D, and E can be combined three at a time in 10 ways, i.e.,

$$
\begin{array}{cc}
ABC & ADE \\
ABD & BCD \\
ABE & BCE
\end{array}
$$

$$\begin{array}{ll} ACD & BDE \\ ACE & CDE \end{array}$$

The *permutation* factorial function determines in how many ways n objects can be brought together and arranged r at a time. Permutation contemplates not only combining the objects but arranging them in as many ways as possible. The permutation function is expressed as

$$_nP_r = \frac{n!}{(n-r)!}$$

where n = number of objects

r = how many objects are to be combined and arranged at one time

1. As an example, to determine how many ways objects A, B, and C can be arranged two at a time, let $n = 3$ and $r = 2$. Then

$$_nP_r = \frac{n!}{(n-r)!} = \frac{3!}{(3-2)!} = \frac{3!}{1!} = \frac{(3)(2)(1)}{(1)} = 6$$

Thus, objects A, B, and C can be brought together and arranged two at a time in six ways, i.e.,

$$\begin{array}{lll} AB & AC & BC \\ BA & CA & CB \end{array}$$

2. To determine how many ways four objects A, B, C, and D can be arranged three at a time, let $n = 4$ and $r = 3$. Then

$$_nP_r = \frac{n!}{(n-r)!} = \frac{4!}{(4-3)!} = \frac{4!}{1!} = \frac{(4)(3)(2)(1)}{(1)} = 24$$

Thus, objects A, B, C, and D can be brought together and arranged three at a time in 24 ways, i.e.,

$$\begin{array}{llll} ABC & ABD & ACD & BCD \\ BCA & BDA & CDA & CDB \\ CAB & DAB & DAC & DBC \\ BAC & BAD & CAD & CBD \\ CBA & DBA & DCA & DCB \\ ACB & ADB & ADC & BDC \end{array}$$

Logarithms

The term *logarithm* stems from the combination of the Greek words *logos*, for reason, and *arithmus*, for number. In mathematical disciplines, a logarithm is an exponent which is applied to a fixed, designated base number to produce a stated number.

There are two types of logarithms used in logistics engineering:

1. *Common logarithm.* The base number of the common logarithm is 10. The mathematical symbol for the common logarithm is log.
2. *Natural logarithm.* The base number of the natural logarithm is e. (The numeric value of e is 2.718281828.) The mathematical symbol for the natural logarithm is ln.

In logistics engineering, the common logarithm (log) has its primary application in maintainability, which entails analysis of maintenance task times and similar human work parameters. The natural logarithm (ln) has its primary application in reliability, inasmuch as failure patterns of electronic and electromechanical equipment tend to reflect the exponential distribution function.

Illustration of the common logarithm

1. To determine the common logarithm (log) of a stated number, such as 100, the problem can be stated in terms of the base number 10, i.e., What exponent applied to the base number 10 will produce 100? or

$$10^? = 100$$

The solution is $10^2 = 100$; therefore

$$\log 100 = 2$$

2. To determine log 275, the problem may be stated

$$10^? = 275$$
$$\log 275 = 2.439332694$$

which means that

$$10^{2.439332694} = 275$$

3. To determine log 10, the problem may be stated

$$10^? = 10$$
$$\log 10 = 1$$

Ten is the base number for the common logarithm. By rules of exponential notation, any number raised to the first power (an exponent of 1) produces the original number itself; i.e., $10^1 = 10$.

4. To determine log 1.5, the problem may be stated

$$10^? = 1.5$$
$$\log 1.5 = 0.176091259$$

therefore

$$10^{0.176091259} = 1.5$$

When the stated number is a positive number less than the base number 10, the logarithm will always be a positive number less than 1.0.

5. To determine log 1.0, the problem may be stated

$$10^? = 1.0$$
$$\log 1.0 = 0$$

therefore

$$10^0 = 1.0$$

By rules of exponential notation, any number with zero as an exponent will always equal 1, irrespective of the magnitude of the base number.

6. To determine log 0.5, the problem may be stated

$$10^? = 0.5$$
$$\log 0.5 = -0.301029995$$

therefore

$$10^{-0.301029995} = 0.5$$

By laws of exponential notation, the logarithm of a number less than 1.0 will always be negative; e.g.,

$$10^{-0.301029995} = \frac{1}{10^{0.301029995}} = \frac{1}{2} = 0.5$$

Illustration of the natural logarithm

1. To determine the natural logarithm (ln) of a stated number, e.g., 100, the problem can be stated in terms of the base number e, i.e., the exponent that must be applied to the base number e to produce 100, or

$$e^? = 100$$

(Note that e can also be stated as the numeric value 2.718281828.) The solution is $e^{4.605170186} = 100$; therefore

$$\ln 100 = 4.605170186$$

2. To determine $\ln e$, the problem may be stated

$$e^? = e$$
$$\ln e = 1$$

e is the base number for the natural logarithm. By rules of exponential notation, any number raised to the first power equals the original number itself; i.e.,

$$e^1 = e$$

3. To determine $\ln 1.0$, the problem may be stated

$$e^? = 1.0$$
$$\ln 1.0 = 0$$
$$e^0 = 1.0$$

By rules of exponential notation, any number raised to a zero exponent always equals 1.

4. To determine $\ln 1.5$, the problem may be stated

$$e^? = 1.5$$
$$\ln 1.5 = 0.405465108$$

If the stated number is a positive number less than the base number e, the natural logarithm will always be a positive value less than 1.0.

5. To determine $\ln 0.5$, the problem may be stated

$$e^? = 0.5$$
$$\ln 0.5 = -0.69314718$$

therefore

$$e^{-0.69314718} = 0.5$$

By rules of exponential notation, the natural logarithm of a stated number less than 1.0 will always be negative; e.g.,

$$e^{-0.69314718} - \frac{1}{e^{0.69314718}} = \frac{1}{2} = 0.5$$

Special rules governing logarithms

First rule.

$$\log (xy) = \log x + \log y$$

As an example, let $x = 20$ and $y = 10$.

$$\log [(20)(10)] = \log 20 + \log 10 = 1.301029996 + 1.0 = 2.301029996$$

To verify the example,

$$\log [(20)(10)] = \log 200 = 2.301029996$$

Second rule.

$$\log \frac{x}{y} = \log x - \log y$$

As an example, let $x = 20$ and $y = 10$.

$$\log \frac{20}{10} = \log 20 - \log 10 = 1.301029996 - 1.0 = 0.301029996$$

To verify the example,

$$\log \frac{(20)}{(10)} = \log 2 = 0.301029996$$

Third rule.

$$\log (y^x) = (x)(\log y)$$

As an example, let $y = 20$ and $x = 3$.

$$\log (20^3) = 3 \log 20 = (3)(1.301029996) = 3.903089987$$

To verify the example,

$$\log (20^3) = \log 8000 = 3.903089987$$

Note: These rules apply equally to the natural logarithm (ln).

Antilogarithms

For both common and natural logarithms, the antilogarithm is the result obtained by applying the logarithm as an exponent to its base number; for example, using the common logarithm,

$$10^{\log} = ? \leftarrow \text{antilogarithm}$$

If log 100 = 2, then antilog 2 = 100.

$$10^2 = 100 \leftarrow \text{antilogarithm}$$

To use an example applying natural logarithms,

$$e^{\ln} = ?$$

If ln 100 = 4.605170186, then antiln 4.605170186 = 100

$$e^{4.605170186} = 100$$

Both the common logarithm and its antilogarithm are applied in the log normal analysis methodology used to evaluate maintenance task times and related maintainability parameters. This illustrates the practical utility of these principles.

Principles of Probability

The mathematics of probability is frequently called *stochastic* mathematics. Probability mathematics is the mathematics of chance, the likelihood of an event randomly occurring. The quantitative value of probability is normally cited as a percentage, but it is mathematically computed in decimal form; e.g., a probability of 97 percent is mathematically expressed as 0.97.

The following are the basic rules governing probability.

Probabilities are real numbers on the interval from 0 to 1.0.

If an event is certain to occur, its probability is 1.0.

If an event is certain *not* to occur, its probability is 0.

Events A and B are considered *mutually exclusive* if both cannot occur at the same time.

Event A is said to be *independent* of event B if the probability of occurrence of event A is the same whether or not event B has occurred.

If two events are mutually exclusive within a defined set of events for which the sum of individual probabilities is 1.0, the probability that one or the other will occur equals the sum of their individual probabilities.

The sum of the probability that an event will occur and the probability that it will not occur is equal to 1.0.

The probability of two independent events occurring simultaneously is the product of the individual event probabilities.

Probability theory

Basic probability theory is explained by the *addition theorem*, the *multiplication theorem*, and the *conditional theorem*.

1. *Addition theorem.* The addition theorem states that the probability of occurrence of one or another of a series of mutually exclusive events is the sum of the probabilities of their individual occurrences.

2. *Multiplication theorem.* The multiplication theorem states that the probability of simultaneous or successive occurrence of two or more independent events is the mathematical *product* of the probabilities of the individual events.

3. *Conditional theorem.* The conditional theorem states that the probability of occurrence of two *dependent* events is the probability of the first event multiplied by the probability of the second event, *given that the first event has occurred.*

Illustration of probability concepts

Coin toss. A coin has two sides, heads and tails. If one flips the coin, there is a 50 percent probability that the coin will land flat with heads showing; there is likewise a 50 percent probability that tails will show. The sum of the individual probabilities of all possible outcomes on a single toss of the coin is 0.50 (heads) + 0.50 (tails) = 1.00 (or 100 percent).

Playing cards. In a standard deck of playing cards there is one ace of spades among the 52 cards. The probability of drawing the ace of spades on a single draw is $1/52$, or 0.019230769. In the same deck there are 13 hearts. The probability of drawing a heart on a single draw is $13/52$ or 0.25. The probability of drawing a king of hearts can be viewed as the probability of two simultaneous events, which is expressed as a product of the individual probabilities. There are 4 kings and 13 hearts in a 52 card deck. The probability of drawing a king of hearts in a single draw is

$$\left(\frac{4}{52}\right)\left(\frac{13}{52}\right) = \frac{1}{52} \quad \text{or} \quad 0.019230769$$

This is validated further by the fact that there is one king of hearts in a 52 card deck, so that the probability of a single draw is $\frac{1}{52}$, or 0.019230769.

To find the probability of drawing *either* a king *or* a heart in a single draw, one must exclude the duplicated probability of the king of hearts:

$$\frac{4}{52} + \frac{13}{52} - \frac{1}{52} = \frac{16}{52} \quad \text{or} \quad 0.307692307$$

Dice.   Each die in a pair of dice is a cube with six sides. Each side of the cube shows a different number of dots, ranging from one through six. With a toss of a die there are six possible outcomes; i.e., the side of the die facing up may display any number of dots from one through six. The probability associated with any outcome of a single toss of a die is $\frac{1}{6}$, or 0.166666666. The total of the individual probabilities of the six possible outcomes is 1.0, as portrayed by Fig. 1.2.

Most games of chance involve rolling a pair of dice to achieve a combination adding up to a specific number (e.g., a roll of 7 on the first toss is usually considered a winner). This involves the simultaneous occurrence of two independent events governed by two separate dice. Because each die has six possible outcomes, the number of possible outcomes involving a pair of dice is 6 × 6, or 36. Each combination of die faces therefore has a $\frac{1}{36}$ or 0.027777777 probability of occurrence on a single roll of a pair of dice.

In some cases, more than one combination of die faces can produce a stated number. For the number 5, for example, there are four combinations that will provide this number, as portrayed in Fig. 1.3. The probability of rolling 5 on a single toss of the dice is therefore

$$\frac{1}{36} + \frac{1}{36} + \frac{1}{36} + \frac{1}{36} = \frac{4}{36} \quad \text{or} \quad 0.111111111$$

A single roll of a pair of dice has 36 possible individual outcomes, representing the numbers 2 through 12. The sum of the 36 individual probabilities is 1.0. The probability pyramid in Fig. 1.4 portrays the possible combinations, numbers, and cumulative probabilities associated with each number.

Matching socks.   Assume that a drawer contains three white socks and two black socks and that one needs to determine the probability of drawing two white socks in two successive attempts (i.e., a person in an

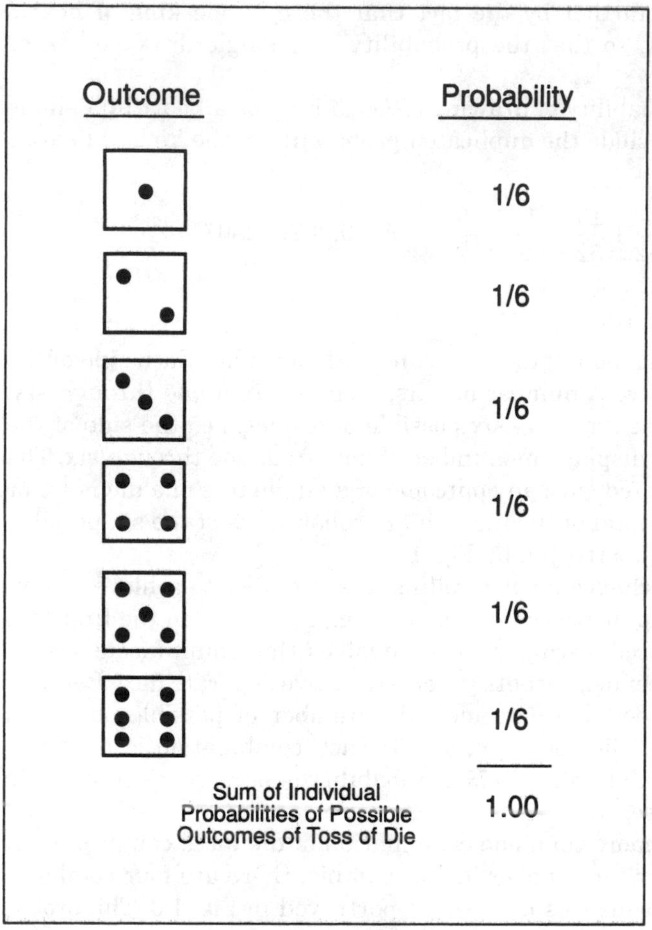

Outcome	Probability
	1/6
	1/6
	1/6
	1/6
	1/6
	1/6
Sum of Individual Probabilities of Possible Outcomes of Toss of Die	1.00

Figure 1.2 Possible outcomes for a single toss of a die.

unlit bedroom is trying to draw a matched pair to wear to work). This problem typifies the conditional theorem and is determined by the probability of selecting the first white sock on the first attempt times the probability of subsequently selecting a second white sock, given that a white sock was successfully selected on the first attempt; i.e.,

$$P_{w1} = \text{probability of first white} = \frac{3 \text{ white socks}}{3 \text{ whites} + 2 \text{ blacks}} = \frac{3}{5}$$

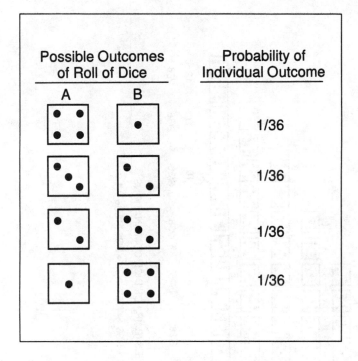

Figure 1.3   Dice combinations to produce the number five.

Now there are 2 white socks and 2 black socks remaining.

$$P_{w2} = \text{probability of second white} = \frac{2}{4}$$

$$P(w1 \cdot w2) = \text{probability of 2 successful successive attempts}$$

$$= (P_{w1})(P_{w2}) = \left(\frac{3}{5}\right)\left(\frac{2}{4}\right) = \frac{6}{20} = 0.3$$

Probability density functions

In logistics systems engineering, probability methodologies are used primarily for predictions of system performance characteristics and to incorporate logistics considerations based on these predictions into system design. These analytical tools harness empirical data developed from real-world operational experience. Empirical data by their nature tend to reflect random occurrences of events that affect the system (e.g., failures of equipment). The vehicles for translating recorded empirical

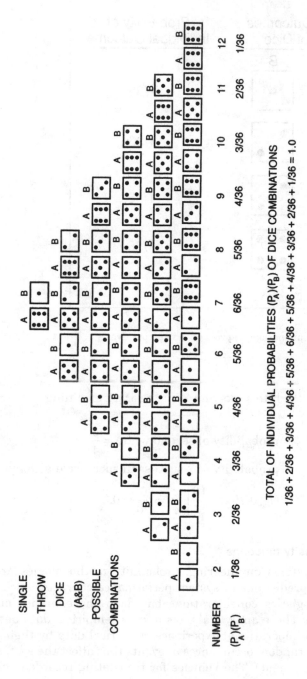

Figure 1.4 Probability pyramid: single-throw dice combinations.

18

data into probablistic predictions are provided by probability density functions (also referred to as probability distribution functions). Different types of systems tend to have individual peculiarities and inherently unique attributes; consequently, data resulting from recorded observations of the behavior of particular systems will not necessarily conform to a specific mathematical function. The techniques of regression analysis are applied to scatter data drawn from recorded observations in order to determine the mathematical function which would provide the best fit graph line for the plotted data points. The best-fit mathematical function selected should be the one that best accommodates the trend, pattern, and dispersion of the scatter data points.

A probability density model portrays the distribution of individual probabilities over all possible outcomes. It also provides the means of determining the likelihood of occurrence of all possible values within a defined population or sample, e.g., prediction of system performance characteristics. Variables described in terms of a probability distribution are referred to as *random variables*. The specific value of a random variable is determined by the mathematical function in the probability density model. When the pattern describing the probability distribution is expressed as a function of the individual variables, the function is called a probability density function.

The essential parameters of classic probability density functions are the *mean*, the *standard deviation*, and the *variance*.

Mean. The mean is the average of all the observed values in a defined sample or population.

Standard deviation. The standard deviation (also called the root-mean-square) is a factor calculated from a standard formula which indicates the central tendency of observed values, i.e., the degree to which data points tend to cluster around the mean.

Variance. The variance is the mathematical square of the standard deviation and is an indicator of the dispersion of the observed values relative to the mean.

Development of probability density functions.   The basic probability distribution formulas which the student of logistics should know are the binomial distribution, the *Poisson distribution*, the *exponential distribution*, and the *normal distribution*. These are the probability density functions typically used to massage data to construct probability density models. Probability functions and models portray the individual probability of each specific value of a variable (or item) occurring in a given population or sample. In addition to the pertinent descriptions of these functions provided below, Table 1.1 provides a summary of the

TABLE 1.1   Summary of Probability Distribution

Probability distribution	Function	Mean	Variance
Normal	$P_x = \dfrac{1}{\sigma\sqrt{2\pi}}\, e^{-(x-\mu)^2/2\sigma^2}$	μ	σ^2
Exponential	$P_x = \dfrac{1}{\mu}\, e^{-x/\mu}$	μ	μ^2
Poisson	$P_x = \dfrac{\mu^x e^{-\mu}}{x!}$	μ	μ
Binomial	$P_x = \dfrac{n!}{x!\,(n-x)!}\, p^x q^{n-x}$	np	npq

probability density functions; their corresponding probability density models are given in Table 1.2.

Binomial distribution.   The binomial distribution is applicable in cases where it is necessary to determine the probability of exactly x occurrences in n trials of a specific event that has a constant probability in a single trial. The probability of exactly x occurrences in n trials of an event that has a single-trial probability of occurrence of p is expressed

$$P_x = \frac{n!}{x!(n-x)!}\, p^x q^{n-x} \qquad 0 \le x \le n$$

where P_x = probability of exactly x occurrences
    x = exact number of occurrences in question
    n = number of trials
    p = probability of event occurring on a single trial
    $q = 1 - p$, or probability of event not occurring in a single trial

The *mean* for the binomial distribution is expressed as np. The *variance* is expressed as npq.

Application of the binomial distribution requires the following assumptions:

1. There is a fixed number of trials n.
2. The probability of success p for the specific event is the same for each trial.
3. All trials are independent.

A classic illustration of the binomial distribution is the toss of a coin, one side of which is heads, the other tails. For example, to determine the

TABLE 1.2 Probability Density Models

Distribution	Form	Probability function
Normal		$y = \dfrac{1}{\sigma\sqrt{2\pi}}\,e^{-(x-\mu)^2/2\sigma^2}$ μ = population average σ = population standard deviation e = 2.718 π = 3.141
Exponential		$y = \dfrac{1}{\mu}\,e^{-x/\mu}$ μ = population average
Poisson		$y = \dfrac{\mu^x e^{-\mu}}{x!}$
Binomial		$y = \dfrac{n!}{r!(n-r)!}\,p^r q^{n-r}$ n = number of trials r = number of occurrences p = probability of occurrence $q = 1 - p$

probability of *exactly* three heads in five tosses of the coin, $n = 5$ (number of trials), $x = 3$ (exactly three heads), $p = 0.5$ (probability of heads on a single toss), and $q = 1 - 0.5 = 0.5$ [probability of tails (other than heads) on a single trial]. If

$$P_x = \frac{n!}{x!(n-x)!}\,p^x q^{n-x}$$

then

$$P_{x=3} = \frac{5!}{3!(5-3)!}(0.5^3)(0.5^2)$$

$$= \frac{120}{(6)(2)}(0.125)(0.25)$$

$$= \frac{120}{12}(0.03125)$$

$$= (10)(0.03125)$$

$$= 0.3125$$

This is the probability of exactly 3 heads in 5 tosses of the coin.

If the student were to address the problem of *up to* 3 heads instead of *exactly* 3 heads, it would be necessary to calculate and determine the sum of the probabilities of zero heads ($P_{x=0}$), 1 head ($P_{x=1}$), and 2 heads ($P_{x=2}$), as well as 3 heads ($P_{x=3}$).

Poisson distribution. The Poisson distribution is an approximation of the binomial distribution to be applied when the number of trials n is large and the probability p of a specific event occurring on a single trial is small. The general governing criteria for the Poisson distribution function stipulate that n shall be at least 100 and the product np less than 10.

For the Poisson distribution function, the mean and variance are expressed as μ, where $\mu = np$. The function for the Poisson distribution is expressed as

$$P_x = \frac{\mu^x e^{-\mu}}{x!}$$

or

$$P_x = \frac{(np)^x e^{-np}}{x!}$$

on condition that

$$0 \le x \le \infty \qquad n \ge 100 \qquad np \le 10 \text{ or } \mu \le 10$$

where P_x = probability of exactly x occurrences
  x = exact number of occurrences in question
  n = number of trials
  p = probability of occurrence of specific event in a single trial
  $\mu = np$
  $e = 2.7182818$

A quality control situation illustrates use of the Poisson distribution function. Assume that a sample of 100 items is selected for analysis from a population of items known to be 2 percent defective. The probability of a quality control inspector obtaining exactly 4 defectives in the sample is calculated as follows:

$$P_x = \frac{(np)^x e^{-np}}{x!}$$

When $n = 100$ (number of trials), $x = 4$ (exact number of defective items), and $p = 0.02$ (2 percent known defective in item population),

$$P_{x=4} = \frac{(100 \times 0.02)^4 e^{-(100 \times 0.02)}}{4!}$$

$$= \frac{(2^4)(2.7182818^{-2})}{24}$$

$$= \left(\frac{16}{24}\right)\left(\frac{1}{2.7182818^2}\right) = \left(\frac{2}{3}\right)\left(\frac{1}{7.3890559}\right)$$

$$= (0.66666)(0.135335286)$$

$$= 0.090222622$$

There is a 9.0 percent probability of selecting exactly 4 defectives in a sample of 100 from a population known to be 2 percent defective.

To determine the probability of selecting *up to* 4 defective items instead of *exactly* 4 defective items, it is necessary to calculate and determine the sum of the probabilities for zero defectives ($P_{x=0}$), 1 defective ($P_{x=1}$), 2 defectives ($P_{x=2}$), and 3 defectives ($P_{x=3}$), as well as 4 defectives ($P_{x=4}$).

Exponential distribution. The probability associated with the exponential distribution typically involves a variable defined by temporal (time-based) intervals. The exponential distribution function is

$$P_x = \frac{1}{\mu}e^{-x/\mu} \qquad 0 \le x \le \infty$$

where x = variable, defined by time interval units
  μ = mean
  e = 2.7182818

For the exponential distribution function, the mean is expressed as μ and the variance as μ^2. The exponential distribution determines the probability of *exactly* x time units occurring, given that μ is the *mean* time value of the population.

The general form of the exponential distribution is illustrated in Fig. 1.5. The following define the characteristics of the mean and extreme parametric variable values of the exponential distribution function. When x = 0,

$$P_{x=0} = \frac{1}{\mu}e^{-x/\mu} = \left(\frac{1}{\mu}\right)e^{-0/\mu}$$

Note that $e^{-0/\mu} = e^0 = 1$. Therefore

$$P_{x=0} = \frac{1}{\mu}(1) = \frac{1}{\mu}$$

When x = ∞ (undefined, infinite value),

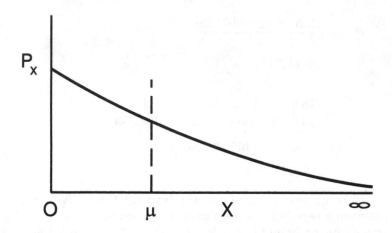

Figure 1.5   The general form of exponential distribution. $P_x = 1/\mu(e^{-x/\mu})$

$$P_{x=\infty} = \frac{1}{\mu}e^{-\infty/\mu} = \left(\frac{1}{\mu}\right)\left(\frac{1}{e^{\infty/\mu}}\right)$$

Note that

$$\left(\frac{1}{e^{\infty/\mu}}\right) = \left(\frac{1}{e^{\infty}}\right) = \left(\frac{1}{\infty}\right)$$

and the reciprocal of ∞ is zero. Therefore

$$P_{x=\infty} = \left(\frac{1}{\mu}\right)(0) = 0$$

P_x asymptotically approaches zero when x approaches infinity (∞), as shown in Fig. 1.5.

When $x = \mu$

$$P_{x=\mu} = \frac{1}{\mu}e^{-\mu/\mu} = \left(\frac{1}{\mu}\right)e^{-1} = \left(\frac{1}{\mu}\right)\left(\frac{1}{e}\right) = \left(\frac{1}{\mu}\right)\left(\frac{1}{2.7182818}\right) = \left(\frac{1}{\mu}\right)(0.3678)$$

$$P_{x=\mu} = \frac{0.3678}{\mu}$$

Fig. 1.6 identifies the extreme variable and mean values on the exponential distribution curve.

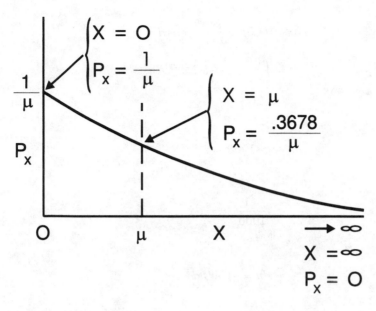

Figure 1.6 Significant parametric points on the exponential distribution curve. $P_x = 1/\mu(e^{-x/\mu})$

Determining P_x when $0 \leq x \leq \mu$ is done by using integral calculus techniques to develop a density function. In essence, this approach is to determine the probability that x will be no more than the mean value μ.

The integration process determines the area under the curve between 0 and ∞ and hence provides the probability density function. To illustrate this concept, consider the case of a cathode ray tube (CRT) used in a personal computer. Its life is known to be exponentially distributed with a mean μ of 1000 h. The probability of the variable life u of the CRT *not* exceeding 1000 would be expressed as $P(u \leq 1000)$. There would be a proportional area under the exponential curve over the range of $u = 0$ to $u = 1000$, as shown in the computation below.

Two special rules for integration are applicable to this case:

1. The integral of a constant times a function is the constant times the integral of the function: $\int_{kf}(x)dx = k\int f(x)dx$.
2. $\int e^x dx = e^x + c$.
   For this problem, u = mean life *variable*

$$p(u \leq 1000) = \int_0^{1000} f(u)\, du$$

$$= \int_0^{1000} \frac{1}{1000} e^{-u/1000} du$$

Let

$$\frac{-u}{1000} = x$$

and let

$$\frac{-1}{1000} du = dx$$

therefore

$$du = -1000\, dx$$

Substituting x,

$$P(u \leq 1000) = \int_0^{1000} \frac{1}{1000} e^x (-1000\, dx)$$

$$= \frac{1}{1000} \int_0^{1000} e^x (-1000\, dx)$$

$$= \frac{-1000}{1000} \int_0^{1000} e^x\, dx$$

$$= (-1) \int_0^{1000} e^x\, dx$$

$$= (-1)(e^x) \Big|_0^{1000}$$

Substitute $\dfrac{-u}{1000}$ for x:

$$P(u \leq 1000) = -e^{-u/1000} \Big|_0^{1000}$$

$$= -e^{-1000/1000} - (-e^{0/1000})$$

$$= -e^{-1} + e^0$$

$$= -e^{-1} + 1$$

$$= 1 - e^{-1}$$

$$= 0.632$$

The probability of the mean life u being equal to or less than 1000 h (≤ 1000) is 0.632. Consequently, the probability of the CRT life exceeding 1000 h is

$$1 - 0.632 = 0.368$$

This exercise confirms that 63.2 percent of the area under the curve of the exponential distribution is to the left of the mean μ and 36.8 percent of the area under the curve is to the right of the mean.

Normal distribution. The normal, or gaussian, distribution is reflected by the normal curve, sometimes referred to as the bell curve. The critical attributes of the normal distribution function are the *mean, standard deviation*, and *variance*.

The mathematical symbols for the mean, standard deviation, and variance differ depending on whether they are used to describe a *sample* of items or the *total population* of items.

	Sample attributes (English characters)	Total population (Greek characters)
Mean	$\bar{x}$	μ
Standard deviation	s	σ
Variance	s^2	σ^2
Number of observations	n	N

For purposes of discussing normal distribution concepts in this text, the population (Greek) symbols will be used. Discussion of sample characteristics will use the sample (English) symbology.

The normal distribution function is expressed as

$$P_x = \frac{1}{\sigma\sqrt{2\pi}} e^{[-(x-\mu)^2]/2\sigma^2}$$

where P_x = probability of exactly x occurring
  σ = population standard deviation
  μ = population mean
  e = 2.7182818
  π = 3.1415927

Like all probability functions, the normal distribution function determines the probability of exactly x occurrences of an event. The standard deviation is an indicator of the tendency of observed values in a population to cluster around the mean value. It is a mathematical factor calculated on the basis of critical attributes applied to a standard formula, which will be addressed later in this section.

The significance of the standard deviation is in the manner in which it defines density intervals under the normal curve. Fig. 1.7 portrays a classic normal curve and its constituent density intervals based on the standard deviation.

As shown in Fig. 1.7, 99.73 percent (or almost 100 percent) of all observed values are between -3σ and $+3\sigma$, 95.45 percent are between -2σ and $+2\sigma$, and 68.26 percent are between -1σ and $+1\sigma$. These attributes also describe area density under the curve: 99.73 percent of the area under the curve is between 3σ and $+3\sigma$, 95.45 percent of the area is between -2σ and $+2\sigma$, and 68.26 percent of the area is between -1σ and $+1\sigma$.

Examples of normal distribution probability functions. Suppose the mean of a population is 5.0 and the standard deviation is 1.0. To determine the probability that an observation would have the value of exactly 5.0, let $x = 5.0$ (variable), $\mu = 5.0$ (mean), and $\sigma = 1.0$ (standard deviation).

$$P_x = \frac{1}{\sigma\sqrt{2\pi}}e^{[-(x-\mu)^2]/2\sigma^2}$$

$$P_{x=5} = \frac{1}{(1.0)\sqrt{2\times 3.1415927}}e^{[-(5-5)^2]/2(1)^2}$$

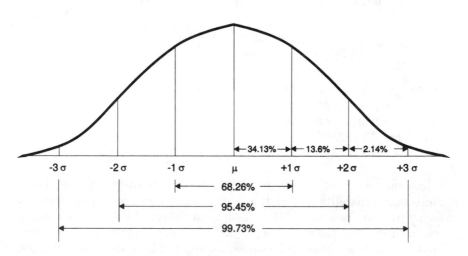

Figure 1.7   Normal curve distribution.

$$= \frac{1}{(1.0)\sqrt{6.283185307}} e^{-0/2}$$

$$= \frac{1}{2.506628275} (1)$$

$$= 0.39894228$$

Fig. 1.8 portrays the above calculated point on the normal distribution curve.

Sample—definition and analysis

1. *Purpose.* The purpose of a sample is to provide the analyst with randomly selected and observed values of a prescribed number of units from a total population from which to determine within acceptable boundaries of mathematical confidence the attributes of the total population. The primary, definitive attributes of such a sample are:

- n is the number of observations.
- $\bar{x}$ is the mean, or average value.
- The median is the middle or midpoint value.
- The mode is the most frequently observed value.
- The range is the highest observed value minus the lowest observed value.
- A frequency polygon: is a distribution curve constructed by grouping the observed values according to class value intervals. The frequency polygon of a sample is equivalent in principle to the normal distribution curve portraying the total population.
- s is the standard deviation.
- s^2 is the variance.

2. *Analytical Techniques.* Analysis of a sample to determine the primary attributes entails three sequential tasks.

1. Tabulate the observations and determine the number of observations, the mean, the median, the mode, and the range.
2. Develop a histogram on the basis of class intervals and plot a frequency polygon.
3. Pursuant to the results of tasks (1) and (2), calculate standard deviation and variance.

The elemental steps of the three sequential tasks are best explained by using an illustrative sample.

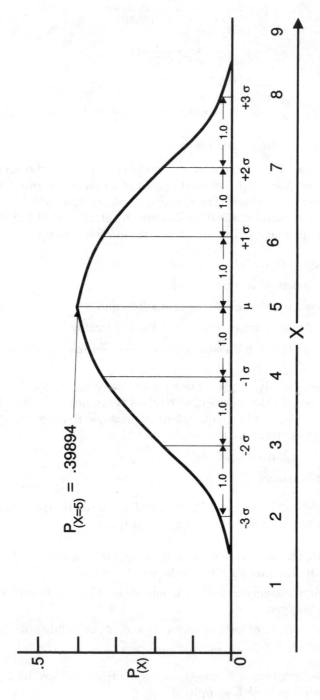

Figure 1.8 Normal curve probability plot for X = 5, σ = 1.0 and μ = 5.0.

As a case example, a random sample of 29 units provided the following observed values:

42	28	40	50	60	26
47	50	56	54	30	61
50	49	74	58	48	53
38	44	69	63	49	52
39	46	59	67	37	

- *Task 1*. Refer to Table 1.3.

    Step 1. List the individual observations in the x_i column (i.e., x_1 through x_{29}; x_i is a generic citation of any ith observation). For example, in the citation of x_4 the subscript "4" denotes that this is the fourth documented observation; therefore, $i = 4$. This also documents the sample size: $n = 29$.

    Step 2. In the observed value (x_i) column, list the individual value attributable to each observation; e.g., for the first observation x_1, the observed value is 42.

    Step 3. Compute the sum of all the observed values listed in the observed value (x_i) column. For this sample,

$$\sum x_i = 1439$$

    Step 4. Calculate the mean by dividing the sum of all observed values by the number of observed values. For this sample,

$$\bar{x} = \frac{\Sigma x_i}{n} = \frac{1439}{29} = 49.62$$

    Step 5. Construct an additional column in which the observed values are rearrayed in ascending order of value. For this sample, the ascending order is from 26 to 74.

    Step 6. Determine the value of the midpoint observation in the rearrayed sequence of observation values. For this sample, the midpoint value for the 29 observations is the 15th observation in ascending sequence. Therefore,

$$\text{median} = 50$$

    Step 7. Determine the most frequently occurring observation value. For this sample, the most frequently observed value is 50, which occurs three times. Therefore,

$$\text{mode} = 50$$

(1) x_i	(2) Observed value (x_i)	(5) Rearray x_i values for median	Sample attributes
x_1	42	(8) 26	Number of Observations:
x_2	47	28	
x_3	50	30	(1) $n = 29$
x_4	38	37	
x_5	39	38	
x_6	28	39	Mean:
x_7	50	40	
x_8	49	42	(4) $\bar{x} = \dfrac{\Sigma x_i}{n}$
x_9	44	44	
x_{10}	46	46	
x_{11}	40	47	
x_{12}	56	48	
x_{13}	74	49	
x_{14}	69	49	$= \dfrac{1439}{29}$
x_{15}	59	(7) (6) 50 (Median)	
x_{16}	50	(7) 50	
x_{17}	54	(7) 50	
x_{18}	58	52	$= 49.62$
x_{19}	63	53	
x_{20}	67	54	
x_{21}	60	56	(5) (6) Median = 50
x_{22}	30	58	
x_{23}	48	59	
x_{24}	49	60	(7) Mode = 50
x_{25}	37	61	
x_{26}	26	63	
x_{27}	61	67	(8) Range:
x_{28}	53	69	$74 - 26 = 48$
x_{29}	52	(8) 74	
(3) $\Sigma x_i =$	1439		

Step 8. Determine the difference between the highest observed value and the lowest observed value. For this sample, the highest value, 74, minus the lowest value, 26, results in a difference of 48. Therefore,

$$\text{range} = 48$$

■ *Task 2.* Refer to Figs. 1.9 to 1.11.

Step 1. Refer to Fig. 1.9. The first step in constructing a frequency polygon is to develop a histogram. Construct a graph on which to plot

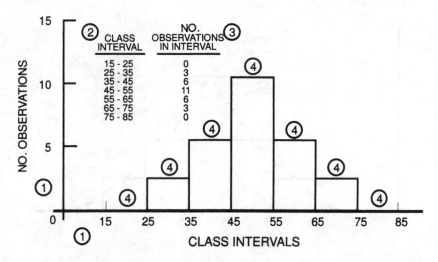

Figure 1.9 Histogram class interval definitions.

class intervals of observation values (horizontal scale) and number of observations (vertical scale).

Step 2. Within the sample range of 26 to 74, determine the class intervals which best group the observed values. In addition, provide for an interval below the lowest value (26) and an interval above the highest value (74), within which there would logically be zero observations. The purpose of the two extreme intervals is to provide "anchors" for the frequency polygon, as discussed later. For this sample there are seven defined class intervals: 15–25, 25–35, 35–45, 45–55, 55–65, 65–75, and 75–85. Note that the class intervals for a sample are subject to the judgment of the analyst; for this sample, seven intervals were considered logical.

Step 3. On the basis of the sample values in Table 1.3, determine the number of observations which would fall within each class interval. For this sample,

Class interval	Number of observations	Class interval	Number of observations
15–25	0	55–65	6
25–35	3	65–75	3
35–45	6	75–85	0
45–55	11		

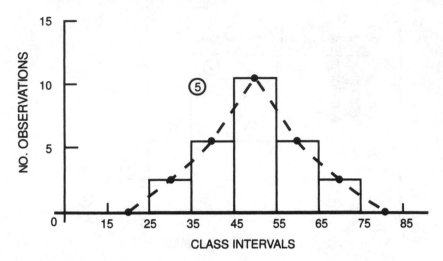

Figure 1.10 Construction of frequency polygon from midpoints of class intervals.

Step 4. Construct vertical bar-chart segments based on the assignment of the observed values to the defined class intervals. The individual bars are bounded by the class intervals on the horizontal scale and the number of observations for the individual class interval on the vertical scale. The resultant profile is the histogram for the sample, as depicted by Fig. 1.9.

Step 5. Refer to Fig. 1.10. Connect the midpoints of the histogram bars developed during the preceding steps. This illustrates the evolution of the frequency polygon.

Step 6. Refer to Fig. 1.11. Complete the frequency polygon by closing the dotted line from Step 5. For this sample, the frequency polygon portrays a distribution curve based on the scatter data over the seven class intervals defined from $x = 15$ to $x = 85$.

■ *Task 3*. Refer to Table 1.4.

Step 1. Construct the x_i column in the same manner as in task 1, step 1.

Step 2. List the individual ith values in the same manner as in task 1, Step 2.

Step 3. Construct the x column. Enter the sample mean $\bar{x}$ of 49.62, as calculated in task 1, on the columnar row element corresponding to each x_i entry, as shown in Table 1.4.

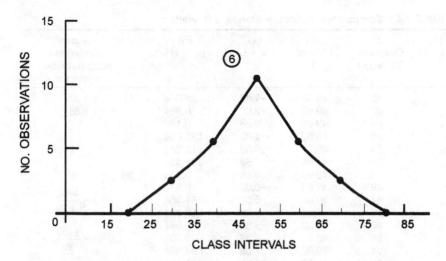

Figure 1.11 Completed frequency polygon.

Step 4. Complete the $x_i - \bar{x}$ column. In each columnar row element, list the result of subtracting the mean value $\bar{x}$ from the corresponding ith observation value, as shown in Table 1.4. For example, for x_1, the first observation,

x_1 (observed value)	42.00
Minus $\bar{x}$ (mean)	−49.62
Results in $x_1 - \bar{x} =$	−7.62

For x_{12}, the 12th observation,

x_{12} (observed value)	56.00
Minus $\bar{x}$ (mean)	−49.62
Results in $x_{12} - \bar{x}$	+6.38

It is emphasized that the accuracy of the calculations in step 4 is critical to execution of step 5.

Step 5. Construct the $(x_1 - \bar{x})^2$ column. The individual differences based on $x_1 - \bar{x}$, calculated in step 4, are each squared. For example, for x_1,

$$(x_1 - \bar{x})_2 = (42 - 49.62)^2 = (-7.62)^2 = 58.064$$

TABLE 1.4   Computation of Sample Standard Deviation

(1) x_i	(2) Observed Value (x_i)	(3) Mean $\bar{x}$	(4) $x_i - \bar{x}$	(5) $(x_i - \bar{x})^2$	Standard deviation computational sequence
x_1	42	49.62	−7.62	58.064	(1) $n = 29$
x_2	47	49.62	−2.62	6.864	$n - 1 = 28$
x_3	50	49.62	0.38	0.144	
x_4	38	49.62	−11.62	135.024	(2)
x_5	39	49.62	−10.62	112.784	$\Sigma x_i = 1439$
x_6	28	49.62	−21.62	467.424	
x_7	50	49.62	0.38	0.144	(3)
x_8	49	49.62	−0.62	0.384	
x_9	44	49.62	−5.62	31.584	$\bar{x} = \dfrac{\Sigma x_i}{n}$
x_{10}	46	49.62	−3.62	13.104	
x_{11}	40	49.62	−9.62	92.544	
x_{12}	56	49.62	6.38	40.704	$= \dfrac{1439}{29}$
x_{13}	74	49.62	24.38	594.384	
x_{14}	69	49.62	19.38	375.584	
x_{15}	59	49.62	9.38	87.984	$= 49.62$
x_{16}	50	49.62	0.38	0.144	
x_{17}	54	49.62	4.38	19.184	(4) (5)
x_{18}	58	49.62	8.38	70.224	
x_{19}	63	49.62	13.38	179.024	$\Sigma(x_i - \bar{x})^2 = 3946.796$
x_{20}	67	49.62	17.38	302.064	
x_{21}	60	49.62	10.38	107.744	
x_{22}	30	49.62	−19.62	384.944	(6)
x_{23}	48	49.62	−1.62	2.624	$s = \sqrt{\dfrac{\Sigma(x_i - \bar{x})^2}{n - 1}}$
x_{24}	49	49.62	−0.62	0.384	
x_{25}	37	49.62	−12.62	159.264	$= \sqrt{\dfrac{3946.796}{28}}$
x_{26}	26	49.62	−23.62	557.904	
x_{27}	61	49.62	11.38	129.504	
x_{28}	53	49.62	3.38	11.424	$= \sqrt{140.957}$
x_{29}	52	49.62	2.38	5.644	$s = 11.8725$
	1439			3946.796	(7) $s^2 = (11.8725)^2$
					$s^2 = 140.9563$

Note: Standard deviation formula (sample size $\leq$ 30), $s = \sqrt{\dfrac{\Sigma(x_i - \bar{x})^2}{n - 1}}$

For x_{12},

$$(x_{12} - \bar{x})^2 = (56 - 49.62)^2 = (6.38)^2 = 40.704$$

After completing this procedure, determine the sum of the individual squares. This results in

$$\sum_{i=1}^{i=29} (x_i - \bar{x})^2 = 3946.796$$

Step 6. Determine the sample standard deviation by use of the small-sample formula

$$s = \sqrt{\frac{\Sigma(x_i - \bar{x})^2}{n-1}}$$

Note: For sample sizes of 30 or less, $n - 1$ is used as the denominator under the square root radical. For sample sizes greater than 30, n is used.

For this sample,

$$\sum_{i=1}^{i=29}(x_i - \bar{x})^2 = 3946.796$$

$$n-1 = 29-1 = 28$$

Therefore

$$s = \sqrt{\frac{3946.796}{28}}$$

$$= \sqrt{140.957}$$

$$= 11.8725$$

Step 7. To determine the variance of this sample, square the standard deviation, i.e.,

$$s^2 = (11.8725)^2$$

$$= 140.9563$$

Population standard deviational variance. The formulas for determining the standard deviation and variance of the total population—or universe—of observations are based on the same methodologies applicable to large samples:

1. Standard deviation:

$$\sigma = \sqrt{\frac{\sum_{i=1}^{i=N}(x_i - \mu)^2}{N}}$$

where σ = population standard deviation
N = number of observations in total population
μ = population mean, or average
x_i = individual ith observation values

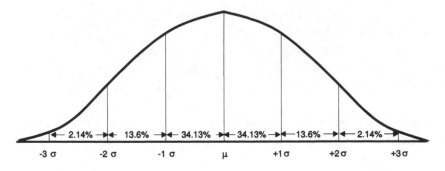

Figure 1.12 Normal curve attributes. $\sigma = \sqrt{\sum_{i=1}^{i=n}(x_i - \mu)^2 / N}$

2. Variance:

$$\text{Population variance} = \sigma^2$$

The purpose of sample analysis is to develop a realistic portrayal of the attributes of the total population. The number of observations N on the total population, the population mean μ, and the population standard deviation σ would produce a histogram conforming to the theoretical normal distribution curve and class interval densities described by Fig. 1.12.

Bayes' Theorem:

Bayes' theorem is based on mathematical inference. It is the scientific basis for quantitatively drawing conclusions about effects based on statistical measurements of the causes of a characteristic in a given population. The constituent elements of bayesian mathematics are (1) the probability of an established condition or inherent characteristic of a defined statistical group, and (2) the probability of occurrence of an event involving an item from that defined statistical group.

The generic formula for Bayes' theorem is

$$P(B_i \mid A) =$$

$$\frac{[P(B_i)][P(A_i \mid B_i)]}{[P(B_1) \cdot P(A_1 \mid B_1)] + [P(B_2) \cdot P(A_2 \mid B_2)] + \cdots + [P(B_k) \cdot P(A_k \mid B_k)]}$$

$$i = 1, 2, \ldots, k$$

where $P(B_i)$ = probability of the existing condition or characteristic in the ith member of the group.

$P(A_i | B_i)$ = probability of designated event A_i occurring for ith member subject to condition B_i.

$P(B_i | A)$ = ratio of probability of occurrence of the event for the ith item relative to the sum of probabilities for all items associated with conditions B_i.

Illustration of Bayes' theorem

Case one. A plant facility is expected to produce items of which 25 percent will be defective and 75 percent not defective. In process testing procedures normally reject 99 percent of the defective items. The same test procedures typically reject 7.0 percent of the nondefective items. Based on these data, determine the percentage of rejected items which are actually defective. Given

$$A_1 = \text{rejected items from defective group}$$
$$B_1 = \text{defective items}$$
$$A_2 = \text{rejected items from nondefective group}$$
$$B_2 = \text{nondefective items}$$

$$P(A_1 | B_1) = 0.99 \qquad P(A_2 | B_2) = 0.07$$
$$P(B_1) = 0.25 \qquad P(B_2) = 0.75$$

construct a logic tree, as portrayed in Fig. 1.13a.

To determine the ratio of defective items rejected to total items rejected,

$$P(B_1 | A) = \frac{P(B_1) \cdot P(A_1 | B_1)}{P(B_1) \cdot P(A_1 | B_1) + P(B_2) \cdot P(A_2 | B_2)}$$

$$= \frac{(0.25)(0.99)}{(0.25)(0.99) + (0.75)(0.07)}$$

$$= \frac{0.3475}{0.2475 + .0525} = \frac{0.2475}{0.3}$$

$$= 0.825, \text{ or } 82.5 \text{ percent}$$

Case two. The production output of three product departments in a plant is as follows:

Department	Percent of total plant output
1	25
2	40
3	35

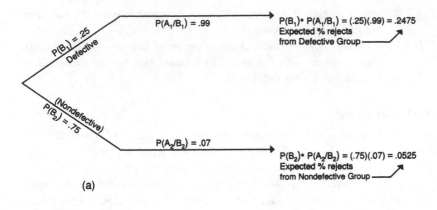

(a)

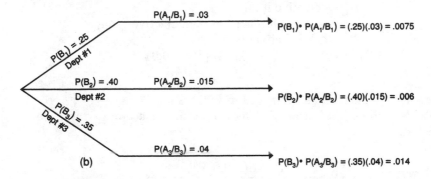

(b)

Figure 1.13   Bayes' theorem logic tree. (a) Case one. (b) Case Two.

For Department 1, the quality control rejection rate is 3.0 percent; for Department 2, the rejection rate is 1.5 percent; and for Department 3, the rejection rate is 4.0 percent. Determine the percentage of rejected products that are produced by Department 1.

	Output*	Rejection rate*
Department 1	$P(B_1) = 0.25$	$P(A_1 \mid B_1) = 0.03$
Department 2	$P(B_2) = 0.40$	$P(A_2 \mid B_2) = 0.015$
Department 3	$P(B_3) = 0.35$	$P(A_3 \mid B_3) = 0.04$

Construct a logic tree as portrayed in Fig. 1.13b. To determine the ratio of rejected units from Department 1 to total units rejected,

$$P(B_1 \mid A) = \frac{P(B_1) \cdot P(A_1 \mid B_1)}{P(B_1) \cdot P(A_1 \mid B_1) + P(B_2) \cdot P(A_2 \mid B_2) + P(B_3) \cdot P(A_3 \mid B_3)}$$

$$= \frac{(0.25)(0.03)}{(0.25)(0.03) + (0.40)(0.015) + (0.35)(0.04)} = \frac{0.0075}{0.0275}$$

$$= 0.2727, \text{ or } 27.3 \text{ percent}$$

2

Reliability

Overview

Reliability with respect to systems and equipment is defined as the probability that a system will perform its intended function for a specified interval under stated conditions. Reliability projects the probability of a system's success in fulfilling its functional requirement *after* the system has been activated and operationally engaged. The constituent elements governing reliability are

Operating cycle. The number of time units or operational cycles (as for engines or winches) a system is designed to successfully complete without malfunction or failure.

Average operating intervals. The average number of time units or operational cycles a system achieves between failures.

Failure frequency. The number of failures per time unit or operational cycle.

Reliability is a significant contributory factor to maintainability. The failure rate dictates the frequency of unscheduled corrective maintenance (or repair) of a system affected by random malfunctions. Low reliability indicates frequent failures, which in turn dictate more frequent corrective maintenance, which in turn mandates increased maintenance support in the form of facilities, skilled technicians, tools, and supporting stocks of spare components and repair piece-parts. Increased system reliability based on high quality components can greatly extend the intervals of operation between failure and eliminate or minimize corrective maintenance support requirements.

Principles of Reliability

Single-point reliability

The reliability of a single operational entity (e.g., a system, subsystem, or assembly) is determined by an algorithm derived from the exponential distribution function. This principle is called *single-point reliability*. As reliability $\underline{R}$ expresses the probability that a system will perform as required for a defined interval of time under specified conditions, the reliability function $\underline{R}_t$ can be expressed as

$$\underline{R}_t = 1 - F_t$$

where $\underline{R}_t$ = probability that the system will successfully perform as required over the interval of time t.

F_t = probability that system will fail during the interval of time t.

If the random variable t has a density function of f_t then the expression for reliability is

$$\underline{R}_t = 1 - F_t = \int_t^\infty f_{(t)d}$$

Based on the exponential probability density function,

$$f_t = \frac{1}{\mu} e^{-1/\mu}$$

where μ = mean life, or MTBF

t = specified operating interval

e = base number of natural logarithm (ln), 2.7182818

Note: MTBF, mean time between failures, is the same parameter as *mean life*. In mathematical logistical formulas, MTBF is most frequently used.

The reliability for time t is

$$\underline{R}_t = \int_t^\infty \frac{1}{\mu} e^{-1/\mu} dt = e^{-t/\mu}$$

The reciprocal of μ or MTBF denotes the average number of failures per time unit. Failure rate is expressed by the Greek letter lambda (λ); i.e.,

$$\lambda = \frac{1}{\mu} = \frac{1}{\text{MTBF}}$$

Therefore

$$\underline{R}_t = e^{(-1/\mu)(t)} = e^{-1/\mathrm{MTBF})(t)} = e^{-\lambda t}$$

Thus, in the reliability engineering discipline, single-point reliability is expressed by

$$\underline{R} = e^{-\lambda t}$$

where $\lambda = \frac{1}{\mathrm{MTBF}}$ = failure rate
t = operating cycle
$e = 2.7182818$

Example of single-point reliability. Compute the reliability or probability of success of an assembly with a *mean life* of 10,000 h and a planned operating cycle of 5000 h.

Based on these characteristics,

$$\lambda = \frac{1}{\mathrm{MTBF}} = \frac{1}{10,000} = 0.0001$$

$$t = 5000$$

$$\begin{aligned}
\underline{R} &= e^{-\lambda t} = e^{-(0.0001)(5000)} \\
&= e^{-0.5} \\
&= 0.60653
\end{aligned}$$

The probability of the assembly successfully accomplishing its intended function for a 5000-h operating cycle is 0.60653, or about 60 percent.

Fig. 2.1 portrays the single-point reliability function when the normalized time is expressed in units of t/MTBF.

Example of reliability of a series configuration. A series configuration of modules is treated as follows: First, add the individual modular failure rates. Second, incorporate the resulting sum of the individual failure rates in the single-point reliability formula as a single failure rate factor. This stipulates that

$$\underline{R}_{\mathrm{series}} = e^{-(\lambda_1 + \lambda_2 + \cdots \lambda_n)(t)}$$

where n = number of modules in series
t = operating cycle

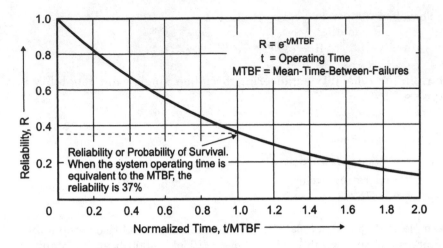

Figure 2.1   Exponential reliability function.

Assume that a system has an operating cycle of 1000 h with three modules in series, and

$$\text{Module 1 failure rate} = \lambda_1 = 0.0001$$
$$\text{Module 2 failure rate} = \lambda_2 = 0.0002$$
$$\text{Module 3 failure rate} = \underline{\lambda_3 = 0.0003}$$
$$0.0006$$

(The total failure rate is the sum of the individual modular failure rates.)

$$\underline{R}_{\text{series}} = e^{-(\lambda_{\text{series}})(t)}$$

where $\lambda_{\text{series}} = 0.0006$
$\quad\quad\quad t = 1000 \text{ h}$

$$\underline{R}_{\text{series}} = e^{-(0.0001 + 0.0002 + 0.0003)(1000)}$$
$$= e^{-(0.0006)(1000)} = e^{-0.6}$$
$$= 0.5488$$

Reliability of components

Constituent modular components of an immediately higher assembly have interconnective relationships defined in the form of a series, parallel, or combination of series and parallel arrangement. Such modular relationships are portrayed by reliability block diagrams (RBDs).

Reliability of components in series.   The reliability of modules connected in series is the mathematical product of the individual modular reliability values; i.e.,

$$\underline{R}_{\text{series}} = (\underline{R}_1)(\underline{R}_2)(\underline{R}_3) \cdots (\underline{R}_n)$$

where n = total number of modules in series

Example   Three modules in a series have individual reliability values of 0.90, 0.80, and 0.70, respectively; i.e.,

$$\text{Module 1: } \underline{R}_1 = 0.9$$
$$\text{Module 2: } \underline{R}_2 = 0.8$$
$$\text{Module 3: } \underline{R}_3 = 0.7$$

The reliability block diagram describing this three module series relationship is shown in Fig. 2.2.

The reliability of this series is computed by determining the product of $\underline{R}_1$, $\underline{R}_2$ and $\underline{R}_3$:

$$\begin{aligned}\underline{R}_{\text{series}} &= (\underline{R}_1)(\underline{R}_2)(\underline{R}_3) \\ &= (0.9)(0.8)(0.7) \\ &= 0.504\end{aligned}$$

Reliability of components in parallel.   The reliability of two modular components in a parallel relationship is expressed as

$$\underline{R}_{\text{parallel}} = \underline{R}_1 + \underline{R}_2 - (\underline{R}_1)(\underline{R}_2)$$

Example   An assembly has two modules in a parallel (redundancy) relationship. The first module has reliability of 0.9, and the second module has reliability of 0.8. The reliability block diagram for this example is shown in Fig. 2.3.

The reliability for this relationship is computed as follows:

$$\underline{R}_{\text{parallel}} = \underline{R}_1 + \underline{R}_2 - (\underline{R}_1)(\underline{R}_2)$$

where $\underline{R}_1$ = reliability for module 1 = 0.9
$\underline{R}_2$ = reliability for module 2 = 0.8

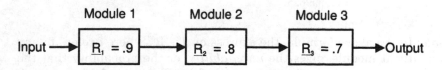

Figure 2.2   Reliability block diagram for series modules.

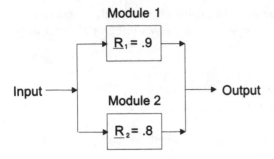

Figure 2.3   Parallel reliability block diagram.

$$\underline{R} = 0.9 + 0.8 - (0.9)(0.8) = 1.7 - 0.72$$
$$= 0.98$$

The reliability of more than two modules is better determined by an algorithmic method that can be used for an unlimited number of modules in parallel relationships.

$$\underline{R}_{\text{parallel}} = 1 - (1 - \underline{R}_1)(1 - \underline{R}_2)(1 - \underline{R}_3)\cdots(1 - \underline{R}_n)$$

where n = the total number of modules in parallel relationship

Example   An assembly has five modules in redundant or parallel relationship, as follows:

Module 1: $\underline{R}_1$ = 0.95
Module 2: $\underline{R}_2$ = 0.9
Module 3: $\underline{R}_3$ = 0.85
Module 4: $\underline{R}_4$ = 0.8
Module 5: $\underline{R}_5$ = 0.75

The reliability is computed as follows:

$$\underline{R} = 1 - (1 - 0.95)(1 - 0.9)(1 - 0.85)(1 - 0.8)(1 - 0.75)$$
$$= 1 - (0.05)(0.1)(0.15)(0.2)(0.25)$$
$$= 1 - 0.0000375$$
$$= 0.9999625$$

Note: In the above formula, the expression $1 - \underline{R}_i$, where i refers to any constituent module, gives the *unreliability*, or the probability that the specific module will not function successfully. The logic of this approach is to subtract from 1.0 (100 percent probability of success) the probability that all modules in the assembly will fail. In the example, the probability that all modules will fail is 0.0000375, which is the product of the individual modular failure probabilities. Hence, subtracting 0.0000375 from 1.0 results in 0.9999625, the probability that at least one of the redundant modules in the assembly will perform successfully.

Reliability of components in combined parallel and series relationships.

Determining the reliability of interconnected modules in a network that includes both parallel and series relationships entails calculations based on both of the aforementioned mathematical techniques. Given the basic quantitative principles, each situation must be based on individual case calculations.

Example Determine the reliability of an assembly defined by the reliability block diagram in Fig. 2.4, showing eight interconnected modules in combined relationships.

The object of this type of computation is to convert the combined network into an equivalent series relationship, for which the resultant reliability is determined by the product of the individual constituent reliabilities. This is accomplished by grouping the modular relationships into manageable segments. To apply this principle to the example,

1. Modules 1, 2, and 3, with respective reliability values of R_1, R_2 and R_3, are grouped into a segment designated as segment A, with segment reliability R_A.
2. Module 4, as a single point module, would be treated as segment B, where $R_4 = R_B$
3. Modules 5, 6, 7, and 8, with respective reliability values of R_5, R_6, R_7, and R_8, would be grouped in segment C, with segment reliability R_C.

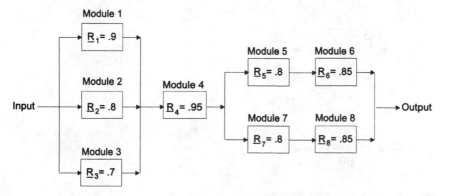

Figure 2.4   Interconnectivity reliability block diagram.

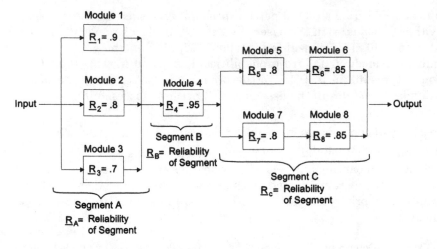

Figure 2.5   Segmental zoning of constituent modules.

This segmental zoning portrayed by Fig. 2.5. The methodology is as follows:

1. Determine R_A using the multiple unit parallel relationship formula:

$$\begin{aligned}
R_A &= 1 - (1 - R_1)(1 - R_2)(1 - R_3) \\
&= 1 - (1 - 0.9)(1 - 0.8)(1 - 0.7) \\
&= 1 - (0.1)(0.2)(0.3) \\
&= 1 - 0.006 \\
&= 0.994
\end{aligned}$$

2. Because R_4 is the single-point reliability for segment B, a single module,

$$R_B = R_4 = 0.95$$

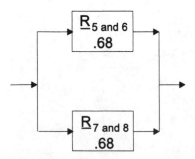

Figure 2.6   Reliability block diagram
for segment C.

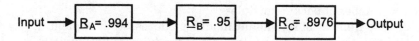

Figure 2.7 Equivalent series network.

3. Determine $\underline{R}_C$ by treating segment C as a parallel arrangement of two individual series relationships, $\underline{R}_5$ with $\underline{R}_6$ and $\underline{R}_7$ with First, calculate the reliability of each series arrangement:

$$\underline{R}_{5\text{ and }6} = (\underline{R}_5)(\underline{R}_6) = (0.8)(0.85) = 0.68$$
$$\underline{R}_{7\text{ and }8} = (\underline{R}_7)(\underline{R}_8) = (0.8)(0.85) = 0.68$$

After this calculation, the equivalent RBD for segment C would reflect the equivalent parallel circuit shown by Fig. 2.6. The new $\underline{R}_{5\text{ and }6}$ and $\underline{R}_{7\text{ and }8}$ equivalent relationship is now treated as a parallel arrangement to determine $\underline{R}_C$:

$$\underline{R}_C = \underline{R}_{5\text{ and }6} + \underline{R}_{7\text{ and }8} - (\underline{R}_{5\text{ and }6})(\underline{R}_{7\text{ and }8})$$
$$= 0.68 + 0.68 - (0.68)(0.68)$$
$$\underline{R}_C = 0.8976$$

As a result of the three sequential steps, the original combined relationship is ultimately defined by an equivalent series network, as portrayed in Fig. 2.7. To calculate the reliability of the assembly based on Fig. 2.7,

$$\underline{R}_{\text{assembly}} = (\underline{R}_A)(\underline{R}_B)(\underline{R}_C)$$
$$= (0.994)(0.95)(0.8976)$$
$$= 0.84760368$$

Reliability of multiple operating modules with multiple redundancy

The probability of success of a system configuration consisting of more than one operating interconnected module backed up by more than one interchangeable redundant module is derived from the Poisson distribution function:

$$f(x)\frac{(n\lambda t)^x e^{n\lambda t}}{(x)!}$$

The Poisson formula can be tailored to address all on-line operational modules and redundancies, as follows:

$$\underline{R}_{\text{system}} = \sum_{i=0}^{i=s} \frac{(n\lambda t)^{x_i} e^{-n\lambda t}}{(x_i)!}$$

where s = number of backup modules

$\quad x_i$ = the number of redundant modules addressed in sequential groupings from 0 modules up to s modules as backup (e.g., $i = 0, 1, 2, \ldots, s$)

$\quad n$ = number of operating modules

$\quad \lambda$ = failure rate

$\quad t$ = operating cycle

The expanded expression for this formula is

$$R_{\text{system}} = e^{-n\lambda t} + (n\lambda t)e^{-n\lambda t} \frac{(n\lambda t)^2 e^{-n\lambda t}}{2!}$$

$$= \frac{(n\lambda t)^3 e^{-n\lambda t}}{3!} + \cdots + \frac{(n\lambda t)^s e^{-n\lambda t}}{s!}$$

Example Determine the reliability of a system with two operational modules backed up by three redundant modules. All modules are interchangeable. Each module has a mean time between failures (MTBF) of 5000 h. The operating cycle is 1000 h. Therefore,

Operating cycle: $t = 1000$

Operating modules: $n = 2$

Number of backup modules: $s = 3$

Modular failure rate: $\lambda = \dfrac{1}{\text{MTBF}} = \dfrac{1}{5000} = 0.0002$

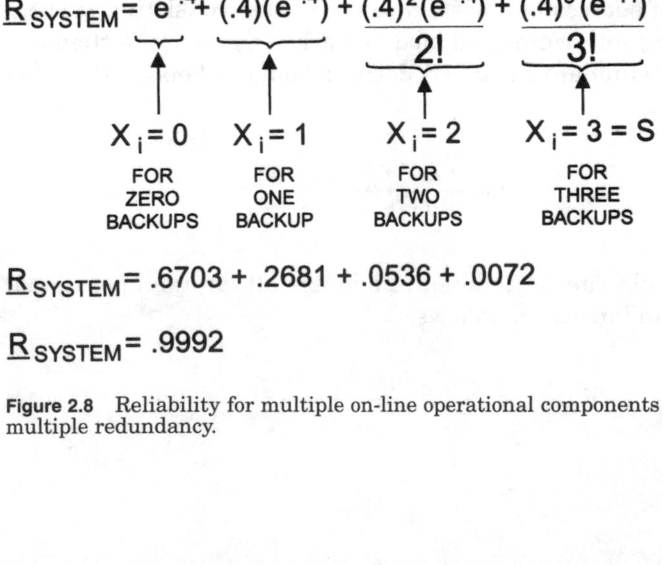

$$\underline{R}_{\text{SYSTEM}} = \underbrace{e^{-.4}}_{} + \underbrace{(.4)(e^{-.4})}_{} + \underbrace{\frac{(.4)^2(e^{-.4})}{2!}}_{} + \underbrace{\frac{(.4)^3(e^{-.4})}{3!}}_{}$$

$$X_i = 0 \qquad X_i = 1 \qquad X_i = 2 \qquad X_i = 3 = S$$

FOR	FOR	FOR	FOR
ZERO	ONE	TWO	THREE
BACKUPS	BACKUP	BACKUPS	BACKUPS

$\underline{R}_{\text{SYSTEM}} = .6703 + .2681 + .0536 + .0072$

$\underline{R}_{\text{SYSTEM}} = .9992$

Figure 2.8 Reliability for multiple on-line operational components with multiple redundancy.

The system reliability can preliminarily be parameterized; i.e.,

$$\underline{R}_{system} = \sum_{i=0}^{i=3} \frac{(n\lambda t)^{xi} e^{-n\lambda t}}{(x_i)!}$$

Solution   Because $n\lambda t$ is repeated throughout the equation, it is advisable to initially determine this factor:

$$n\lambda t = (2)(0.0002)(1000) = 0.4$$

The expanded equation incorporating the elements of this example is shown in Fig. 2.8.
   Based on this formulation,

$$\underline{R}_{system} = 0.6703 + 0.2681 + 0.0536 + 0.0072$$
$$= 0.9992$$

Note: It is emphasized that to address the reliability potential of all the redundant modules, it is necessary to address sequentially the probability effects of each ith module grouping starting with $i = 0$ until i is equal to the total number of redundant modules, i.e., $i = s$.

System Life Cycle Reliability

Phases of system reliability

The principle of the negative exponential distribution function, from which the single-point reliability formula is derived, presumes a relatively constant failure rate during system operations. This presumption is based on the system's having begun to mature and performance efficiency having stabilized.

It is important to recognize that when a system is initially produced, tested, and distributed for operation, there is typically a higher frequency of failures as a result of defects in manufacturing, defects in design, component irregularities, etc., which have to be debugged during the introductory period. This initial phase of the system life cycle is called the "burn-in" or "infant mortality" period.

Susequent to the burn-in period, the system stabilizes in terms of failure frequencies and performance. However, when the system reaches a certain age, there is a "wear-out" period during which the failure rate increases as a result of such influences as cumulative component stress, age of assemblies, and material deterioration.

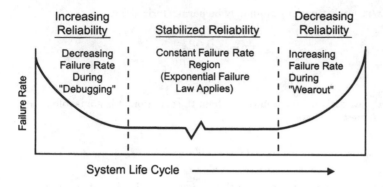

Figure 2.9   The bathtub curve.

The bathtub curve

The composite effect of the system aging phenomenon is described by the life-cycle system failure rate or bathtub curve, as shown in Fig. 2.9.

The typical failure rate curve in Fig. 2.9 shows a downward trend when the failure rate decreases, which indicates an *increase* in system reliability. The curve flattens when the reliability stabilizes and the failure rate is constant. During the wear-out period, the failure rate will tend to increase, indicating a decrease in reliability, as the system approaches termination.

Trade-Off of Key Performance Parameters within Specified Reliability Targets

It is likely that during development of a system or equipment there will be changes in some of the performance factors that determine the systems level reliability (R) value. Generally, system level reliability is specified as key performance parameter (KPP), which is mathematically based on the failure rate (λ) and operating cycle (t). If there is an adjustment dictated in either of these factors, revision in the other factor will be required in order to fulfill the target system reliability. The following explains the algorithmic approach to analysis and determination of this trade-off process.

$$R = e^{-\lambda t}$$

where R = reliability value
e = base number of natural logarithm
λ = failure rate
t = operating cycle

Derivation of λ and t Factor Algorithms from Rules of Natural Logarithms. (The reader is referred to special rules governing logarithms in Chapter 1).

$$\ln y^x = (x)(\ln y)$$

Therefore:

$$\ln R = (-\lambda t)(\ln e)$$

Note: $\ln e = 1$

Therefore:

$$\ln R = (-\lambda t)(1)$$

$$\ln R = (-\lambda t)$$

Therefore:

$$-\lambda = \frac{\ln R}{t} \qquad -t = \frac{\ln R}{\lambda}$$

$$\lambda = -\frac{\ln R}{t} \qquad t = -\frac{\ln R}{\lambda}$$

Example of application

A system has specified 0.95 reliability, based on a failure rate of 0.00005 (or 5×10^{-5}), with a required operating cycle of 1,000 hours, i.e.

$$R = 0.95$$
$$\lambda = 0.00005 \text{ (or MTBF of 20,000 h)}$$
$$t = 1.000 \text{ h}$$

The operating cycle must be increased to 1,500 h. Determine the revised failure rate (λ) necessary to accommodate the increased operating cycle in order to maintain the specified reliability of 0.95:

$$\text{Revised } \lambda = -\frac{\ln R}{t}$$

$$\text{Revised } \lambda = -\frac{\ln 0.95}{1500}$$

$$\text{Revised } \lambda = -\frac{(-.051293294)}{1500} = 0.000034196 \text{ (or } 3.4196 \times 10^{-5})$$

Note: Failure rate of 0.000034196 translates to revised MTBF of @ 29,243 h.

3

Maintainability

Overview

Definition of maintainability

Maintainability, $\underline{M}$, is the measure of the ability of a system to be restored to a specified level of operational readiness within defined intervals with the use of prescribed personnel, facility, and equipment resources.

Professional practitioners of logistics tend to describe maintainability in terms of critical system performance specifications. The maintainability parameters most frequently cited in system specifications are

1. *Mean-corrective-maintenance time,* $\overline{M}_{ct}$ the indicator of repairability at the organizational (retail) level of customer utility

2. *Mean preventive maintenance time,* $\overline{M}_{pt}$ the indicator of scheduled, planned maintenance time at the organizational level

Systems contracts within the purview of the U.S. Department of Defense typically stipulate the organizational-level repair measurement parameter, $\overline{M}_{ct}$, as the governing maintainability criterion; $\overline{M}_{ct}$ is also the companion systems engineering parameter to mean time between failures (MTBF).

The inherent availability A_i of a system (discussed in Chap. 4) is predominantly a system contract specification governed by the trade-off between $\overline{M}_{ct}$ as the maintainability parameter, and MTBF, as the reliability parameter. It is further noted that the term mean time to repair

(MTTR), normally cited in the lexicon of reliability engineering, is identical to and interchangeable with the term $\overline{M}_{ct}$. Many logistics engineering algorithms will incorporate MTTR in the same application as $\overline{M}_{ct}$.

Approach to understanding maintainability

This chapter addresses the fundamentals of maintainability by focusing on the basic quantitative parameters:

1. Corrective maintenance is based on
   - Mean corrective maintenance time ($\overline{M}_{ct}$)
   - Mean time between failures (MTBF)
   - Failure rate (λ)

2. Preventive maintenance is measured by
   - Mean preventive maintenance time ($\overline{M}_{pt}$)
   - Mean time between preventive maintenance (MTBM$_{pt}$)
   - Preventive maintenance frequency (f_{pt})

3. Logistics/administrative delay is governed by
   - Mean logistics delay time ($\overline{M}_{LD}$)
   - Mean time between logistics delay (MTBL)
   - Logistics delay frequency (f_{LD})

4. General maintenance, considering both corrective and preventive maintenance elements, is measured by
   - Mean active maintenance time ($\overline{M}$)
   - Mean time between maintenance (MTBM)
   - Maintenance downtime (MDT)

   The essential ingredients of maintainability analysis are

   Average task or event times, e.g., $\overline{M}_{ct}$ or $\overline{M}_{pt}$
   Average intervals between tasks or events, e.g., MTBF or MTBM
   Frequencies of occurrence, e.g., λ or f_{pt}

These generic maintainability attributes are algorithmically manipulated to produce the various indicators of maintenance capability.

In keeping with tutorial logic, the parameters describing corrective maintenance, preventive maintenance, logistics delay, and general main-

TABLE 3.1 Interchangeable Maintainability Terminology

Term most frequently used	Alternative terms used
Corrective maintenance (M_{ct})	Unscheduled maintenance (M_{unsch})
	Repair
Mean corrective maintenance time ($\overline{M}_{ct}$)	Mean time to repair (MTTR)
	Mean unscheduled maintenance time ($\overline{M}_{unsch}$)
Failure rate (λ)	Corrective maintenance frequency (f_{ct})
	Unscheduled maintenance frequency (f_{unsch})
Mean time between failures (MTBF)	Mean time between corrective maintenance (MTBM$_{ct}$)
	Mean time between unscheduled maintenance (MTBM$_{unsch}$)
Preventive maintenance (M_{pt})	Scheduled maintenance (M_{sch})
Mean preventive maintenance time ($\overline{M}_{pt}$)	Mean scheduled maintenance time ($\overline{M}_{sch}$)
Preventive maintenance frequency (f_{pt})	Scheduled maintenance frequency (f_{sch})
Mean time between preventive maintenance (MTBM$_{pt}$)	Mean time between scheduled maintenance (MTBM$_{sch}$)

tenance will be addressed in sequence to establish their elemental significance in the measurement of maintainability and determination of system availability.

Interchangeable maintainability terminology

While the student of logistics should be aware of those terms most frequently utilized in maintainability algorithms, it is also important to be cognizant of the proliferation of other terms that have become part of the language of logistics and may be used interchangeably in maintainability analysis. Table 3.1 provides a summary of the primary maintainability terms and their corresponding alternatives.

Corrective Maintenance

Concept of corrective maintenance

Corrective maintenance M_{ct} relates to the repair and restoration of a system required as a result of a random, unplanned failure or a disabling malfunction of the system. Hence, *corrective maintenance* is synonymous with *repair*. Logistics managers evaluate corrective maintenance requirements through assessment of empirical data developed from prior failure and repair experience with an item, monitoring of current corrective maintenance activity, and prediction of future corrective maintenance requirements through analysis of repair history as reinforced by quantitative methodologies. The classic corrective maintenance cycle runs from the point of failure detection through verification of restoration, as shown in Fig. 3.1.

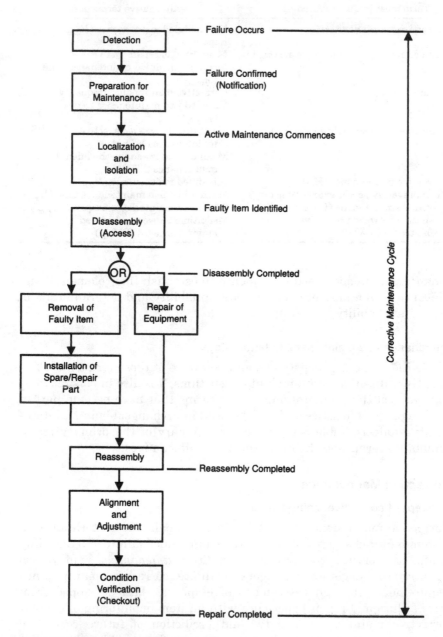

Figure 3.1 Corrective maintenance cycle.

Note that M_{ct} is the symbolic expression for corrective maintenance as a discipline, whereas $\overline{M}_{ct}$ is a mathematical factor referring to mean corrective maintenance time.

Mean corrective maintenance time

Mean corrective maintenance time $\overline{M}_{ct}$ is the average time required to accomplish repair actions within the defined steps of corrective maintenance which occurs at random intervals. Mean corrective maintenance time is measured by computing the mean time required to accomplish a series of repetitive maintenance actions executing the same function; $\overline{M}_{ct}$ is determined by averaging the individual task times. The following example illustrates this.

Observation i	Observed task time, h
$i = 1$	$M_{ct1} = 1.5$
$i = 2$	$M_{ct2} = 1.2$
$i = 3$	$M_{ct3} = 1.6$
$i = 4$	$M_{ct4} = 1.2$
$i = 5$	$M_{ct5} = \underline{1.5}$
	7.0

$$\sum_{i=1}^{i=5} M_{ct}$$

$$n = 5$$

$$\overline{M}_{ct} = \frac{\sum_{i=1}^{i=5} M_{ct}}{n} = \frac{7.0}{5}$$

$$= 1.4 \text{ h}$$

Repair rate. The *repair rate* μ, commonly utilized in evaluation of the inherent availability (A_i) tradeoff, represents the frequency of repair actions per period:

$$\mu = \text{repair rate}$$

$$= \frac{1}{\overline{M}_{ct}}$$

With respect to the example above,

$$\mu = \frac{1}{1.4 \text{ h}} = 0.714 \text{ repair actions/h}$$

Mean time between corrective maintenance

Mean time between corrective maintenance (MTBM_{ct}) is synonymous with mean time between failures (MTBF). This figure of merit expresses the average interval between repair actions. It is determined by the ratio of total operating hours to the number of failures, i.e., the reciprocal of the sum of the individual modular failure rates in a system.

Example (Note that operating time excludes downtime for repair.)

Module number	Interval between failure/repair, h
1	$\text{MTBM}_{ct1} = 1000$ (also MTBF_1)
2	$\text{MTBM}_{ct2} = 2000$ (also MTBF_2)
3	$\text{MTBM}_{ct3} = 3000$ (also MTBF_3)
4	$\text{MTBM}_{ct4} = 4000$ (also MTBF_4)

$$
\begin{aligned}
\text{MTBM}_{ct,\text{system}} &= \frac{1}{\dfrac{1}{\text{MTBM}_{ct1}} + \dfrac{1}{\text{MTBM}_{ct2}} + \dfrac{1}{\text{MTBM}_{ct3}} + \dfrac{1}{\text{MTBM}_{ct4}}} \\[2mm]
&= \frac{1}{\dfrac{1}{1000} + \dfrac{1}{2000} + \dfrac{1}{3000} + \dfrac{1}{4000}} \\[2mm]
&= \frac{1}{0.001 + 0.0005 + 0.00033 + 0.00025} \\[2mm]
&= \frac{1}{0.00208} = 480.77 \text{ h}
\end{aligned}
$$

Frequency of corrective maintenance

Corrective maintenance frequency is equivalent to the failure rate λ, and is determined on the basis of the same mathematical principle.

Example   If $MTBM_{ct} = MTBF = 2500$ h, then

$$\text{Corrective maintenance frequency} = \frac{1}{MTBM_{ct}} = \frac{1}{MTBF} = \lambda$$

The corrective maintenance frequency, also known as the failure rate, is

$$\lambda = \frac{1}{MTBF} = \frac{1}{2500} = 0.0004$$

Determining $\overline{M}_{ct}$ based on measurement of constituent component attributes

This approach involves calculating the $\overline{M}_{ct}$ of a higher-level assembly by evaluating the repair experience of its subsidiary physical elements. The mean corrective maintenance time $\overline{M}_{ct}$ of a system is calculated based on the $\overline{M}_{ct}$ values of the constituent subsystems:

$$\overline{M}_{ct} = \frac{\dfrac{1}{MTBF_1}(\overline{M}_{ct1}) + \dfrac{1}{MTBF_2}(\overline{M}_{ct2}) + \cdots + \dfrac{1}{MTBF_n}(\overline{M}_{ctn})}{\dfrac{1}{MTBF_1} + \dfrac{1}{MTBF_2} + \cdots + \dfrac{1}{MTBF_n}}$$

$$M_{ct,\text{system}} = \frac{\lambda_1(\overline{M}_{ct1}) + \lambda_2(\overline{M}_{ct2}) + \cdots + \lambda_n(\overline{M}_{ctn})}{\lambda_1 + \lambda_2 \cdots \lambda_n}$$

where n represents the final subsystem addressed in sequence.

Example   Calculate the $\overline{M}_{ct}$ of a system which consists of 3 subsystems as follows:

Subsystem	MTBF	$\overline{M}_{ct}$
1	1000	3
2	1500	4
3	2000	5

$$\overline{M}_{ct,\text{system}} = \frac{\dfrac{1}{1000}(3) + \dfrac{1}{1500}(4) + \dfrac{1}{2000}(5)}{\dfrac{1}{1000} + \dfrac{1}{1500} + \dfrac{1}{2000}}$$

$$= \frac{(0.001)(3) + (0.00067)(4) + (0.0005)(5)}{0.0001 + 0.00067 + 0.0005}$$

$$= \frac{0.003 + 0.00268 + 0.0025}{0.00217}$$

$$= 3.77$$

Preventive Maintenance

Concept of preventive maintenance

Preventive maintenance M_{pt} is planned, scheduled maintenance. A system is periodically removed from service for routine calibration, lubrication, cleaning, inspection, etc. A logistics manager establishes the preventive maintenance intervals and, therefore, the scheduled maintenance frequency.

Mean preventive maintenance time

Mean preventive maintenance time $\overline{M}_{pt}$ is determined by the logistics manager based on task times for individual scheduled maintenance actions. The principles of preventive maintenance apply equally to all constituent physical elements, higher-level assemblies, and the system itself.

For example, if the system manager establishes a program for routine servicing that entails 3.5 calendar hours at specified operational intervals, the $\overline{M}_{pt}$ is 3.5. It is noted that this factor is clock time, not total number of maintenance technician hours (e.g., the $\overline{M}_{pt}$ of 3.5 h could require two technicians to contribute 7 labor hours to the task).

Mean time between preventive maintenance actions

The mean time between preventive maintenance (MTBM$_{pt}$) is identical to the average of the planned intervals of operating hours between scheduled servicing. For higher-level assemblies, consideration must be given to the subassemblies and components.

Example If a logistics manager determines that an assembly requires routine lubrication every 1000 h of operation, inspection every 4000 h of operation, and calibration at 10,000 h intervals, the average scheduled maintenance interval is the reciprocal of the sum of the reciprocals of the individual intervals; therefore, if MTBM$_{pt1}$ = lubrication interval, MTBM$_{pt2}$ = inspection interval, and MTBM$_{pt3}$ = calibration interval, then for the assembly

$$\text{MTBM}_{pt} = \frac{1}{\dfrac{1}{\text{MTBM}_{pt1}} + \dfrac{1}{\text{MTBM}_{pt2}} + \dfrac{1}{\text{MTBM}_{pt3}}}$$

$$= \frac{1}{\dfrac{1}{1000} + \dfrac{1}{4000} + \dfrac{1}{10,000}}$$

$$= \frac{1}{0.001 + 0.00025 + 0.0001}$$

$$= \frac{1}{0.00135}$$

$$= 740.74 \text{ h}$$

Frequency of preventive maintenance actions

The scheduled maintenance frequency f_{pt} is the reciprocal of MTBM_{pt}. Using the results from the example above,

$$\text{System } f_{pt} = \frac{1}{\text{MTBM}_{pt}} = \frac{1}{740.74}$$

$$= 0.00135$$

The term f_{pt} is the M_{pt} companion factor to the M_{ct} failure rate λ.

Determining $\overline{M}_{pt}$ based on measurement of constituent component attributes.

As in the case of $\overline{M}_{ct}$ calculation, the calculation of $\overline{M}_{pt}$ for a higher-level assembly is based on calculation of scheduled maintenance planned for the subsidiary elements. The mean preventive maintenance time of a system is calculated based on the $\overline{M}_{pt}$ values of the constituent subsystems:

$$\overline{M}_{ct,\text{system}} = \frac{\dfrac{1}{\text{MTBM}_{pt1}}(\overline{M}_{pt1}) + \dfrac{1}{\text{MTBM}_{pt2}}(\overline{M}_{pt2}) + \cdots + \dfrac{1}{\text{MTBM}_{ptn}}(\overline{M}_{ptn})}{\dfrac{1}{\text{MTBM}_{pt1}} + \dfrac{1}{\text{MTBM}_{pt2}} + \cdots + \dfrac{1}{\text{MTBM}_{ptn}} +}$$

$$= \frac{f_{pt1}(\overline{M}_{pt1}) + f_{pt2}(\overline{M}_{pt2}) + \cdots + f_{ptn}(\overline{M}_{ptn})}{f_{pt1} + f_{pt2} + \cdots + f_{ptn}}$$

where n represents the final subsystem addressed in sequence.

Example Calculate the $\overline{M}_{pt}$ of a system which consists of three subsystems as follows:

Subsystem	MTBM_{pt}	$\overline{M}_{pt}$
1	2000	1.5
2	2500	2.0
3	2800	2.5

$$\overline{M}_{pt,\text{system}} = \frac{\dfrac{1}{2000}(1.5) + \dfrac{1}{2500}(2.0) + \dfrac{1}{2800}(2.5)}{\dfrac{1}{2000} + \dfrac{1}{2500} + \dfrac{1}{2800}}$$

$$= \frac{0.0005(1.5) + 0.0004(2.00) + 0.000357(2.5)}{0.0005 + 0.0004 + 0.000357}$$

$$= \frac{0.00075 + 0.0008 + 0.0008925}{0.001257}$$

$$= 1.943$$

Logistics Delay

Concepts governing logistics delay

Logistics delay pertains to the time other than actual repair time and scheduled maintenance time which contributes to the downtime of the equipment. Logistics delay as a quantitative parameter is a contributory element affecting maintenance downtime (MDT) and system operational availability A_o. The definition and techniques of measurement of logistics delay and its effect on logistics readiness are determined by the logistics manager for the system through identification and evaluation of the types of delay which have predominant effects on the maintenance turnaround time. Logistics delay is the result of all delay factors that are not attributable to actual maintenance actions, including such elements as

Administrative actions

Ordering and shipping time; e.g., for spare parts

Requisitioning of necessary skills

Research of technical data

Review and decision time

This is only a partial list of the multiplicity of factors that contribute to logistics delays.

It is incumbent upon the logistics manager to analyze the maintenance activity to determine the types of events that contribute critically to delay time. The approach in this text is predicated upon mathematical methodologies, based on measurement of individual elements of delay. In some cases, a logistics manager may aggregate the results of analyses and stipulate a single default factor for all systems within the manager's purview; this default factor would thereby be a standard factorial value applicable to all systems under the manager's control. The mathematical approach is essentially micro-managerial; the default approach represents the macro-managerial measurement concept. Either technique is applied in accordance with managerial judgment.

Mean logistics delay time

Mean logistics delay time $\overline{M}_{LD}$ is a quantitative reflection of the average elapsed time caused by delays. It stems from a series of observations of the elapsed time of individual events attributable to specific types of delay. The following case illustrates.

Example During a system operational period, the following events were recorded documenting administrative action time for determination of work center assignment. To determine the mean logistics delay time, define the attributes:

	Observation	Delay time recorded, h
$i = 1$	(1st event)	0.75
$i = 2$	(2nd event)	1.25
$i = 3$	(3rd event)	0.5
$i = 4$	(4th event)	1.75
$i = 5$	(5th event)	1.5
	$\Sigma_{i=1}^{i=5} \text{LOGDelay}$	5.75

When $n = 5$ (number of observed events)

$$\overline{M}_{LD,\text{admin}} = \frac{\sum_{i=1}^{i=5} \text{LOGDelay}}{n} = \frac{5.75}{5}$$

$$= 1.15 \text{ h}$$

The average administrative action delay during the operational period is 1.15 h.

Mean time between logistics delay

Mean time between logistics delay (MTBL) is the average interval between occurrences of an event causing delay. For example, if a system operating period encompasses 15,000 h and data search was contributory to 15 occurrences of delay, the MTBL is calculated as follows:

$$n = 15 \text{ (number of occurrences)}$$
$$\text{Operating period} = 15{,}000 \text{ h}$$

$$\text{MTBL}_{\text{data search}} = \frac{\text{operating period}}{n} = \frac{15{,}000 \text{ h}}{15}$$

$$= 1000 \text{ h}$$

Frequency of logistics delay events

The logistics delay frequency f_{LD} indicates the average rate of occurrence per operating hour of a specific type of delay. It is computed by determining the reciprocal of MTBL, the average interval between occurrences of the delay event. For example, using the results of the case above, f_{LD} is computed as follows:

$$f_{LD,\text{data search}} = \frac{1}{\text{MTBL}_{\text{data search}}} = \frac{1}{1000 \text{ h}}$$

$$= 0.001 \text{ h}$$

The average rate of occurrence of delays attributable to data search is 0.001 event per operating hour.

Determination of summary $\overline{M}_{LD}$ of a system based on contributory types of delay

The summary $\overline{M}_{LD}$ is based on calculation of the average delay time, taking into consideration the various individual causal factors of delay. The summary $\overline{M}_{LD}$ is therefore calculated from the $\overline{M}_{LD}$ of the individual contributory elements and the frequencies of their occurrence.

The generic formula for summary mean logistics delay time is

$$\overline{M}_{LD} = \frac{f_{LD1}(\overline{M}_{LD1}) + f_{LD2}(\overline{M}_{LD2}) + \cdots + f_{LDn}(\overline{M}_{LDn})}{f_{LD1} + f_{LD2} + \cdots + f_{LDn}}$$

where n is the last type of delay addressed in sequence. *Note*: The frequency f_{LD} of any type of delay is equal to the reciprocal of the mean time between occurrences of that delay:

$$f_{LD} = \frac{1}{\text{MTBL}}$$

The example provided in Fig. 3.2 illustrates a methodology for developing summary $\overline{M}_{LD}$.

Problem: Calculate the summary $\overline{M}_{LD}$ for a system based on the following empirical data on delays during a total of 10,000 h operation.

Given:

Type of delay	Number of occurrences	$\overline{M}_{LD}$
Data search (DatSch)	$n_{DatSch} = 190$	$\overline{M}_{LD,DatSch} = 2\,h$
Requisition skills (ReqSkls)	$n_{ReqSkls} = 150$	$\overline{M}_{LD,ReqSkls} = 3\,h$
Orders & shipping time for parts (O&ST)	$n_{O\&ST} = 50$	$\overline{M}_{LD,O\&ST} = 220\,h$

Step 1: Restructure data for MTBL and f_{LD}:

Type of delay	$\text{MTBL} = \dfrac{\text{total operating hours}}{\text{number of occurrences}}$	$f_{LD} = \dfrac{1}{\text{MTBL}}$
Data search	$\text{MTBL}_{DatSch} = \dfrac{10,000}{190} = 52.63$	$f_{LD,DatSch} = \dfrac{1}{52.63} = 0.019$
Requisition skills	$\text{MTBL}_{ReqSkls} = \dfrac{10,000}{150} = 66.67$	$f_{LD,ReqSkls} = \dfrac{1}{66.67} = 0.015$
Orders & shipping	$\text{MTBL}_{O\&ST} = \dfrac{10,000}{50} = 200$	$f_{LD,O\&ST} = \dfrac{1}{200} = 0.005$

Step 2: Apply $\overline{M}_{LD}$ algorithm·

Summary

$$\overline{M}_{LD} = \frac{(f_{LD,DatSch})(\overline{M}_{LD,DatSch}) + (f_{LD,ReqSkls})(\overline{M}_{LD,ReqSkls}) + (f_{LD,O\&ST})(\overline{M}_{LD,O\&ST})}{f_{LDDatSch} + f_{LDReqSkls} + f_{LDO\&ST}}$$

$$= \frac{(0.019)(2) + (0.015)(3) + (0.005)(220)}{0.019 + 0.015 + 0.005} = \frac{0.038 + 0.045 + 1.1}{0.039}$$

$$= 30.33\,h$$

Figure 3.2   Example of computation of summary $\overline{M}_{LD}$.

Aggregate Maintainability Indicators

Consolidation of measurement of constituent activities

The assessment of the total scope of maintenance effectiveness of a system requires the analytical amalgamation of the subsidiary maintainability indicators. The constituent activities and events defining the total maintenance purview are included in corrective maintenance, preventive maintenance, and logistics delay. Factors based on measurement of these individual areas contribute to the development of the aggregate indicators of maintainability, i.e.,

> Mean active maintenance time
> Mean time between maintenance
> Maintenance downtime

These critical parameters are mathematically derived from algorithmic application of the measures of effectiveness related to corrective maintenance, preventive maintenance, and logistics delay.

Mean active maintenance time

Mean active maintenance time $\overline{M}$ is a function of both corrective maintenance and preventive maintenance. This parameter reflects the average maintenance task time, taking into consideration mean corrective maintenance time $\overline{M}_{ct}$ and mean preventive maintenance time $\overline{M}_{pt}$. The equation for mean active maintenance time is

$$\overline{M} = \frac{\left(\dfrac{1}{\text{MTBM}_{ct}}\right)(\overline{M}_{ct}) + \left(\dfrac{1}{\text{MTBM}_{pt}}\right)(\overline{M}_{pt})}{\dfrac{1}{\text{MTBM}_{ct}} + \dfrac{1}{\text{MTBM}_{pt}}}$$

or

$$\overline{M} = \frac{(\lambda)(\overline{M}_{ct}) + (f_{pt})(\overline{M}_{pt})}{\lambda + f_{pt}}$$

Note: The failure rate λ is treated as the corrective maintenance frequency.

Example Determine the $\overline{M}$ of a system with the following attributes:

$$\begin{aligned}
\overline{M}_{ct} &= 3 \text{ h} \\
\text{MTBM}_{ct} &= 1000 \text{ h} \\
\overline{M}_{pt} &= 2 \text{ h} \\
\text{MTBM}_{p} &= 5000 \text{ h}
\end{aligned}$$

Therefore,

$$\overline{M} = \frac{\left(\dfrac{1}{\text{MTBM}_{ct}}\right)(\overline{M}_{ct}) + \left(\dfrac{1}{\text{MTBM}_{pt}}\right)(\overline{M}_{pt})}{\dfrac{1}{\text{MTBM}_{ct}} + \dfrac{1}{\text{MTBM}_{pt}}}$$

$$= \frac{\left(\dfrac{1}{1000}\right)(3) + \left(\dfrac{1}{5000}\right)(2)}{\dfrac{1}{1000} + \dfrac{1}{5000}}$$

$$= \frac{(0.001)(3) + (0.0002)(2)}{0.001 + 0.0002}$$

$$= \frac{0.003 + 0.0004}{0.0012} = \frac{0.0034}{0.0012}$$

$$= 2.833 \text{ h}$$

Mean time between maintenance

Mean time between maintenance (MTBM) is the average interval between maintenance actions, taking into consideration both mean time between corrective maintenance $MTBM_{ct}$ and mean time between preventive maintenance $MTBM_{pt}$. The algorithm for mean time between maintenance is

$$\text{MTDM} - \frac{1}{\dfrac{1}{\text{MTBM}_{ct}} + \dfrac{1}{\text{MTBM}_{pt}}}$$

or

$$\text{MTBM} = \frac{1}{\lambda + f_{pt}}$$

Note: The failure rate λ is treated as the corrective maintenance frequency.

Example If the scheduled maintenance interval of a system is 1200 operating hours and the failure analyses indicate an MTBF of 500 h, determine the MTBM of the system.

$$\text{MTBM} = \frac{1}{\dfrac{1}{\text{MTBM}_{ct}} + \dfrac{1}{\text{MTBM}_{pt}}}$$

$$= \frac{1}{\dfrac{1}{500} + \dfrac{1}{1200}}$$

$$= \frac{1}{0.002 + 0.000833} = \frac{1}{0.002833}$$

$$= 352.94 \text{ h}$$

Mean time between maintenance (MTBM) is both a maintainability and a reliability parameter. It is also a contributory factor to operational availability A_o and achieved availability (A_a).

Maintenance downtime

Maintenance downtime MDT represents the total system downtime attributable to maintenance actions and maintenance-related events. It takes into consideration the contributory effects of corrective maintenance, preventive maintenance, and logistics delay periods. This methodology incorporates the intervals, mean task or event times, and frequencies of occurrence in these critical areas of effectiveness measurement. The algorithm for maintenance downtime MDT) is

$$MDT = \frac{\dfrac{1}{MTBF}(\bar{M}_{ct}) + \dfrac{1}{MTBM_{pt}}(\bar{M}_{pt}) + \dfrac{1}{MTBL}(\bar{M}_{LD})}{\dfrac{1}{MTBF} + \dfrac{1}{MTBM_{pt}} + \dfrac{1}{MTBL}}$$

or

$$MDT = \frac{\lambda(\bar{M}_{ct}) + f_{tp}(\bar{M}_{pt}) + f_{LD}(\bar{M}_{LD})}{\lambda + f_{pt} + f_{LD}}$$

Example Calculate the MDT of a system with the following maintainability elements

MTBF $= 1200$	$\lambda = 0.00083$	$M_{ct} = 2$
$MTBM_{pt} = 2500$	$f_{pt} = 0.0004$	$M_{pt} = 3$
MTBL $= 1500$	$f_{LD} = 0.00066$	MLD $= 100$

$$\begin{aligned} MDT_{system} &= \frac{\lambda(\bar{M}_{ct}) + f_{pt}(\bar{M}_{pt}) + f_{LD}(\bar{M}_{LD})}{\lambda + f_{pt} + f_{LD}} \\ &= \frac{(0.00083)(2) + (0.0004)(3) + (0.00066)(100)}{0.00083 + 0.0004 + 0.00066} \\ &= \frac{0.00166 + 0.0012 + 0.066}{0.00189} \\ &= 36.434 \end{aligned}$$

MDT is the predominant factor affecting operational availability A_o.

4

Availability

Overview

Definition of availability

Availability is the measure of the readiness of a system to fulfill its assigned function. Whereas reliability defines the probability of a system's successfully completing its operational mission after having been engaged, availability, as a preengagement indicator, measures the capability of a system to be committed to operation. Availability is sensitive to and determined by the trade-off between system reliability and maintainability.

Availability is the measurement of the relationship of the system uptime to the total active time of the system. The uptime determines the average time the system is available. The *total active time* is the uptime plus the downtime. Fig. 4.1 illustrates this concept. When a system is up, it is ready for commitment to operation. When the system is down, it is Lot ready for operational engagement because of required maintenance and the attendant technical and administrative delays.

Availability parameters

The availability of a system is expressed by three individual figures of merit.

Inherent availability A_i
Achieved availability A_a
Operational availability A_o

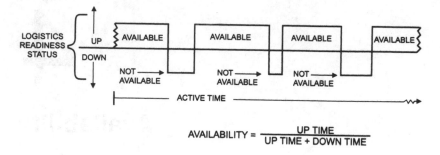

$$\text{AVAILABILITY} = \frac{\text{UP TIME}}{\text{UP TIME} + \text{DOWN TIME}}$$

Figure 4.1   System availability concept.

The three methods of measuring availability correlate to the evaluation of a system from engineering design through its operational phases. Inherent availability is a system design parameter, achieved availability applies to system test and evaluation, and operational availability measures system readiness after deployment in its actual operational environment.

Inherent Availability

Inherent availability A_i defines the readiness of a system when evaluated under the conditions of an ideal environment. This presumes readily available logistics resources (e.g., maintenance physical facilities, technical skills, support and test equipment, spares and repair parts). The mathematical computation of inherent availability includes the corrective maintenance intervals, or MTBF, and mean corrective maintenance time $\overline{M}_{ct}$. It excludes preventive maintenance time and logistics delays. Inherent availability thresholds are typically incorporated in system specifications invoked as contractual requirements.

A_i algorithm

The inherent availability formula is

$$A_i = \frac{\text{MTBF}}{\text{MTBF} + \overline{M}_{ct}}$$

where MTBF = mean time between failure
$\overline{M}_{ct}$ = mean corrective maintenance time

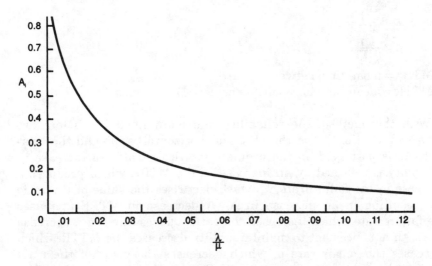

Figure 4.2   Inherent availability as a function of λ/μ.

Example   Determine A_i for a system with mean time between failure of 1000 operating hours and mean corrective maintenance time of 2.5 h.
If MTBF = 1000 and $\overline{M}_{ct} = 2.5$, then

$$A_i = \frac{\text{MTBF}}{\text{MTBF} + \overline{M}_{ct}} = \frac{1000}{1000 + 2.5}$$

$$= \frac{1000}{1002.5}$$

$$= 0.9975$$

Inherent availability: tradeoff techniques

Inherent availability, A_i, serves as a criterion for sensitivity analysis of reliability and maintainability attributes as well as a parameter governing system readiness design considerations. Inherent availability reflects and is sensitive to variations in factors derived from failure factors and those derived from repair factors. The following are selected applications of A_i trade-off techniques.

A_i as a function of failure rate and repair rate.   Fig. 4.2 illustrates a technique for portraying inherent availability as a function of the failure rate, λ, and the repair rate, μ, i.e.,

$$A_i = \frac{\lambda}{\mu}$$

where $\lambda = \frac{1}{MTBF}$

   $\mu = \frac{1}{MTTR}$

MTBF = mean time between failure
MTTR = mean time to repair (same as $\overline{M}_{ct}$)

With this method, the reliability and maintainability factors are shown as a ratio, λ/μ, on the abscissa (horizontal) scale and the variable, A_i, is portrayed on the ordinate (vertical) scale. An increase in reliability is reflected by an increase in the MTBF, which produces a decrease in the failure rate, λ, which decreases the value of the ratio λ/μ, resulting in an increase in A_i. A decrease in MTBF increases the failure rate, which increases the ratio λ/μ, resulting in a decrease in A_i. An enhancement to maintainability decreases the MTTR, which increases the repair rate μ, which decreases the value of the ratio λ/μ, resulting in an increase in A_i. Degradation in maintainability would increase the MTTR, which would decrease the repair rate, which would increase the value of the ratio λ/μ, resulting in degradation of A_i.

A_i as a function of MTBF and repair rate.   Fig. 4.3 describes a method of plotting A_i as a functional line determined by variation in the values of MTBF and the repair rate μ. With this method, a fixed A_i target or a number of alternative A_i target values are each described by a *curve of indifference*. To plot each A_i curve, a series of points are determined by holding the target A_i constant and computing the repair rate μ by varying the values of MTBF to determine the corresponding values of MTTR.

Each A_i curve of indifference is plotted as follows:

1. Use A_i = MTBF/(MTBF + MTTR) as the governing relationship.
2. Holding A_i constant, calculate the repair rate μ (or 1/MTTR) for varying values of MTBF.
3. Plot the point of intersection of each ordinate scale value of the MTBF and abscissa scale value of the calculated repair rate μ.
4. Connect the series of points to construct the curve of indifference for the designated A_i target value.

Example   Assume the A_i target is 0.99 and MTBF = 70 is applied.

$$A_i = \frac{MTBF}{MTBF + MTTR}$$

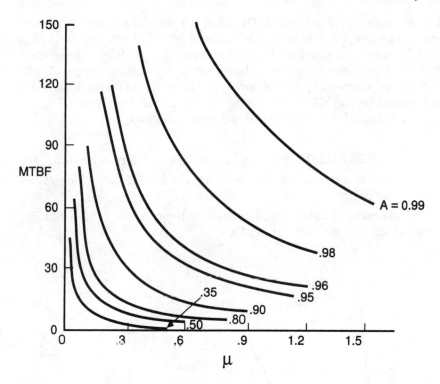

Figure 4.3 Target A_i curve of indifference governed by MTBF and repair rate μ, where $\mu = 1/\text{MTTR}$.

To determine MTTR,

$$0.99 = \frac{70}{70 + \text{MTTR}}$$
$$(0.99)(70 + \text{MTTR}) = 70$$
$$(0.99)(70) + 0.99\text{MTTR} = 70$$
$$0.99\text{MTTR} = 70 - (0.99)(70)$$
$$\text{MTTR} = \frac{70 - 69.3}{0.99} = \frac{0.7}{0.99}$$
$$\text{MTTR} = 0.707$$
$$\mu = \frac{1}{0.707}$$
$$\mu = 1.414$$

Therefore, where MTBF is 70 h, the repair rate μ is 1.414 for a fixed A_i of 0.99.

By determining a series of plot points by varying MTBF values and calculating the corresponding repair rates μ, a curve of indifference for a target A_i of 0.99 would be constructed.

A_i as a function of MTBF and MTTR. Fig. 4.4 portrays a method of plotting A_i as a graphic function of MTBF and MTTR. With this method, a fixed A_i target or a number of alternative A_i target values are described by lines of indifference, each of which emanates from the origin. To plot each A_i line, a series of points is derived by holding the target A_i constant and computing the MTTR by varying the values of the MTBF.

Each A_i line of indifference is developed as follows:

1. Use A_i = MTBF/(MTBF + MTTR) as the governing relationship.
2. Holding A_i at a constant value, calculate the MTTR for varying values of the MTBF.
3. Plot the point of intersection of each ordinate scale value of MTBF and abscissa scale value of MTTR.

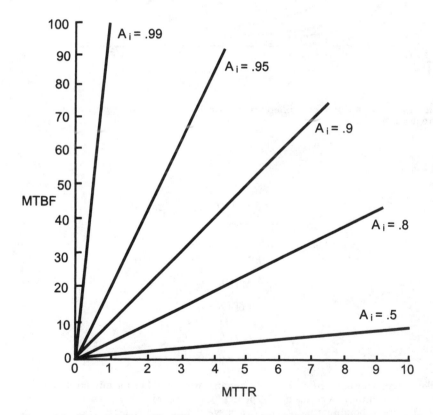

Figure 4.4   Target A_i line of indifference governed by MTBF and MTFR.

4. Connect the series of points to construct the A_1 line of indifference. It is noted that at the origin,

$$A_i = 0$$
$$\text{MTBF} = 0$$
$$\text{MTTR} = 0$$

Example If the A_i target is 0.9 and MTBF = 50 is applied, to determine the corresponding MTTR,

$$0.9 = \frac{50}{50 + \text{MTTR}}$$
$$0.9(50 + \text{MTTR}) = 50$$
$$(0.9)(50) + 0.9\text{MTTR} = 50$$
$$0.9\text{MTTR} = 50 - (0.9)(50)$$
$$\text{MTTR} = \frac{50 - 45}{0.9} = \frac{5}{0.9}$$
$$\text{MTTR} = 5.56$$

By determining a series of points by varying MTBF values and calculating the corresponding MTTR values, a line of indifference for target $A_i = 0.9$ could be constructed.

A_i as a function of failure rate and repair rate. Fig. 4.5 describes a method of plotting A_i as a graphic function of the failure rate λ and the repair rate μ. The approach defined by this method is similar to that for MTBF and MTTR. The essential difference is that the failure rate λ and repair rate μ are, respectively, the reciprocals of MTBF and MTTR. As a consequence of this characteristic, the order of ascending values of the A_i lines in the two methodologies is reversed. Whereas when MTBF and MTTR are used, the A_i line values increase in a counterclockwise direction within the graphic quadrant (see Fig. 4.4), use of failure rate λ and repair rate μ causes the A_i line values to increase in a clockwise direction (see Fig. 4.5).

Each line of indifference is determined as follows:

1. Use $A_i = \text{MTBF}/(\text{MTBF} + \text{MTTR})$ as the governing relationship.
2. Holding A_i constant, compute MTTR using varying values of MTBF.
3. Convert MTBF values to failure rates (that is, $\lambda = 1/\text{MTBF}$). Convert MTTR values to repair rates (that is, $\mu = 1/\text{MTTR}$).
4. Determine the point of intersection of each ordinate scale value for failure rate λ and abscissa scale value for the repair rate μ.

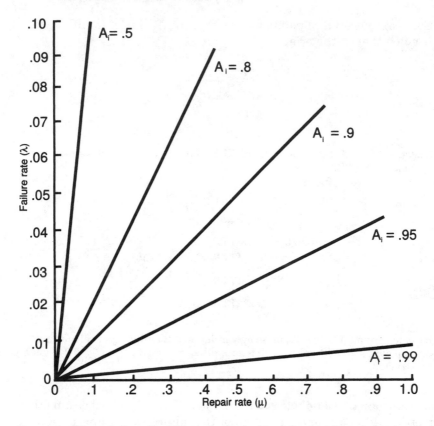

Figure 4.5   Target A_i line of indifference governed by failure rate λ and repair rate μ.

Example   The first step parallels the procedures for the preceding example. For this example, the following factors were determined.

$$A_i = 0.9$$
$$\text{MTBF} = 50$$
$$\text{MTTR} = 5.56$$

Based on the attributes determined by incorporating MTBF and MTTR in the A_i formula, determine the failure rate λ and repair rate μ values as follows:

$$\lambda = \frac{1}{\text{MTBF}} = \frac{1}{50} = 0.02$$

$$\mu = \frac{1}{\text{MTTR}} = \frac{1}{5.56} = 0.18$$

Thus, with respect to an A_i target of 0.9, $\lambda = 0.02$ and $\mu = 0.18$.

By determining a series of plot points derived by varying the MTBF and failure rate values and calculating the corresponding MTTR and repair rate values, a line of indifference for target $A_i = 0.9$ could be constructed.

Reliability-maintainability trade-off. Fig. 4.6 illustrates the method of graphically plotting system reliability, maintainability, and availability (R, M, & A) factors to define parametric constraints stipulated by performance specifications. Typically specified are the target A_i, minimum MTBF, and maximum MTTR. The procedures for constructing reliability (minimum MTBF) and maintainability (maximum MTTR) constraint parameters are:

1. Construct a graph, using the ordinate scale to plot MTTR and the abscissa scale to plot MTBF.
2. Establish target A_i, minimum MTBF, and maximum MTTR.

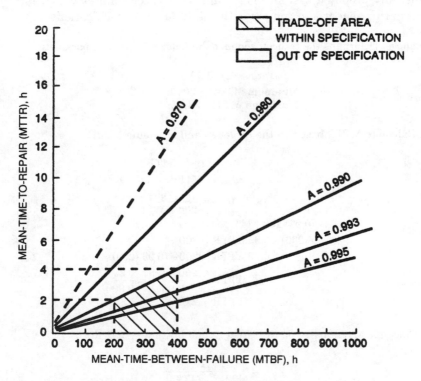

Figure 4.6   Reliability-maintainability trade-off zone of acceptability. Target $A = 0.99$, minimum MTBF = 200 h, and maximum MTTR = 4 h.

3. Calculate the plot points for the A_i, line of indifference by holding target A_i constant and using the A_i formula to
   a. Calculate the MTTR by incorporating the minimum MTBF under conditions of the stated A_i target value.
   b. Calculate the MTBF by incorporating the maximum MTTR under conditions of the stated A_i target value.

4. Draw a line from the intersection of the ordinate and abscissa scales through the two previously constructed plot points. This defines the target A_i line of indifference.

5. Determine the tradeoff area as follows:
   a. Draw a perpendicular line from the minimum MTBF point on the abscissa scale to the target A_i line.
   b. Draw a perpendicular line from the MTBF point on the abscissa scale determined by incorporating the maximum MTTR to the target A_i line.

6. The area of the graph *under* the target A_i line of indifference, above the abscissa scale, and *between* the two perpendicular lines is the trade-off zone. Any proposed system MTBF and MTTR coordinates which fall within this zone would comply with the R, M, & A specifications.

Example   Determine the tradeoff zone for the following specified parameters.

$$\text{Target } A_i = 0.99$$
$$\text{Minimum MTBF} = 200 \text{ h}$$
$$\text{Maximum MTTR} = 4.0 \text{ h}$$

1. Calculate MTTR based on the target A_i and minimum MTBF.

$$A_i = \frac{\text{MTBF}}{\text{MTBF} + \text{MTTR}}$$
$$0.99 = \frac{200}{200 + \text{MTTR}}$$
$$(0.99)(200 + \text{MTTR}) = 200$$
$$(0.99)(200) + 0.99\text{MTTR} = 200$$
$$0.99\text{MTTR} = 200 - (0.99)(200)$$
$$\text{MTTR} = \frac{200 - 198}{0.99} = \frac{2}{0.99}$$
$$\text{MTTR} = 2.02$$

2. Calculate MTBF based on the target A_i and maximum MTTR.

$$A_i = \frac{\text{MTBF}}{\text{MTBF} + \text{MTTR}}$$
$$0.99 = \frac{\text{MTBF}}{\text{MTBF} + 4}$$

$$0.99(\text{MTBF} + 4) = \text{MTBF}$$
$$0.99\text{MTBF} + (0.99)(4) = \text{MTBF}$$
$$(0.99)(4) = \text{MTTR} - 0.99\text{MTBF}$$
$$3.96 = 0.01\text{MTBF}$$
$$\text{MTBF} = \frac{3.96}{0.01}$$
$$\text{MTBF} = 396$$

Thus, the plot points for a target A_i of 0.99, minimum MTBF of 200 h, and maximum MTTR of 4 h are MTBF = 0, MTTR = 0 (origin coordinate); (minimum) MTBF = 200, MTTR = 2.02; and MTBF = 396, (maximum) MTTR = 4. The tradeoff zone is as described in Fig. 4.6.

Achieved Availability

Achieved availability A_a defines system readiness under the same ideal environment prescribed for inherent availability A_i. The quantitative approach for achieved availability differs from that for inherent availability in that

1. Mean preventive maintenance time $\overline{M}_{pt}$ is considered as well as mean corrective maintenance time $\overline{M}_{ct}$ and $\overline{M}_{ct}$ are consolidated into mean active maintenance time $\overline{M}$.
2. Mean time between preventive maintenance MTBM_{pt} is considered in addition to mean time between failure MTBF. MTBM_{pt} and MTBF are consolidated into mean time between maintenance MTBM.

A_a is applicable for use in system test and evaluation.

A_a Algorithm

The achieved availability formula is

$$A_a = \frac{\text{MTBM}}{\text{MTBM} + \overline{M}}$$

where MTBM = mean time between maintenance
$\overline{M}$ = mean active maintenance time

Example Calculate the achieved availability for a system with the following attributes: $\overline{M}_{ct} = 2.4$, $\overline{M}_{pt} = 3.0$, MTBF = 1200, and $\text{MTBM}_{pt} = 2000$.

1. Compute $\overline{M}$.

$$\overline{M} = \frac{\left(\dfrac{1}{\text{MTBF}}\right)(\overline{M}_{ct}) + \left(\dfrac{1}{\text{MTBM}_{pt}}\right)(\overline{M}_{pt})}{\dfrac{1}{\text{MTBF}} + \dfrac{1}{\text{MTBM}_{pt}}}$$

$$= \frac{\left(\dfrac{1}{1200}\right)(2.4) + \left(\dfrac{1}{2000}\right)(3.0)}{\dfrac{1}{1200} + \dfrac{1}{2000}}$$

$$= \frac{(0.0008)(2.4) + (0.0005)(3)}{0.0008 + 0.0005} = \frac{0.00192 + 0.0015}{0.0013}$$

$$= 2.63$$

2. Compute MTBM.

$$\text{MTBM} = \frac{1}{\dfrac{1}{\text{MTBF}} + \dfrac{1}{\text{MTBM}_{pt}}} = \frac{1}{\dfrac{1}{1200} + \dfrac{1}{2000}}$$

$$= \frac{1}{0.0008 + 0.0005} = \frac{1}{0.0013}$$

$$= 769.23$$

3. Compute A_a.

$$A_a = \frac{\text{MTBM}}{\text{MTBM} + \overline{M}}$$

$$= \frac{769.23}{769.23 + 2.63} = \frac{769.23}{771.86}$$

$$= 0.9966$$

Operational Availability

Operational availability A_o measures the readiness of a system in its actual operational environment. The A_o parameter mathematically consolidates mean time between maintenance (MTBM) and maintenance downtime (MDT). It considers the impacts of mean corrective maintenance time $\overline{M}_{ct}$; mean preventive maintenance time $\overline{M}_{pt}$, mean logistics delay time $\overline{M}_{LD}$, mean time between failure MTBF, mean time between preventive maintenance MTBM_{pt}, and mean time between logistics delay MTBL. Operational availability is the functional responsibility of and relates to the logistics readiness goal stipulated by the user of the system.

A_o algorithm

The operational availability formula is

$$A_o = \frac{\text{MTBM}}{\text{MTBM} + \text{MDT}}$$

where MTBM = mean time between maintenance
      MDT = maintenance downtime

Example Determine the operational availability of a system based on the following: $\overline{M}_{ct} = 2.0$, $\overline{M}_{pt} = 3.5$, $\overline{M}_{LD} = 50.0$, $\text{MTBM}_{pt} = 3000$, $f_{pt} = 0.00033$, $\text{MTBM}_{ct} = 1000$, $\lambda = 0.001$, $\text{MTBL} = 1500$, and $f_{LD} = 0.00066$.
Note: The failure rate λ is treated as the corrective maintenance frequency.

1. Calculate MTBM.

$$\text{MTBM} = \frac{1}{\lambda + f_{pt}} = \frac{1}{0.001 + 0.00033}$$

$$= \frac{1}{0.00133}$$

$$= 751.87$$

2. Compute MDT.

$$\text{MDT} = \frac{(\lambda)(\overline{M}_{ct}) + (f_{pt})(\overline{M}_{pt}) + (f_{LD})(\overline{M}_{LD})}{\lambda + f_{pt} + f_{LD}}$$

$$= \frac{(0.001)(2) + (0.00033)(3.5) + (0.00066)(50)}{0.001 + 0.00033 + 0.00066}$$

$$= \frac{0.002 + 0.001155 + 0.033}{0.00199} = \frac{0.036155}{0.00199}$$

$$= 18.168$$

3. Determine A_o.

$$A_o = \frac{\text{MTBM}}{\text{MTBM} + \text{MDT}}$$

$$= \frac{751.87}{751.87 + 18.168} = \frac{751.87}{770.038}$$

$$= 0.976$$

Quality Assurance

Overview

Quality assurance is a systems engineering discipline which embodies the process of quality control; the art of inspection; and the management philosophy, policies, and oversight procedures that will instill understanding, integration of individual objectives, and supportiveness of quality at all levels of the organization. Quality control and inspection are governed by statistical analysis methodologies applied to measurement, examination, testing, and other appropriate methods of comparison of the output of the enterprise with the stipulated product requirements. The focus of management is on sustaining a quality-oriented attitude and performance on the part of everyone in the enterprise who contributes value to the product. The ANSI/ASQ Z1.4-2003 Standard and MIL-STD-105E, each entitled *Sampling Procedures and Tables for Inspection by Attributes*, set forth the basic procedures, tables, and sampling planning guidance employed within the Department of Defense (DoD) and throughout private industry. Although MIL-STD-105E was officially cancelled in February 1995 by the U.S. Air Force, the same technical guidance, supporting tables, and data provided by MIL-STD-105E have been incorporated into the ANSI/ASQ Z1.4-2003 Standard, which was distributed by the American Society for Quality (ASQ) in January 2004. The textual content of ANSI/ASQ Z1.4-2003 is nearly identical to MIL-STD-105E. It was therefore determined by DoD to cancel MIL-STD-105E in order to reduce costs through the elimination of duplication with civilian standards. It is noteworthy that many companies continue to use MIL-STD-105E as the standard for inspection. In view of the significant similarities between the military and civilian standards, the source of the tables and technical guidance provided in this chapter will, where appropriate, be cited as "ANSI/ASQ Z1.4-2003/MIL-STD-105E." The material in this chapter has been developed by synthesis and tailoring of pertinent sections of ANSI/ASQ Z1.4-2003/MIL-STD-105E. Applicable tables from ANSI/ASQ Z1.4-2003/ MIL-STD-105E, as well as case studies, are included as an appendix to this chapter.

Quality assurance applications

This chapter focuses on the basic techniques and quantitative elements of quality assurance. These elements include criteria for sample inspection by attributes, sampling procedures, and sampling plans. The methods described are applicable, but not limited, to the following types of products and activities:

1. *End items*. These are completed products; they may be inspected before or after packaging and packing for shipment or storage.

2. *Components and raw materials*. These are the materials which are shaped, treated, or assembled to form the end items.

3. *Operations*. In many cases, repetitive work performed by machines and operators can be judged to be acceptable or unacceptable.

4. *Materials in process*. Materials may be inspected on a sampling basis to determine their quality after any step along the production line.

5. *Supplies in storage*. Attribute sampling inspection procedures can be used to determine the quality of supplies in storage.

6. *Maintenance operations*. These operations are usually performed on reparable materials to restore them to a serviceable condition.

7. *Data or records*. Whenever large volumes of data are processed (i.e., accounting records, cost data, maintenance tickets, invoices, bills of lading, etc.), attribute sampling inspection procedures can be used as the basis for determining the quality of the data or records, using a measure such as monetary volume, item count, or accuracy.

8. *Administrative procedures*. If the results of administrative procedures can be measured on an attribute basis, attribute sampling procedures can be applied.

Approaches to quality control

The approaches to quality control are reflected by sampling methodologies. The individual methodologies are based upon some aspect of producer and/or consumer protection offered by the various sampling plans. Following are some of the most commonly used methods of indexing or grouping sampling plans:

1. Indifference quality level, where probability of acceptance $P_a = 0.5$
2. Limiting quality (LQ) protection
3. Average outgoing quality limit (AOQL)
4. Acceptable quality level (AQL)

Indifference quality (P_a = 0.5).   Sampling plans based on the indifference quality are commonly called 50 percent plans. The indifference quality is that level of lot quality at which the probability of acceptance P_a is 0.5. The level of lot quality at which $P_a = 0.5$ depends, of course, upon the sampling plan being used for lot acceptance. A single sampling plan for the indifference quality level can be computed very easily by using the following approximate equation:

$$n = \frac{100c + 67}{\text{indifference quality (in percent defective)}}$$

where n = sample size
c = acceptance number

For example, if a product that is 3 percent defective should be accepted with a probability of 50 percent and an acceptance number of 2 defectives is to be used, the sample size is computed as follows:

$$n = \frac{100c + 67}{\text{indifference quality (in percent defective)}}$$

$$= \frac{(100 \times 2 + 67)}{3}$$

$$= \frac{267}{3}$$

$$= 89$$

The single sampling plan would be to draw a sample of 89 units at random from the lot. If 2 or fewer defectives are found, accept the lot. As long as the consumer and supplier do not care about their own specific risks at the quality level that divides tolerable quality from intolerable quality, this is a very simple way to perform sampling inspection.

Limiting quality (LO).   The protection provided to the consumer by a sampling plan is usually described by the term *consumer's risk*. The consumer's risk is the probability of accepting a lot whose quality is at or below a level which the customer can tolerate only a small part of the time. This quality level is called the limiting quality, LQ, or lot tolerance percent defective, LTPD. Sampling plans called LQ sampling plans may be devised to provide a specified LQ protection or consumer's risk protection when the product quality is at the LQ.

A typical example of an LQ sampling plan is one based on a statement by the consumer that he or she is willing to accept a maximum of 6.5 percent defective (LQ = 6.5 percent) no more than 5 percent (consumer's risk = 5 percent) of the time.

Average outgoing quality limit (AOQL).   The average outgoing quality (AOQ) is the average quality of outgoing product, including all accepted lots or batches plus all rejected lots or batches after they have been effectively screened and defectives removed or replaced by nondefectives.

Acceptable quality level (AOL).   The protection provided to the producer (supplier) by a sampling plan is usually given in terms of *producer's risk*. The producer's risk is the probability that a lot of acceptable quality is rejected, and the producer is "protected" when this probability is low. While the producer's risk is of interest, especially to the producer, at all levels of good quality, it is common practice to be especially interested in the producer's risk at the lowest level of acceptable quality. Because of the extensive use and application of the AQL method in government and private industry, this chapter will address the AQL approach so as to provide a threshold for basic understanding and further development of the reader's insight into quality assurance discipline.

Basic terms of quality assurance

Acceptable quality level (AQL)   When a continuous series of lots is considered, the AQL is the quality level which, for the purposes of sampling inspection, is the limit of a satisfactory process average. A sampling plan and an AQL are chosen in accordance with the risk assumed. The AQL represents a designated percent defective (or number of defects per 100 units) for which lots will be accepted most of the time with the sampling procedure being used. The sampling plans are arranged in such a way that the probability of acceptance at the designated AQL value depends upon the sample size; it is generally higher for large samples than for small ones, for a given AQL. Within the international community, the term AQL has also been referred to as the *assured quality level* and *acceptable quality limit*.

Attribute   An attribute is a characteristic or property which is appraised in terms of whether it does or does not meet a given requirement, e.g., too big or not too big; hard enough or not hard enough.

Defect   A defect is any nonconformance of the unit of product with specified requirements.

Defective   A defective is a unit of product which contains one or more defects.

Defects per hundred units   The number of defects per 100 units is 100 times the number of defects contained in any given quantity of units of products (with one or more defects being possible in any unit of product) divided by the total number of units of product:

$$\text{Defects per 100 units} = \frac{\text{number of defects} \times 100}{\text{number of units inspected}}$$

Inspection   Inspection is the process of measuring, examining, testing, or otherwise comparing the unit of product with the requirements.

Inspection by attributes   Inspection by attributes is inspection in which certain characteristics of units of product are classified simply as defects or nondefects. Any unit of product found to have one or more defects is classified as a defective. Under attributes inspection, characteristics of the units of product are considered on the basis of "go or not go"—defective or nondefective, within tolerance or out of tolerance, correct or incorrect, complete or incomplete, etc.

Lot or batch   The term *lot* or *batch* means inspection lot or inspection batch, i.e., a collection of units of product *from which a sample is to be drawn* and inspected. This may differ from the collection of units designated as a lot or batch for other purposes (e.g., production and shipment).

Lot or batch size   The lot or batch size is the number of units of product in a lot or batch.

Percent defective   The percent defective of any given quantity of units of product is 100 times the number of defective units of product contained therein divided by the total number of units of product:

$$\text{Percent defective} = \frac{\text{number of defects} \times 100}{\text{number of units inspected}}$$

Sample   A sample consists of one or more units of product selected at random from a lot or batch without regard to their quality. The number of units of product in the sample is the sample size.

Sample size code letter   A sample size code letter is a factor used along with the AQL to locate a sampling plan on a table of sampling plans.

Sampling plan A sampling plan indicates the number of units of product from each lot or batch which are to be inspected (sample size or series of sample sizes) and the criteria for determining the acceptability of the lot or batch (acceptance and rejection numbers).

Total quality management *(TQM)* Total quality management is a system of management philosophy, policy, and procedures for the organizational engineering, production, logistics, marketing, and service groups, developed to afford full customer satisfaction consistent with sound principles of economy and cost-effectiveness.

Unit of product A unit of product is an entity that is inspected in order to determine its classification as defective or nondefective or to count the number of defects. It may be a single article, a pair, a set, a length, an area, a maintenance procedure, an operation, a volume, a component of an end product, or an end product itself.

Applicability of Statistically Based Quality Control

The quantitative methodologies for quality control and inspection are applicable at any point of the production process—to product components, workstation processes, final product factory acceptance testing, or first article or prototype system testing.

Quality control based on statistical techniques can reinforce an organizational TQM program by providing a safety net to the degree deemed necessary to achieve the confidence level stipulated by management. TQM-oriented organizations can apply statistical quality control methodologies to final evaluation of randomly selected production units or, if they have a high level of confidence in the production process, use alternative methods of oversight. An organization without TQM might rely heavily on a comprehensive quality control and inspection program based on statistical analysis.

Because of customers' sensitivity and adverse reaction to defective products, it is considered advisable to maintain some form of statistically based system of screening out defective units so as to assure continued customer confidence and satisfaction.

Inspection

General

Inspection is the examination or testing of supplies and services (including, when applicable, raw materials, documents, data, components, and intermediate assemblies) to determine whether they conform to technical and contractual requirements. The inspection criteria used to determine whether the quality requirements have been met are stated in appropriate documents, such as purchase descriptions, project descriptions, inspection instructions, technical orders, drawings, technical bulletins, and process specifications.

Amount of inspection

Before deciding how much inspection should be done in a particular situation, it is important to realize that for a given acceptance criterion, the less the inspection that is done, the greater the risk that nonconforming products will be accepted. Also, before a decision on the amount of inspection can be made, how units of product will be submitted for inspection—whether on a lot-by-lot basis or on a unit-by-unit basis—must be determined. Once this has been done, the following factors are among the most important to consider when deciding the amount of inspection:

1. The type of product to be inspected
2. The quality characteristics to be examined for conformance
3. The quality history of the producer
4. The cost of inspection
5. The effect of inspection upon the unit of product (i.e., destructive or nondestructive)

The question of how much inspection should be done is related to the problem of selecting a sampling plan. Because of the interdependence of these two questions, they should be studied and considered together.

One hundred percent inspection.   One hundred percent inspection is the inspection of every unit of product (procedure, data, operations, etc.). In some cases of 100 percent inspection, the accept/reject decision will be made not for the entire lot, but for each unit individually, based upon the results of inspecting the unit for the quality characteristics concerned. For critical quality characteristics, 100 percent inspection or inspection of relatively large samples is usually required to assure the desired quality protection. One hundred percent inspection cannot be specified when inspection is destructive and is not likely to be specified when the individual tests are expensive or take extremely long periods of time—for example, qualification and environmental tests. One hundred percent testing can always be specified for nondestructive tests—that is, for tests in which the characteristic can be measured without damaging the product.

The obvious advantage of 100 percent inspection is that it gives a better indication of product quality than does sampling inspection. However, it does not guarantee detection of all defects, especially when the inspection is done by human inspectors. The direct costs of 100 percent inspection will generally be greater than the costs of sampling inspection. However, the cost of permitting a defect to go undetected may be so great that the cost of 100 percent inspection is justified.

Sampling inspection.   In sampling inspection, a sample consisting of one or more (but not all) units of product is selected at random from the production process output and examined for one or more quality characteristics. Sampling inspection is usually the most practical and economical means of determining the conformance or nonconformance of product to specified quality requirements. Sampling inspection costs are typically lower than 100 percent inspection costs since sampling inspection does not require that every unit of product be inspected for conformance with specified quality requirements. When determining which inspection method is to be employed, the lower costs of sampling inspection must be weighed against the risk of greater cost incurred by permitting defective units of product to be accepted:

$$P_b = \frac{\text{cost to inspect one piece}}{\text{damage done by one defective}}$$

where P_b = breakeven fraction defective

If $P_b < 1.0$, inspection is cost-effective; $P_b > 1.0$, inspection is not cost-effective.

Severity of inspection

The severity of inspection involves the total amount of inspection and the accept/reject criterion specified by the quality assurance provisions established for the unit of product or dictated by quality history. Sampling inspection generally contemplates three degrees of severity:

1. Tightened inspection
2. Normal inspection
3. Reduced inspection

Normal inspection.   Normal inspection is that which is used when there is no evidence that the quality of product being submitted is better or worse than the specified quality level. Normal inspection is usually used at the start of inspection and is continued as long as there is evidence that the product quality is consistent with the specified requirements. Tightened inspection is instituted in accordance with established procedures when it becomes evident that product quality is deteriorating.

Reduced inspection may be instituted in accordance with established procedures when it is evident that product quality is very good.

Tightened inspection. Tightened inspection under a sampling scheme involves a more stringent acceptance criterion than that of the normal inspection plan with which it is used. This requirement is usually met by decreasing the acceptable number of defectives or defects per hundred units produced for the sample. The effect of this decrease is generally to increase the producer's risk while reducing the consumer's risk.

Reduced inspection. Reduced inspection normally involves a smaller sample size than that for normal inspection under the same sampling scheme. The effect of this decrease in sample size is to slightly reduce the producer's risk while significantly increasing the consumer's risk. A proven history of good quality for the product is essential for application of reduced inspection procedures.

Inspection levels

Inspection levels in general provide the quality engineer with a means of selecting one of several sample size code letters for a given lot size. The effect is to offer several sampling plans, each with approximately the same probabilities of acceptance in the region of good quality (AQL or better), but with differing probabilities of acceptance when lot quality is worse than the AQL. Table I of the appendix to this chapter gives three general inspection levels, numbered I, IJ, and III, and four special inspection levels, numbered S-1, S-2, S-3, and S-4. The general levels will be used most often, and it is assumed that level IJ will be used unless one of the other levels is specified. The following example illustrates the effects of inspection levels for a lot size of 600:

Inspection level	Code letter	Sample size (single sampling)
I	G	32
II	J	80
III	K	125

The level of inspection, therefore, has a significant impact on the determination of sample size.

Quality Control Procedures

The basic steps in the quality control and inspection process entail (1) definition of the criteria for nonconformance of the product to specified requirements, (2) formulation of lots or batches, (3) determination of the acceptable quality level (AQL), (4) determination of sampling techniques, (5) development of the sampling plan, and (6) determination of acceptability. Table 5.1 provides a step-by-step summary of the quality control process.

TABLE 5.1   The Quality Control Process

Steps	Explanation
1. Determine lot size	Lot size may be controlled by lot formation criteria contained in procurement documents. Otherwise, establish it by agreement between the responsible authority and the supplier.
2. Determine inspection level	If the item specification does not give the inspection level, use inspection level II.
3. Determine sample size code letter	Found in the appendix, Table I, based on lot size and inspection level.
4. Determine sampling plans	Single sampling is generally selected. Double or multiple sampling may be used.
5. Establish severity of inspection	Normal inspection is generally used at the start of a contract or production.
6. Determine sample size and acceptance number	Assuming single sampling, normal inspection, and a given size code letter, the sample size and acceptance number are found in the appendix, Table IIA (Single Sampling Plan for Normal Inspections).
7. Select sample	The sample, consisting of the number of units of product as determined from the appendix, Table IIA, is selected at random from the lot. Additionally, any obvious defectives that have not been selected for the inspection sample are removed from the lot (but are not included in the sample).
8. Inspect sample	The defectives (or defects) are counted. If this count does not exceed the acceptance number (Ac), the entire lot is accepted. If the count equals or exceeds the rejection number, the lot is rejected.
9. Record inspection results	Compute the estimated process average if this is required by operating procedures. Maintain a record of accept/reject decisions in order that switching rules may be followed.

The ANSI/ASQ Z1.4-2003/MIL-STD-105E tables can be found in the appendix to this chapter.

Nonconformance criteria

The extent of nonconformance of product should be expressed either in terms of percent defective or in terms of defects per hundred units.

Formation and identification of lots or batches

The product should be assembled into identifiable lots, sublots, or batches, or in such other manner as may be prescribed. Each lot or batch should, as far as is practicable, consist of units of product of a single type, grade, class, size, and composition, manufactured under essentially the same conditions and at essentially the same time.

Acceptable quality level (AOL)

AOL use. The AQL, together with the corresponding sample size code letter noted in the applicable inspection table, is used for indexing the sampling plans (see Figs. 5.7 and 5.8, in the appendix to this chapter for an example applicable to single-sample plans).

Choosing AQLs. Different AQLs may be chosen for groups of defects considered collectively and for individual defects. An AQL for a group of defects may be chosen in addition to AQLs for individual defects or for subgroups within that group. AQL values of 10.0 or less may be expressed either in percent defective or in defects per hundred units; those over 10.0 are expressed only in defects per hundred units.

Sample techniques

Representative (stratified) sampling. When appropriate, the number of units in the sample should be selected in proportion to the size of sublots or subbatches, or parts of the lot or batch, identified by some rational criterion. When representative sampling is used, the units from each sublot, subbatch, or part of the lot may be selected at random.

Time of sampling. A sample may be drawn after all the units making up the lot or batch have been assembled, or sample units may be drawn during assembly of the lot or batch, in which case the size of the lot or batch is determined before any sample units are drawn. When the sample units are drawn during assembly of the lot or batch, if the rejection number is reached before the lot is completed, that portion of the lot already completed shall be rejected.

Double or multiple sampling. When double or multiple sampling is to be used, each sample should be selected over the entire lot or batch.

Sampling plans

The sampling plan is determined by the level of inspection (degree of scrutiny); sample size, as established by the inspection level, and lot or batch size pursuant to Table I of the appendix acceptable quality level (AQL); and whether single, double, or multiple sampling will be applied.

Sample selection.   Basic to sampling inspection is the selection of a sample which can reasonably be expected to represent the quality of the parent lot. Hence, the procedure used to select units from a lot must be such that it assures a sample free of bias. The process of selecting a sample meeting this requirement is called *random sampling*.

A sample consists of one or more units of product drawn from a lot or batch. Random sampling is any procedure used to draw units from an inspection lot in such a way that each unit in the lot has an equal chance, without regard to its quality, of being included in the sample.

The amount of information about the process quality gained from examining samples depends mostly upon the absolute size of the samples and only slightly upon the percentage of the lot that is examined. It is sometimes asked, therefore, "Why is the sample size made to depend upon the lot size?" There are three reasons:

1. A sample of small size that has a high probability of representing the quality of a small lot or batch may be too small to represent, with high probability, the quality of a larger lot or batch.

2. When there is more at stake, it is more important to make the right decision. Proper use of the tables leads to the result that lots from a good process are more likely to be accepted as the lot size increases, whereas lots from a bad process are more likely to be rejected.

3. A sample size that can be afforded for a large lot is likely to be uneconomical for a small lot. For example, a sample size of 80 from a lot of 1,000 may be easy to justify economically, whereas a sample of 80 from a lot of 100 would be relatively expensive.

Inspection level.   The inspection level determines the relationship between the lot or batch size and the sample size. The inspection level to be used for any particular requirement is normally prescribed by the quality manager's written procedures. Three inspection levels, I, IJ, and III, for general use are given in Table I of the appendix. Normally, inspection level II is used. However, inspection level I may be used when less discrimination is needed, or level III may be used for greater discrimination. Four additional special levels, S-1, S-2, S-3, and S-4, are given in the same table and may be used where relatively small sizes are necessary and large sampling risks can or must be tolerated.

Sample size code letters.   Sample sizes are designated by code letters. Table I of the appendix is used to find the applicable code letter for the particular lot or batch size and the prescribed inspection level.

Determination of the sampling plan.   The AQL and the code letter shall be used to obtain the sampling plan from Table IJ, III, or IV of the appendix. Parts of these tables, which respectively, provide governing criteria for single-sample, double-sample, and multiple-sample plans, whichever is appropriate, are shown in the appendix to this chapter. A decision as to type of plan (either single, double, or multiple) is usually governed by the comparison between the administrative difficulty and average sample sizes of the available plans.

Determination of acceptability

Percent defective inspection.   To determine the acceptability of a lot or batch under percent defective inspection, the applicable sampling plan should be used in accordance with the following.

Single sampling plan.   The number of sample units inspected shall be equal to the sample size given by the plan. If the number of defectives found in the sample is equal to or less than the acceptance number, the lot or batch shall be considered acceptable. If the number of defectives is equal to or greater than the rejection number, the lot or batch shall be rejected. (See Table IIA.)

Double sampling plan.   A number of sample units equal to the first sample size given by the plan shall be inspected. If the number of defectives found in the first sample is equal to or less than the first acceptance number, the lot or batch shall be considered acceptable. If the number of defectives found in the first sample is equal to or greater than the first rejection number, the lot or batch shall be rejected. If the number of defectives found in the first sample is between the first acceptance and rejection numbers, a second sample of the same size shall be inspected. The number of defectives found in the second sample shall be added to the number found in the first. If the cumulative number of defectives is equal to or less than the second acceptance number, the lot or batch shall be considered acceptable. If the cumulative number of defectives is equal to or greater than the second rejection number, the lot or batch shall be rejected. (See Table IIIA.)

Multiple sampling plan.   Under multiple sampling, the procedure shall be similar to that specified for the double sampling plan, except that the number of successive samples required to reach a decision may be as many as seven. (See Table IVA.)

Defects per hundred units inspection.   To determine the acceptability of a lot or batch under defects per hundred units inspection, the procedure specified for percent defective inspection above shall be used, except that the word "defects" shall be substituted for "defectives."

Operating Characteristic (OC) Curves

Sampling risks

Figure 5.1 portrays the ideal sampling plan, in which there is 100 percent probability of accepting good lots and zero percent probability of accepting bad lots. In such a case, the operating characteristic curve would form a rectangular graph bounded by the horizontal and vertical scales and the acceptance line.

Notwithstanding the attraction of an ideal scenario, there are certain risks inherent in inspection. In addition to errors in human performance, in the case of sampling inspection, a special kind of risk that can be attributed to the "luck of the draw" results in erroneous decisions concerning good and bad lots. In other words, whenever sampling is

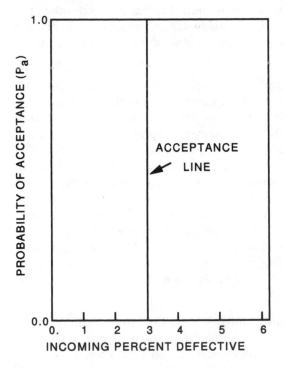

Figure 5.1   An ideal sampling plan performance.

involved, there is always the risk (or chance) that good lots may be rejected and bad lots accepted. In general, the smaller the sample size, the greater the risk of selecting a sample which does not truly reflect the quality of its parent lot, and therefore of making an erroneous accept/reject decision. Since risks are inherent in sampling plans, this relationship should be clearly understood. The problem of these risks may be restated as follows: "Assuming that a lot is some given percent defective, what is the chance (probability) that the lot will be accepted or rejected by the sampling plan?" When the given percent defective is in the region of good quality, both supplier and consumer will be interested in a high probability of lot acceptance, and when the given percent defective is in the region of bad quality, the consumer especially will be interested in a high probability that the lot will be rejected. These probabilities of acceptance and rejection can be determined from the performance curve, or operating characteristic curve, of the sampling plan. The curve for a single sampling plan shown in Fig. 5.2 indicates the chance that lots of varying quality (percent defective) will be accepted. Because of chance variations, samples drawn from lots of identical quality may themselves be very unequal in quality and thus yield very different test results. Some of the test results may be so far from correctly reflecting the quality of the parent lot that the parent lot is either incorrectly rejected (if the lot quality is good) or incorrectly accepted (if the lot quality is bad). The probability that a sampling plan leads to the rejection of a good lot is called the producer's or *alpha* risk. The probability that a sampling plan leads to the acceptance of a bad lot is called the consumer's or *beta* risk.

Attributes of operating characteristic (OC) curves

The protection afforded by a sampling plan—that is, its ability to discriminate between good and bad quality—can be accurately calculated. The fact that the alpha and beta risks can be quantified makes it possible to state these risks statistically (numerically) and predict the quantities rejected on the average over the entire possible range of product quality. Such calculations, based on the mathematical theory of probability, provide the basis for the curve shown in Fig. 5.2. This curve indicates the relationship between the quality of lots submitted for inspection and the probability of acceptance and is identified as the plan's *operating characteristic curve*, or OC curve. OC curves are a graphical means of showing the relationship between the quality of a lot submitted for sampling inspection (usually expressed in percent defective, but sometimes expressed in defects per hundred units) and the probability that the sampling plan will yield a decision to accept the lot (described as the *probability of acceptance*). In preparing the OC

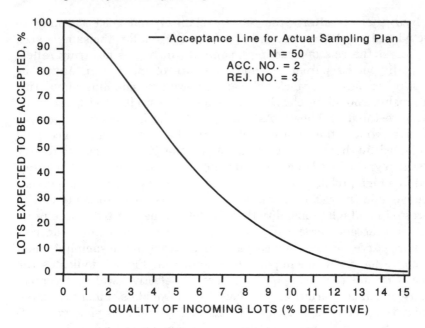

Figure 5.2   A theoretical sampling plan (single sample).

curve, the percent defective of submitted lots, ranging from zero to some conveniently selected value of percent defective (not exceeding 100 percent) or defects per hundred units representing less than perfect quality, is generally shown graphically on the abscissa (horizontal axis). Along the ordinate (vertical axis) of the graph, the percent (or fraction) of lots that may be expected to be accepted by the particular sampling plan are shown, also ranging from 0 to 100 percent (0 to 1, if the scale is in fractions of lots). Obviously, lots which contain 0 percent defective will be accepted 100 percent of the time by any sampling plan, and lots which are 100 percent defective will never be accepted; consequently, the initial and terminal points (highest and lowest) on the graph can be plotted without the need for calculation. The points in between follow a smooth curve and are obtained from mathematical probability computation.

Effects of changes to the sampling plan on the OC curve

A sampling plan and its associated risks are completely defined by the lot size, sample size, and acceptance number. The lot size, except in the case of very small lots, is of relatively little importance in determining the risks associated with any given sampling plan. Thus, sample size and acceptance number are the two important factors which influence the risk pattern of sampling plans. If the risks of a tentative sampling plan are considered unsatisfactory, the question which follows is, "What changes must be made to obtain the desired sampling protection?" This can be answered by considering the effect on the OC curve of changes in the sampling plan.

To understand the effect of such changes, a more detailed study of the OC curve (see Fig. 5.3) is appropriate. From examination of this curve, it is seen that if lots to be inspected are 2 percent defective, approximately 90 percent of the lots are expected to be accepted, whereas if the lots submitted are 8 percent defective, about 10 percent of the lots are expected to be accepted. If 2 percent defective and 8 percent defective represent good- and bad-quality lots, respectively, the good lots will be rejected 10 percent (100 – 90 = 10) of the time (producer's risk) and bad lots will be accepted 10 percent of the time (consumer's risk). This rejection/acceptance frequency will occur by chance. If this frequency is intolerable, appropriate changes in the sampling plan are required.

Changes in sample size. An increase in the sample size results in a steepening of the OC curve, as indicated in Fig. 5.4. The steeper the OC curve, the greater the power of the sampling plan to discriminate between good and bad quality. Figure 5.4 clearly illustrates the effect that increasing the sample size has on the OC curve.

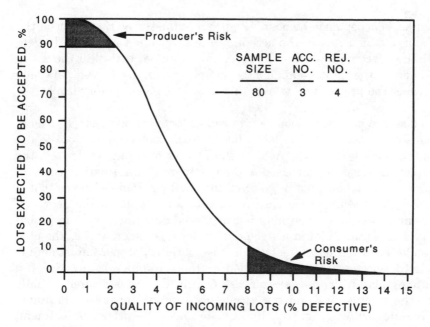

Figure 5.3 OC curve for a typical sampling plan.

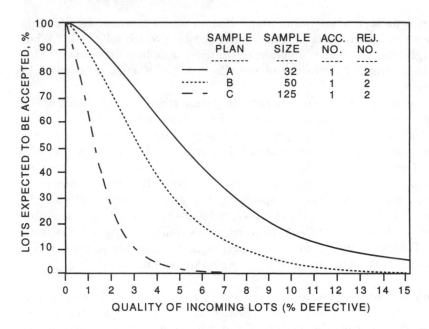

Figure 5.4   Effect on OC curve of changing sample size.

Changes in acceptance number.   Figure 5.5 illustrates the effect on the OC curve of changes in the acceptance and rejection numbers. In general, the effect of increasing the acceptance number is to shift the entire OC curve to the right. Changing the sampling plan in this way generally increases the probability of accepting a lot at a given quality level.

Simultaneous change of sample size and acceptance number.   If a more accurate disposition of the lots whose percent defective is close to the selected quality level (the AQL or the LQ, for example) is desired, the sample size must be increased to provide more discrimination. Also, an acceptance number must be selected that will yield an OC curve that is properly located about the desired quality level. Thus, if the degree of discrimination of a given plan is considered adequate, but the probability of accepting a lot at a given quality level is too great (i.e., the plan is too loose) or too small (i.e., the plan is too tight), proper adjustment is made by selecting the appropriate acceptance number. In practice, if a sampling plan which has certain risk characteristics is desired, usually both the sample size and the acceptance number must be simultaneously adjusted (see Fig. 5.6). In order to make the proper adjustment, however, the effect of each must be understood.

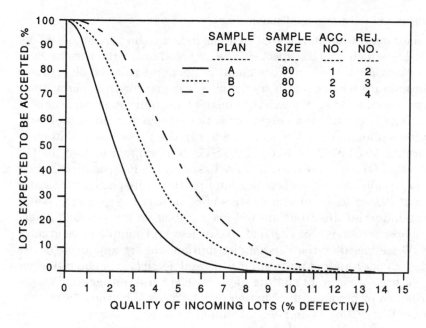

Figure 5.5 Effect on OC curve of changing acceptance number.

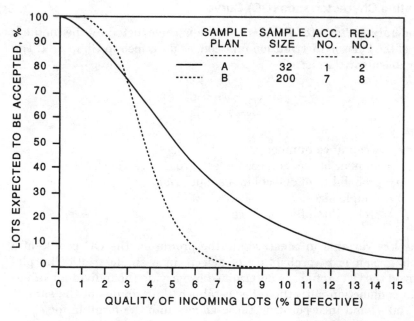

Figure 5.6 Effects on OC curve of simultaneous changes in sample size and acceptance number.

Conditions of OC curve application

Standard operating characteristic curves indicate the percentage of lots or batches which may be expected to be accepted under the various sampling plans for a given process quality. A general rule for algorithmic application prescribes that if the lot size N is greater than eight times the sample size n (i.e., $N > 8n$), the binomial distribution is applicable. If the lot size is greater than eight times the sample and the sample size is greater than 15 (i.e., $N > 8n$ and $n > 15$), the Poisson distribution is applicable. ANSI/ASQ Z1.4-2003/MIL-STD-105E stipulates that the OC curves for AQLs greater than 10.0 are based on the Poisson distribution and are applicable for defects per hundred units inspection; those for AQLs of 10.0 or less and sample sizes of 80 or less, are as a rule, based on the binomial distribution and are applicable for percent defective inspection; those for AQLs of 10.0 or less and sample sizes greater than 80 are based on the Poisson distribution and are applicable either for defects per hundred units or for percent defective. It is noted that for sample sizes of 80 units and the AQL is 10.0 or less, either the Poisson distribution or binomial distribution may be utilized by the analyst; the Poisson distribution is an adequate approximation to the binomial distribution under these conditions.

Methodology for Construction of the Operating Characteristics (OC) Curve

The operating characteristics (OC) curve is constructed by the methodology of the Poisson distribution function, as described in Chapter 1. The applicable algorithm is:

$$P_a = \sum_{r=0}^{r=c} \frac{(np)^r (e^{-np})}{r!}$$

where c = acceptance number
  r = number of occurrences (defects) $(0, 1, 2 \ldots c)$
  p = probability of defects in a single trial
  n = sample size
  P_a = probability of acceptance

The key variable in constructing the profile for the OC profile of a sampling plan is p (probability of defects in a single trial). The plot points on the OC profile are determined by varying the individual values of p in conjunction with the other established attributes of the sample (c, r, and n) and incorporating these factors into the formula for P_a, as described above.

The analyst must first designate a range of the values of p (probability of defects in a single trial). The next step is to compute the value P_a

(probability of acceptance) for each designated value of p by exercising the formula for P_a, as detailed above.

To illustrate the process, the following describes a case example developed on basis of the following values of p:

$$p = 0.005$$
$$p = 0.01$$
$$p = 0.02$$
$$p = 0.03$$
$$p = 0.04$$
$$p = 0.05$$
$$p = 0.06$$
$$p = 0.07$$
$$p = 0.08$$

Next, calculate P_a for each value of p by using the additionally specified values for c (acceptance number), r (number of occurrences or defects), and n (sample size). For purpose of illustration of the mathematical methodology, Figure 5.7 provides an example of the calculation of P_a

Computation of Plot Points for Operating Characteristics (OC) Curve

EXAMPLE

Calculation where $p = 0.005$

$$P_a = \sum_{r=0}^{r=c} \frac{(np^r)(e^{-np})}{r!}$$

where c = acceptance number = 3
r = number of occurrences (defects) = 0, 1, 2, 3
p = probability of defects in a single trial = 0.005
n = sample size = 80
$np = (80)(0.005) = 0.4$
P_a = probability of acceptance (from algorithmic computation)

$r = 0$ $\dfrac{\{(0.4)^0\}(e^{-0.4})}{0!} = \dfrac{(1.0)(0.67032)}{1} = 0.67032$

$r = 1$ $\dfrac{\{(0.4)^1\}(e^{-0.4})}{1!} = \dfrac{(0.4)(0.67032)}{1} = 0.268128$

$r = 2$ $\dfrac{\{(0.4)^2\}(e^{-0.4})}{2!} = \dfrac{(0.16)(0.67032)}{2} = 0.053626$

$r = 3$ $\dfrac{\{(0.4)^3\}(e^{-0.4})}{3!} = \dfrac{(0.064)(0.67032)}{6} = 0.00715$

$$P_a = \sum_{r=0}^{r=3} = 0.999224$$

Figure 5.7 Computation of the plot point for the OC curve where $p = 0.005$.

where $p = 0.005$. Figure 5.8 shows the computation of P_a where $p = 0.04$. The same type of calculation is required for the remaining values of p:

$$p = 0.01$$
$$p = 0.02$$
$$p = 0.03$$
$$p = 0.05$$
$$p = 0.06$$
$$p = 0.07$$
$$p = 0.08$$

The results of the calculations for all individual values of p are incorporated into a summary table that will generate the input plot data for construction of the operating characteristics (OC) curve for the illustrative sample. Table 5.2 provides the tabular summary of all the pertinent factors developed for the OC curve. In construction of the OC curve the percent (%) defective values are identified on the horizontal scale and the resultant values of P_a are plotted on the vertical scale.

Computation of Plot Points for Operating Characteristics (OC) Curve

EXAMPLE

Calculation where $p = 0.04$

$$P_a = \sum_{r=0}^{r=c} \frac{(np^r)(e^{-np})}{r!}$$

where c = acceptance number = 3
$\quad r$ = number of occurrences (defects) = 0, 1, 2, 3
$\quad p$ = probability of defects in a single trial = 0.04
$\quad n$ = sample size = 80
$\quad np$ = (80)(0.005) = 3.2
$\quad P_a$ = probability of acceptance (from algorithmic computation)

$r = 0 \qquad \dfrac{\{(3.2)^0\}(e^{-3.2})}{0!} = \dfrac{(1.0)(0.040762)}{1} = 0.040762$

$r = 1 \qquad \dfrac{\{(3.2)^1\}(e^{-3.2})}{1!} = \dfrac{(3.2)(0.040762)}{1} = 0.1304384$

$r = 2 \qquad \dfrac{\{(3.2)^2\}(e^{-3.2})}{2!} = \dfrac{(10.24)(0.040762)}{2} = 0.208701$

$r = 3 \qquad \dfrac{\{(3.2)^3\}(e^{-3.2})}{3!} = \dfrac{(32.768)(0.040762)}{6} = 0.222615$

$$P_a = \sum_{r=0}^{r=3} = 0.6062516$$

Figure 5.8   Computation of the plot point for the OC curve where $p = 0.04$.

Example Development of factors for operating characteristics (OC) curve

$$r = c \qquad n = 80$$

$$P_s = \sum\nolimits_{r=0}^{r=c} \frac{(np)^r (e^{-np})}{r!}$$

TABLE 5.2 Summary of Factors for Constructing the OC Curve

p	n	np	Vertical Scale Plot Value P_a	Horizontal Scale Plot Value 100p (% Defective)
0.005	80	0.4	0.999224	0.5%
0.01	80	0.8	0.990923	1.0%
0.02	80	1.6	0.921188	2.0%
0.03	80	2.4	0.778723	3.0%
0.04	80	3.2	0.602516	4.0%
0.05	80	4.0	0.433479	5.0%
0.06	80	4.8	0.294240	6.0%
0.07	80	5.6	0.190629	7.0%
0.08	80	6.4	0.118951	8.0%

Figure 5.9 portrays the operating characteristics curve constructed on basis of the plot data displayed in Table 5.2. Inspection of the OC curve displayed in Figure 5.9 indicates that, based on the sample attributes, the probability of acceptance of the total lot at the acceptance number of

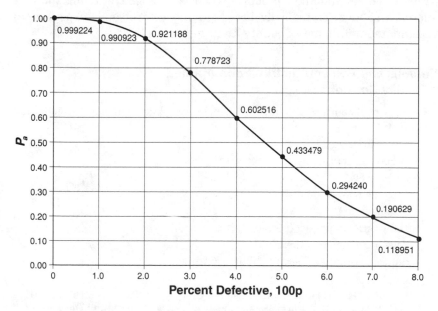

Figure 5.9 Completed operating characteristics (OC) curve with plot prints.

3.0 (percent defective) is about 78% (0.778723); the producer's risk is about 22% (1.0 − 0.778723).

Total Quality Management (TQM)

The focus in a total quality management (TQM)[1] system is on customer satisfaction and on continuing improvement in meeting changing customer needs. In TQM, everyone is responsible for quality, and the customer is the person who receives the output of one's work. TQM encompasses all phases of an organization, from the receptionist to the shipping dock, to ensure that defects do not occur and that quality constantly improves. Under TQM, concerns about quality become a part of everyone's day. The philosophy of TQM predicates that quality cannot be inspected into a product. Thus, TQM concentrates on management of the total process, not the product. Management of the total process requires "quality at the source." This begins with the design of the product and the design of the process for producing that product. Mapping customer needs into specific design requirements is an important TQM objective. Quality at the source also means that the operator and the supplier should monitor their own quality; they are asked and expected to stop production if a job cannot be done right. Operators are also active participants in the quality improvement process. Thus, "respect for people" is an important element of TQM implementation. For purchased parts, quality at the source means quality from the supplier. Suppliers are selected on the basis of demonstrated ability to deliver consistently defect-free goods, and major suppliers are early participants in the design of new products. TQM is an effective companion to a tailored system of statistically based quality control and inspection to ascertain the fulfillment of quality assurance goals.

Binomial and Poisson Distributions Applied to Quality Control

Example of Binomial Distribution.   A single sampling plan has the following characteristics:

Inspection level IJ

Normal inspection

Lot size: 80

AQL: 6.5 percent defectives

Sample size: 13

Acceptance number: 2 (percent)

Rejection number: 3 (percent)

Known percentage defectives in lots: 6 percent

---

[1]Summary of TQM developed from W. Demmy, Steven, and Petrini, "MRP IJ + JIT + TQM + TOC: The Path to World Class Management" *Logistics Spectrum: The Journal of the Society of Logistics Engineers*, 2b(3): 8–13, 1992.

Determine the probability of acceptance of lots with no more than 2 percent defectives as provided by the sampling plan.

As AQL < 10 and lot size = 80, the binomial distribution is applied.

$$P_a = \sum_{r=0}^{r=c} \frac{n!}{(r!)(n-r)} p^r (1-P)^{n-r}$$

where P_a = probability of acceptance
c = 2 (acceptance number)
r = number of occurrences (0, 1, 2)
p = 0.06 (probability of drawing a defective unit in a single trial)
n = 13 (sample size)

Solution

r	Computation	Probability of r occurrences
0	$\left(\dfrac{[13!]}{1!(13-1)!}\right)[(06)^0(1-0.06)^{13-0}] = (1)(0.94^{13})]$	0.447
1	$\left(\dfrac{[13!]}{1!(13-1)!}\right)[(.06)^1(10-.06)^{13-1}] = 13[(0.06)(0.94^{12})]$	0.371
2	$\left(\dfrac{[13!]}{2!(13-2)!}\right)[(0.06)^2(1-0.06)^{13-2}] = 78(0.0036)(0.94^{11})]$	0.142
	P_a (Total)	0.960

The probability of lot acceptance P_a is 0.96 for the stated sampling plan. This can be confirmed by inspection of the governing OC curve portrayed by Table X-E of the appendix.

Example of Poisson Distribution   A single sampling plan is described as follows:

Inspection level III

Lot size: 90

AQL: 15 defects per 100 units

Sample size: 20

Acceptance number: 7

Rejection number: 8

Known number of defects per 100 units: 10

Determine the probability of acceptance P_a of lots with no more than 7 defects per 100 units.

As AQL > 10, the Poisson distribution is applicable.

$$P_a = \sum_{r=0}^{r=c} \frac{(np)^r e^{-np}}{(r!)}$$

where P_a = probability of acceptance
c = 7 (acceptance number)
r = number of occurrences (0, 1, 2, 3, 4, 5, 6, 7)
p = $0.10(\dfrac{10 \text{ defects}}{100 \text{ units}}$; probability of defects in a single trial)
n = 20 (sample size)

Solution

r	Computation	Probability of r occurrences
0	$\dfrac{(20 \times 0.1)^0 (e^{-20 \times 0.1})}{0!} = \dfrac{(1)(e^{-2.0})}{1}$	0.135
1	$\dfrac{(20 \times 0.1)^1 (e^{-20 \times 0.1})}{1!} = \dfrac{(2)(e^{-2.0})}{1}$	0.27
2	$\dfrac{(20 \times 0.1)^2 (e^{-20 \times 0.1})}{2!} = \dfrac{(4)(e^{-2.0})}{2}$	0.27
3	$\dfrac{(20 \times 0.1)^3 (e^{-20 \times 0.1})}{3!} = \dfrac{(8)(e^{-2.0})}{6}$	0.18
4	$\dfrac{(20 \times 0.1)^4 (e^{-20 \times 0.1})}{4!} = \dfrac{(16)(e^{-2.0})}{24}$	0.09
5	$\dfrac{(20 \times 0.1)^5 (e^{-20 \times 0.1})}{5!} = \dfrac{(32)(e^{-2.0})}{120}$	0.036
6	$\dfrac{(20 \times 0.1)^6 (e^{-20 \times 0.1})}{6!} = \dfrac{(64)(e^{-2.0})}{720}$	0.012
7	$\dfrac{(20 \times 0.1)^7 (e^{-20 \times 0.1})}{7!} = \dfrac{(128)(e^{-2.0})}{5040}$	0.003
	P_a (Total)	0.996

The probability of acceptance P_a is .996 for the stated sampling plan. This can be confirmed by inspection of the governing OC curve portrayed by Table X-F of the appendix.

Appendix

Use of Sampling Planning Tables

The student of logistics is well advised to become familiar with the use of the standard sampling planning tables set forth in the appendix, which incorporates the conventional, industrywide procedures for quality control. These tables (shown in the following pages) and the following case examples are provided in the appendix to give a basic understanding of such procedures.

Case no.	Subject
1	Determination of Sample Parameters Based on Prestipulated Quality Criteria
2	Interpretation of Table for Double Sampling Plan
3	Interpretation of Table for Multiple Sampling Plan

TABLE 1 — *Sample size code letters*

Lot or batch size	Special inspection levels				General inspection levels		
	S-1	S-2	S-3	S-4	I	II	III
2 to 8	A	A	A	A	A	A	B
9 to 15	A	A	A	A	A	B	C
16 to 25	A	A	B	B	B	C	D
26 to 50	A	B	B	C	C	D	E
51 to 90	B	B	C	C	C	E	F
91 to 150	B	B	C	D	D	F	G
151 to 280	B	C	D	E	E	G	H
281 to 500	B	C	D	E	F	H	J
501 to 1200	C	C	E	F	G	J	K
1201 to 3200	C	D	E	G	H	K	L
3201 to 10000	C	D	F	G	J	L	M
10001 to 35000	C	D	F	H	K	M	N
35001 to 150000	D	E	G	J	L	N	P
150001 to 500000	D	E	G	J	M	P	Q
500001 and over	D	E	H	K	N	Q	R

TABLE II-A — Single sampling plans for normal inspection (Master table)

Each cell shows "Ac Re" (Acceptance number, Rejection number). ↓ = use first sampling plan below arrow; ↑ = use first sampling plan above arrow.

Sample size code letter	Sample size	*Acceptable Quality Levels (normal inspection)* 0.010	0.015	0.025	0.040	0.065	0.10	0.15	0.25	0.40	0.65	1.0	1.5	2.5	4.0	6.5	10	15	25	40	65	100	150	250	400	650	1000
A	2	↓	↓	↓	↓	↓	↓	↓	↓	↓	↓	↓	↓	↓	↓	↓	↓	0 1	1 2	2 3	3 4	5 6	7 8	10 11	14 15	21 22	30 31
B	3	↓	↓	↓	↓	↓	↓	↓	↓	↓	↓	↓	↓	↓	↓	↓	0 1	1 2	2 3	3 4	5 6	7 8	10 11	14 15	21 22	30 31	44 45
C	5	↓	↓	↓	↓	↓	↓	↓	↓	↓	↓	↓	↓	↓	↓	0 1	1 2	2 3	3 4	5 6	7 8	10 11	14 15	21 22	30 31	44 45	↑
D	8	↓	↓	↓	↓	↓	↓	↓	↓	↓	↓	↓	↓	↓	0 1	1 2	2 3	3 4	5 6	7 8	10 11	14 15	21 22	30 31	44 45	↑	↑
E	13	↓	↓	↓	↓	↓	↓	↓	↓	↓	↓	↓	↓	0 1	1 2	2 3	3 4	5 6	7 8	10 11	14 15	21 22	30 31	44 45	↑	↑	↑
F	20	↓	↓	↓	↓	↓	↓	↓	↓	↓	↓	↓	0 1	1 2	2 3	3 4	5 6	7 8	10 11	14 15	21 22	30 31	44 45	↑	↑	↑	↑
G	32	↓	↓	↓	↓	↓	↓	↓	↓	↓	↓	0 1	1 2	2 3	3 4	5 6	7 8	10 11	14 15	21 22	30 31	44 45	↑	↑	↑	↑	↑
H	50	↓	↓	↓	↓	↓	↓	↓	↓	↓	0 1	1 2	2 3	3 4	5 6	7 8	10 11	14 15	21 22	30 31	44 45	↑	↑	↑	↑	↑	↑
J	80	↓	↓	↓	↓	↓	↓	↓	↓	0 1	1 2	2 3	3 4	5 6	7 8	10 11	14 15	21 22	30 31	44 45	↑	↑	↑	↑	↑	↑	↑
K	125	↓	↓	↓	↓	↓	↓	↓	0 1	1 2	2 3	3 4	5 6	7 8	10 11	14 15	21 22	30 31	44 45	↑	↑	↑	↑	↑	↑	↑	↑
L	200	↓	↓	↓	↓	↓	↓	0 1	1 2	2 3	3 4	5 6	7 8	10 11	14 15	21 22	30 31	44 45	↑	↑	↑	↑	↑	↑	↑	↑	↑
M	315	↓	↓	↓	↓	↓	0 1	1 2	2 3	3 4	5 6	7 8	10 11	14 15	21 22	30 31	44 45	↑	↑	↑	↑	↑	↑	↑	↑	↑	↑
N	500	↓	↓	↓	↓	0 1	1 2	2 3	3 4	5 6	7 8	10 11	14 15	21 22	30 31	44 45	↑	↑	↑	↑	↑	↑	↑	↑	↑	↑	↑
P	800	↓	↓	↓	0 1	1 2	2 3	3 4	5 6	7 8	10 11	14 15	21 22	30 31	44 45	↑	↑	↑	↑	↑	↑	↑	↑	↑	↑	↑	↑
Q	1250	↓	↓	0 1	1 2	2 3	3 4	5 6	7 8	10 11	14 15	21 22	30 31	44 45	↑	↑	↑	↑	↑	↑	↑	↑	↑	↑	↑	↑	↑
R	2000	↓	0 1	1 2	2 3	3 4	5 6	7 8	10 11	14 15	21 22	30 31	44 45	↑	↑	↑	↑	↑	↑	↑	↑	↑	↑	↑	↑	↑	↑

↓ = Use first sampling plan below arrow. If sample size equals, or exceeds, lot or batch size, do 100 percent inspection.

↑ = Use first sampling plan above arrow.

Ac = Acceptance number.

Re = Rejection number.

115

TABLE III-A—Double sampling plans for normal inspection (Master table)

Acceptable Quality Levels (normal inspection)

Sample size code letter	Sample	Cumulative sample size	0.010	0.015	0.025	0.040	0.065	0.10	0.15	0.25	0.40	0.65	1.0	1.5	2.5	4.0	6.5	10	15	25	40	65	100	150	250	400	650	1000

(Table body consists of a grid of Ac/Re acceptance and rejection numbers with arrows indicating "use first sampling plan below/above arrow." Detailed numeric entries not legibly transcribable.)

⇩ = Use first sampling plan below arrow. If sample size equals or exceeds lot or batch size, do 100 percent inspection.

⇧ = Use first sampling plan above arrow.

Ac = Acceptance number

Re = Rejection number

* = Use corresponding single sampling plan (or alternatively, use double sampling plan below, where available).

116

TABLE IV-A—Multiple sampling plans for normal inspection (Master table)

Sample size code letter	Sample	Sample size	Cumulative sample size	0.010 Ac Re	0.015 Ac Re	0.025 Ac Re	0.040 Ac Re	0.065 Ac Re	0.10 Ac Re	0.15 Ac Re	0.25 Ac Re	0.40 Ac Re	0.65 Ac Re	1.0 Ac Re	1.5 Ac Re	2.5 Ac Re	4.0 Ac Re	6.5 Ac Re	10 Ac Re	15 Ac Re	25 Ac Re	40 Ac Re	65 Ac Re	100 Ac Re	150 Ac Re	250 Ac Re	400 Ac Re	650 Ac Re	1000 Ac Re
A B C																												
D	First	2	2														* 2	0 2	0 3	1 4	2 5	3 6	4 7	6 9	7 10	9 13	11 16	16 22
	Second	2	4														2 3	0 3	1 4	2 5	3 7	6 10	7 11	8 13	14 17	19 25	27 34	40 49
	Third	2	6														0 3	0 3	2 6	3 7	6 10	8 13	11 15	17 20	19 27	25 29	34 40	49 58
	Fourth	2	8														0 3	1 4	3 7	5 10	8 13	12 17	17 20	21 23	27 34	36 40	53 58	77 78
	Fifth	2	10														1 3	2 4	3 6	5 8	7 11	11 15	17 20	25 29	33 45	40 47	65 68	
	Sixth	2	12														1 3	3 5	4 6	7 9	10 12	14 17	21 23	31 33	47 54	77 78		
	Seventh	2	14														2 3	4 5	6 7	9 10	13 14	18 19	25 26	37 38				
E	First	3	3													* 2	0 4	0 5	1 7	2 9	4 12	6 16						
	Second	3	6													2 3	0 3	1 5	1 6	3 8	4 10	7 14	11 19	17 27				
	Third	3	9													0 3	1 4	2 6	3 8	6 10	8 13	13 19	19 27	29 39				
	Fourth	3	12													0 3	1 4	2 5	3 7	5 10	8 13	12 17	19 25	27 34	40 49			
	Fifth	3	15													1 3	2 4	3 6	5 8	7 11	11 15	17 20	25 29	36 40	53 58			
	Sixth	3	18													1 3	3 5	4 6	7 9	10 12	14 17	21 23	31 33	45 47	65 68			
	Seventh	3	21													2 3	4 5	6 7	9 10	13 14	18 19	25 26	37 38	53 54	77 78			
F	First	5	5											*		* 2	0 4	0 5	1 7	2 9								
	Second	5	10													2 3	0 3	1 5	1 6	3 8	4 10	7 14						
	Third	5	15													0 3	1 4	2 6	3 8	6 10	8 13	13 19						
	Fourth	5	20													0 3	1 4	2 5	3 7	5 10	8 13	12 17	19 25					
	Fifth	5	25													1 3	2 4	3 6	5 8	7 11	11 15	17 20	25 29					
	Sixth	5	30													1 3	3 5	4 6	7 9	10 12	14 17	21 23	31 33					
	Seventh	5	35													2 3	4 5	6 7	9 10	13 14	18 19	25 26	37 38					
G	First	8	8											*		* 2	0 4	0 5	1 7	2 9								
	Second	8	16													2 3	0 3	1 5	1 6	3 8	4 10	7 14						
	Third	8	24													0 3	1 4	2 6	3 8	6 10	8 13	13 19						
	Fourth	8	32													0 3	1 4	2 5	3 7	5 10	8 13	12 17	19 25					
	Fifth	8	40													1 3	2 4	3 6	5 8	7 11	11 15	17 20	25 29					
	Sixth	8	48													1 3	3 5	4 6	7 9	10 12	14 17	21 23	31 33					
	Seventh	8	56													2 3	4 5	6 7	9 10	13 14	18 19	25 26	37 38					
H	First	13	13										*			* 2	0 4	0 5	1 7	2 9								
	Second	13	26													2 3	0 3	1 5	1 6	3 8	4 10	7 14						
	Third	13	39													0 3	1 4	2 6	3 8	6 10	8 13	13 19						
	Fourth	13	52													0 3	1 4	2 5	3 7	5 10	8 13	12 17	19 25					
	Fifth	13	65													1 3	2 4	3 6	5 8	7 11	11 15	17 20	25 29					
	Sixth	13	78													1 3	3 5	4 6	7 9	10 12	14 17	21 23	31 33					
	Seventh	13	91													2 3	4 5	6 7	9 10	13 14	18 19	25 26	37 38					
J	First	20	20									*				* 2	0 4	0 5	1 7	2 9								
	Second	20	40													2 3	0 3	1 5	1 6	3 8	4 10	7 14						
	Third	20	60													0 3	1 4	2 6	3 8	6 10	8 13	13 19						
	Fourth	20	80													0 3	1 4	2 5	3 7	5 10	8 13	12 17	19 25					
	Fifth	20	100													1 3	2 4	3 6	5 8	7 11	11 15	17 20	25 29					
	Sixth	20	120													1 3	3 5	4 6	7 9	10 12	14 17	21 23	31 33					
	Seventh	20	140													2 3	4 5	6 7	9 10	13 14	18 19	25 26	37 38					

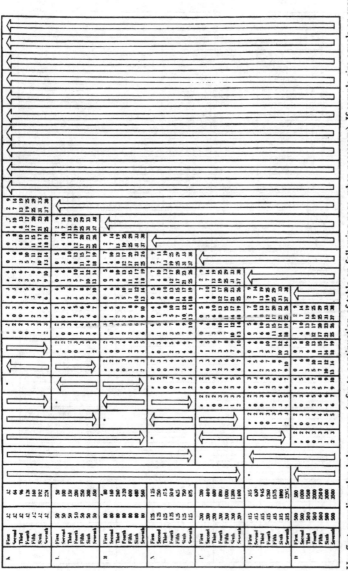

⇨ = Use first sampling plan below arrow (refer to continuation of table on following page, when necessary). If sample size equals or exceeds lot or batch size, do 100 percent inspection.

⇦ = Use first sampling plan above arrow.

Ac = Acceptance number

Re = Rejection number

* = Use corresponding single sampling plan (or alternatively, use multiple sampling plan below, where available).

++ = Use corresponding double sampling plan (or alternatively, use multiple sampling plan below, where available).

" = Acceptance not permitted at this sample size.

TABLE X-E—Tables for sample size code letter: E

CHART E - OPERATING CHARACTERISTIC CURVES FOR SINGLE SAMPLING PLANS

(Curves for double and multiple sampling are matched as closely as practicable)

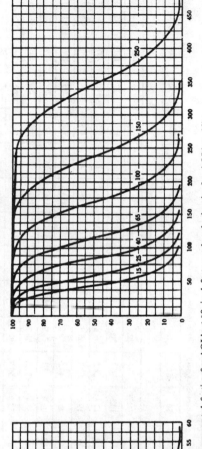

PERCENT OF LOTS EXPECTED TO BE ACCEPTED (P_a)

QUALITY OF SUBMITTED LOTS (p, in percent defective for AQL's ≤ 10; in defects per hundred units for AQL's > 10)

Note: Figures on curves are Acceptable Quality Levels (AQL's) for normal inspection.

119

TABLE X-E-1 - Tabulated Values for Operating Characteristic Curves for Single Sampling Plans

Acceptable Quality Levels (normal inspection)

P_a	p (in percent defective)				p (in defects per hundred units)															
	1.0	4.0	6.5	10	1.0	4.0	6.5	10	15	25	X	40	X	65	X	100	X	150	X	250
99.0	0.077	1.19	3.63	7.00	0.078	1.15	3.35	6.33	13.7	22.4	27.0	36.7	46.9	57.5	79.6	96.7	132	150	219	238
95.0	0.394	2.81	6.63	11.3	0.395	2.73	6.29	10.5	20.1	30.6	36.1	47.5	59.2	71.1	95.7	115	153	173	246	266
90.0	0.807	4.16	8.80	14.2	0.806	4.09	8.48	13.4	24.2	35.8	41.8	54.0	66.5	79.2	105	125	165	185	261	282
75.0	2.19	7.41	13.4	19.9	2.22	7.39	13.3	19.5	32.5	45.8	52.6	66.3	80.2	94.1	122	144	187	208	288	310
50.0	5.19	12.6	20.0	27.5	5.33	12.9	20.6	28.2	43.6	59.0	66.7	82.1	97.5	113	144	168	213	236	321	344
25.0	10.1	19.4	28.0	36.2	10.7	20.7	30.2	39.3	57.1	74.5	83.1	100	117	134	167	192	241	266	355	379
10.0	16.2	26.8	36.0	44.4	17.7	29.9	40.9	51.4	71.3	90.5	100	119	137	155	190	217	269	295	388	414
5.0	20.6	31.6	41.0	49.5	23.0	36.5	48.4	59.6	80.9	101	111	130	150	168	205	233	286	313	409	435
1.0	29.8	41.5	50.6	58.7	35.4	51.1	64.7	77.3	101	123	134	155	176	196	235	264	321	349	450	477
Acceptable Quality Levels (tightened inspection)	1.5	6.5	10	X	1.5	6.5	10	15	25	X	40	X	65	X	100	X	150	X	250	X

Note: Binomial distribution used for percent defective computations; Poisson for defects per hundred units.

120

TABLE X-F—Tables for sample size code letter: F

CHART F - OPERATING CHARACTERISTIC CURVES FOR SINGLE SAMPLING PLANS

(Curves for double and multiple sampling are matched as closely as practicable)

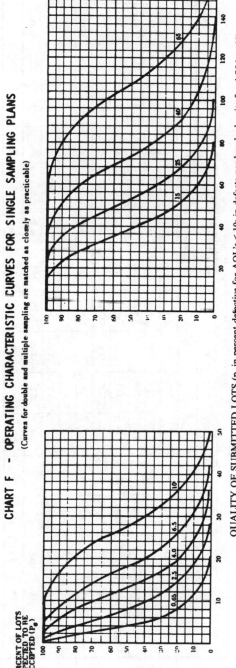

QUALITY OF SUBMITTED LOTS (p, in percent defective for AQL's ≤ 10; in defects per hundred units for AQL's > 10)

Note: Figures on curves are Acceptable Quality Levels (AQL's) for normal inspection.

121

TABLE X-F-1 - Tabulated Values for Operating Characteristic Curves for Single Sampling Plans

Acceptable Quality Levels (normal inspection)

p (in percent defective)

P_a	0.65	2.5	4.0	6.5	10
99.0	0.050	0.75	2.25	4.31	9.75
95.0	0.256	1.80	4.22	7.13	14.0
90.0	0.525	2.69	5.64	9.03	16.6
75.0	1.43	4.81	8.70	12.8	21.6
50.0	3.41	8.25	13.1	18.1	27.9
25.0	6.70	12.9	18.7	24.2	34.8
10.0	10.9	18.1	24.5	30.4	41.5
5.0	13.9	21.6	28.3	34.4	45.6
1.0	20.6	28.9	35.6	42.0	53.4
(tightened AQL)	1.0	4.0	6.5	10	X

p (in defects per hundred units)

P_a	0.65	2.5	4.0	6.5	10	15	X	25	X	40	X	65
99.0	0.051	0.75	2.18	4.12	8.92	14.5	17.5	23.9	30.5	37.4	51.7	62.9
95.0	0.257	1.78	4.09	6.83	13.1	19.9	23.5	30.8	38.5	46.2	62.2	74.5
90.0	0.527	2.66	5.51	8.73	15.8	23.3	27.2	35.1	43.2	51.5	68.4	81.2
75.0	1.44	4.81	8.68	12.7	21.1	29.8	34.2	43.1	52.1	61.2	79.5	93.4
50.0	3.47	8.39	13.4	18.4	28.4	38.3	43.3	53.3	63.3	73.3	93.3	108
25.0	6.93	13.5	19.6	25.5	37.1	48.4	54.0	65.1	76.1	87.0	109	125
10.0	11.5	19.5	26.6	33.4	46.4	58.9	65.0	77.0	88.9	101	124	141
5.0	15.0	23.7	31.5	38.8	52.6	65.7	72.2	84.8	97.2	109	133	151
1.0	23.0	33.2	42.0	50.2	65.5	80.0	87.0	101	114	127	153	172
(tightened AQL)	1.0	4.0	6.5	10	15	X	25	X	40	X	65	X

Acceptable Quality Levels (tightened inspection)

Note: Binomial distribution used for percent defective computations; Poisson for defects per hundred units.

TABLE X-H-I. TABULATED VALUES FOR OPERATING CHARACTERISTIC CURVES FOR SINGLE SAMPLING PLANS

Acceptable Quality Levels (normal inspection)

p (in percent defective)

P_a	0.25	1.0	1.5	2.5	4.0	6.5	X	10
99.0	0.0201	0.300	0.886	1.68	3.69	6.07	7.36	10.1
95.0	0.103	0.715	1.66	2.78	5.36	8.22	9.72	12.9
90.0	0.210	1.07	2.22	3.53	6.43	9.54	11.2	14.5
75.0	0.574	1.92	3.46	5.10	8.51	12.0	13.8	17.5
50.0	1.38	3.33	5.31	7.29	11.3	15.2	17.2	21.2
25.0	2.73	5.29	7.69	10.0	14.5	18.8	21.0	25.2
10.0	4.50	7.56	10.3	12.9	17.8	22.4	24.7	29.1
5.0	5.82	9.14	12.1	14.8	19.9	24.7	27.0	31.6
1.0	8.00	12.6	15.8	18.7	24.2	29.2	31.7	36.3
Tightened AQL	0.40	1.5	2.5	4.0	6.5	X	10	X

p (in defects per hundred units)

P_a	0.25	1.0	1.5	2.5	4.0	6.5	X	10	X	15	X	25
99.0	0.0201	0.297	0.872	1.65	3.57	5.81	7.01	9.54	12.2	15.0	20.7	25.1
95.0	0.103	0.711	1.64	2.73	5.23	7.96	9.39	12.3	15.4	18.5	24.9	29.8
90.0	0.211	1.06	2.20	3.49	6.30	9.31	10.9	14.0	17.3	20.6	27.3	32.5
75.0	0.575	1.92	3.45	5.07	8.44	11.9	13.7	17.2	20.8	24.5	31.8	37.4
50.0	1.39	3.36	5.35	7.34	11.3	15.3	17.3	21.3	25.3	29.3	37.3	43.3
25.0	2.77	5.39	7.84	10.2	14.8	19.4	21.6	26.0	30.4	34.8	43.5	49.9
10.0	4.61	7.78	10.6	13.4	18.5	23.5	26.0	30.8	35.6	40.3	49.5	56.4
5.0	5.99	9.49	12.6	15.5	21.0	26.3	28.9	33.9	38.9	43.8	53.4	60.5
1.0	9.21	13.3	16.8	20.1	26.2	32.0	34.8	40.3	45.6	50.9	61.2	68.7
Tightened AQL	0.40	1.5	2.5	4.0	6.5	X	10	X	15	X	25	X

Acceptable Quality Levels (tightened inspection)

Case 1: Determination of sample parameters based on prestipulated quality criteria

For a planned production run, the production manager has stated the following lot specifications for the product quality control program:

> Lot size: 500
>
> Acceptable quality level (AQL): 15
>
> Limiting quality (LQ) level: 40
>
> Inspection level: IJ
>
> Inspection severity: Normal
>
> Sampling plan: Single

Note: Based on an AQL of 15, defects per hundred units is applicable.

Pursuant to the stated lot quality specifications, the following sampling plan data may be determined.

A. Sample size code letter

B. Sample size

C. Acceptance number

D. Rejection number

E. Estimated producer's risk at 15 defects/100 units for quality of submitted lots

F. Estimated consumer's risk at 40 defect/100 units for quality of submitted lots

By use of the sample planning tables of the appendix, the essential sample data are developed as follows:

Question	Answer	Appendix Source
A. Sample size code letter?	H	Table I (Fig. 5.7)
B. Sample size?	50	Table IJ (Fig. 5.8)
C. Acceptance number?	14	Table IJ (Fig. 5.8)
D. Rejection number?	15	Table IJ (Fig. 5.8)
E. Estimated producer's risk at 15 Defects/100 units for quality of submitted lots?	1%*	Table X-H (Fig. 5.9)
F. Estimated consumer's risk at 40 defects/100 units for quality of submitted lots?	10%	Table X-H (Fig. 5.9)

*Note: P_a = 99%. Therefore, $1 - 0.99 = 0.01$, or 1%.

The circled letters on Figs. 5-10, 5-11, and 5-12 indicate the data points which correlate with the alphabetized questions.

Lot or batch size	Special inspection levels				General inspection levels		
	S-1	S-2	S-3	S-4	I	II	III
2 to 8	A	A	A	A	A	A	B
9 to 15	A	A	A	A	A	B	C
16 to 25	A	A	B	B	B	C	D
26 to 50	A	B	B	C	C	D	E
51 to 90	B	B	C	C	C	E	F
91 to 150	B	B	C	D	D	F	G
151 to 280	B	C	D	E	E	G	H
281 to 500	B	C	D	E	F	H	J
501 to 1200	C	C	E	F	G	J	K
1201 to 3200	C	D	E	G	H	K	L
3201 to 10000	C	D	F	G	J	L	M
10001 to 35000	C	D	F	H	K	M	N
35001 to 150000	D	E	G	J	L	N	P
150001 to 500000	D	E	G	J	M	P	Q
500001 and over	D	E	H	K	N	Q	R

Figure 5.10 Sample size code letters, Case 1 letter

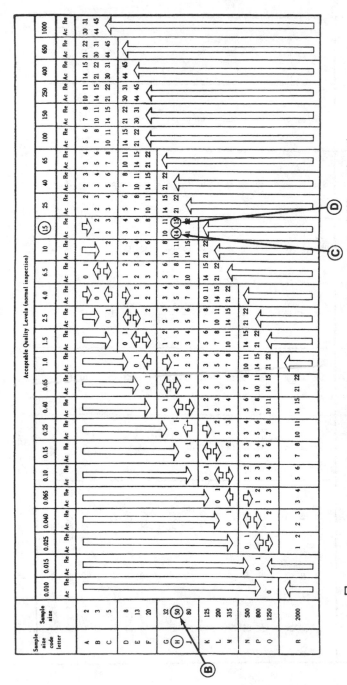

Figure 5.11 Single sampling plans for normal inspection, Case 1

⇩ = Use first sampling plan below arrow. If sample size equals, or exceeds, lot or batch size, do 100 percent inspection.

⇧ = Use first sampling plan above arrow.

Ac = Acceptance number.

Re = Rejection number.

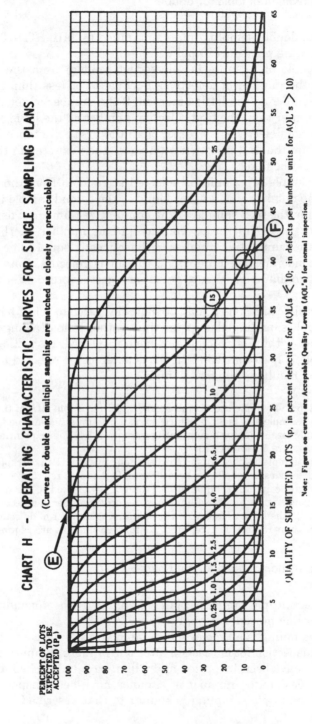

CHART H - OPERATING CHARACTERISTIC CURVES FOR SINGLE SAMPLING PLANS

(Curves for double and multiple sampling are matched as closely as practicable)

PERCENT OF LOTS EXPECTED TO BE ACCEPTED (P_a)

QUALITY OF SUBMITTED LOTS (p, in percent defective for AQL(s ≤ 10; in defects per hundred units for AQL's > 10)

Note: Figures on curves are Acceptable Quality Levels (AQL's) for normal inspection.

Figure 5.12 Single Tables for sample size code, Case 1

127

Case 2: Interpretation of table for double sampling plan

For use of the double sampling plan as typified by MIL-STD-105 Table IIIA, the following procedures apply.

A first sample of n_1 units is selected at random from the lot and inspected. If the number of defectives is equal to or less than the first acceptance number c_1, the lot is accepted. If the number of defectives is equal to or greater than the first rejection number r_1, the lot is rejected. If the number of defectives is greater than the first acceptance number c_1 and less than the first rejection number r_1, the next step in the sampling plan must be taken.

A second sample of n_2 units is selected at random from the lot and inspected. The number of defectives found in the second sample is added to that found in the first sample. If the cumulative number of defectives is equal to or less than the second acceptance number c_2, the lot is accepted. If the cumulative number of defectives is equal to or greater than the second rejection number r_2, the lot is rejected. Note that r_2 equals $c_2 + 1$, so that a decision to accept or reject is forced after the second sample.

Under certain conditions it may be more desirable to select both samples for a double sampling plan at one time, rather than drawing the second sample after the first sample has been inspected. Inspection of the second sample is not required if the lot is accepted or rejected based on the inspection results of the first sample.

Example   Refer to sample code H and an AQL of 10 in Table IIIA of the appendix. The first sample size is 32, at which point the cumulative sample is 32. The second sample size is 32, for a cumulative sample of 64. The sequential accept/reject numbers are 5/9 and 12/13.

If the first sample of 32 results in 5 or less defectives, the lot is accepted; if the number of defectives is equal to or greater than 9, the lot is rejected. If the number of defectives is greater than 5 but less than 9, proceed to the second sample.

If the second sample of 32 results in a *cumulative* number of defects equal to or less than 12, the lot is accepted. If the *cumulative* number of defects is equal to or greater than 13, the lot is rejected. No further samples are required.

Case 3: Interpretation of table for multiple sampling plan

For use of the multiple sampling plan as typified by the appendix, Table IVA, the following procedures apply.

Multiple sampling is a type of sampling in which a decision to accept or reject an inspection lot is reached after one or more samples from the inspection lot have been inspected and will always be reached after not more than a designated number of samples have been inspected. The procedure for multiple sampling is similar to that described for double sampling, except that the number of successive samples required to reach a decision to accept or reject the lot may be more than two.

Under multiple sampling up to seven sequential samples may be taken.

Example Refer to sample code F and an AQL of 15 in Table IVA of the appendix. The sequential samples and corresponding accept/reject numbers are given as

Sample sequence	Sample size	Cumulative sample	Cumulative accept/reject numbers
1	5	5	0/4
2	5	10	1/6
3	5	15	3/8
4	5	20	5/10
5	5	25	7/11
6	5	30	10/12
7	5	35	13/14

Decision Sequence-Multiple Sampling

1 If the first sample results in 0 defectives, accept the lot. If the number of defectives is equal to or greater than 4, reject the lot. If the number of defectives is greater than 0 but less than 4, proceed to the second sample.

2. If the second sample results in a cumulative number of defectives of 1, accept the lot. If the cumulative number of defectives is equal to or greater than 6, reject the lot. If the cumulative number of defectives is greater than 1 but less than 6, proceed to the third sample.

3. If the third sample results in a cumulative number of defectives equal to or less than 3, accept the lot. If the cumulative number of defectives is equal to or greater than 8, reject the lot. If the cumulative number of defectives is greater than 3 but less than 8, proceed to the fourth sample.

4. If the fourth sample results in a cumulative number of defectives equal to or less than 5, accept the lot. If the cumulative number of defectives is equal to or greater than 10, reject the lot. If the cumulative number of defectives is greater than 5 but less than 10, proceed to the fifth sample.

5. If the fifth sample results in a cumulative number of defectives equal to or less than 7, accept the lot. If the cumulative number of defectives is equal to or greater than 11, reject the lot. If the cumulative number of defectives is greater than 7 but less than 11, proceed to the sixth sample.

6. If the sixth sample results in a cumulative number of defectives equal to or less than 10, accept the lot. If the cumulative number of defectives is equal to or greater 12, reject the lot. If the cumulative number of defectives is greater than 10 but less than 12, proceed to the seventh sample.

7. If the seventh sample results in a cumulative number of defectives equal to or less than 13, accept the lot. If the cumulative number of defectives is equal to or greater than 14, reject the lot. No further sampling is required.

6

Human Factors Engineering

General

Overview

Human factors engineering (HFE) is the science of analysis and optimization of human characteristics and capabilities for integration with machine characteristics and capabilities in order to provide the most effective integrated human-machine system capable of accomplishing specified functions. Within the context of human factors engineering, one may consider a human-machine system as a combination of one or more human beings and one or more physical components interacting through a flow of inputs to achieve some desired output. Within the human factors purview, a machine may be any type of physical device, equipment, facility, etc., that people use in executing some activity to achieve a desired purpose. The integrated human-machine system may range in complexity from a person with a hammer to a sprawling telephone network with battalions of operating personnel.

The field of human factors, also know as *ergonomics* (a term of European origin), is not self-contained, but frequently reflects a confluence of elements of related disciplines, such as safety engineering, maintainability engineering, industrial engineering, physiology, psychology, sociology, anthropology, and anthropometry.

Human factors engineering considers human intellect, mentality, expectations, behavioral traits, senses, physical strengths, and anthropometric attributes in the design of things people use in their daily activities and environments. Human factors design has a major impact on the interface of the maintainers and operators with the system and the efficacy of human-machine integration. The perspective of this chapter will, therefore, emphasize those aspects of human factors that are germane to logistics engineering.

Definition of human factors

The frequently elusive definition of *human factors* is best addressed in three stages.

1. *Focus.* The central focus of human factors is the consideration of human beings in carrying out such functions as the design and creation of man-made objects, products, equipment, facilities, and environments that people use; the development of procedures for performing work and other human activities; the provision of services to people; and the evaluation of the things people use in terms of their suitability for people.

2. *Objectives.* The objectives of human factors in these functions are twofold: to enhance the effectiveness and efficiency with which work and other human activities are carried out, and to maintain or enhance certain desirable human values (e.g., health, safety, satisfaction). The second objective is essentially one of human welfare and well-being.

3. *Approach.* The central approach of human factors is the systematic application of relevant information about human abilities, characteristics, behavior, and motivation in the execution of such functions.

Variables affecting human factors analysis

Human factors analysis is governed by the influence of two groups of variables: *independent* variables and *dependent* variables.

Independent variables are classified into three types:

1. Task-related variables, which include equipment variables (e.g., display panels and drive controls) and procedural variables (e.g., work-rest cycles, and sequential task elements)

2. Environmental variables, such as noise, temperature, and illumination

3. Subjective variables, such as human height, weight, age, visual acuity, hearing, and intellectual capacity

Dependent variables are derived from the application of independent variables. They typically reflect the consequences and effects of the independent variables, expressed as measurements of performance, reaction time, and similar output values.

Human factors in the design process

Incorporation of human factors in the design of operating devices, such as machinery, domestic appliances, facilities, and hand tools, involves sequential phases that must be achieved, specifically

1. Statement of functions
2. Delineation of functions
3. Allocation of functions to humans and machines
4. Development of capabilities—i.e., development of human resources, development of machine capabilities, and development of human-machine interfaces

For complex systems (e.g., an automobile assembly plant or a new generation of passenger aircraft), these phases of the process are extensive and highly organized. For simple systems (e.g., an eggbeater or scissors), the process tends to be more informal, with some procedural elements omitted.

Human factors and logistics

Considerations of human factors ripple throughout the total infrastructure of logistics disciplines. They significantly affect the functional applications of maintainability, diagnostics and testing, and materials handling. These elements are manifest primarily in the identification and description of human resource skills requirements, development of maintenance manuals and operators' instructions, and development of logistics training courses.

HFE and maintainability design. Maintainability is described in terms of average maintenance task times. System disassembly, onequipment maintenance, and reassembly task times are governed by the types of fasteners used to secure the components that make up the system, the tools utilized, and the accessibility of the serviceable components. The location of the maintenancesignificant components should (1) accommodate human strength, digital utility, and physical dimensions and (2) be positioned so as to preclude exposure of human body parts to impact and injury while performing service tasks.

Diagnostics and test equipment. Fault localization and isolation routines must be user-friendly. Informative indicators designed within display configurations that require human auditory, visual, tactual (touch), or olfactory (smell) interpretation must be clear and conveniently placed.

Materials packaging, handling, and storage. Packaging of products for off-the-shelf (retail) issue or dealer (wholesale) distribution and palletizing for warehouse storage and handling must accommodate the parameters of human muscular strength and endurance.

Human-Machine Systems

Classification of humanmachine systems

Human-machine systems are defined by (1) whether they are closed-loop or open-loop systems and (2) whether they are manual, mechanical, or automated systems. A closed-loop system is a continuous system which performs an operation subject to continuous control and which requires continuous feedback to ensure successful performance (e.g., an automobile). Such feedback provides information about any errors and adjustments that must be made to continue operation. An open-loop system requires no further control after being energized and made operational. Once its mission is initiated, no further control can be exercised.

Manual vs. machine control systems.   In addition to being categorized as closed-loop or open-loop, systems can be considered in three classes: manual, mechanical, and automated.

1. *Manual systems.* A manual system consists of hand tools and other aids which are coupled with the human operator, who controls the operation. Operators of such systems use their own physical energy as a power source, transmit to and receive from their tools a great deal of information, typically operate at their own speed, and can readily exploit their ability to act as a variable and flexible system.

2. *Mechanical systems.* Mechanical systems consist of integrated physical parts, such as various types of powered components. They are generally designed to perform their functions with little variation. The power is normally provided by the machine, and the operator's function is typically one of control, through the use of control devices.

3. *Automated systems.* A fully automated system performs all operational functions, including sensing, information processing, decision making, and action. Such a system needs to be fully programmed so that it can take appropriate action for all possible contingencies that are sensed. Most automatic systems are of a closed-loop nature. The primary human functions in such systems are those of monitoring, programming, and maintenance.

Human considerations in HFE

The governing human considerations in the design of human-machine systems include strength and endurance; speed and accuracy; the senses of touch, sight, hearing, and smell; and body movements.

Human strength and endurance.   Strength is the maximal force muscles can exert isometrically in a single, focused effort; it is the muscular capacity to apply force under static conditions.

There are certain human variables which influence human strength and endurance. The most frequently observed of these variables are age, sex, body build, and exercise.

Age: Strength reaches a maximum by the middle to late twenties and declines slowly but continuously from then on, until at about age 65 strength is about 75 percent of that exerted in youth.

Sex: On average, women's strength is about two-thirds that of men.

Body build: Although body build is related to strength and endurance, the relationships are complicated; for example, athletic-looking individuals generally are stronger than others, but less powerfully built persons may be more efficient. For rapidly fatiguing, severe exercise, slender subjects are best and obese subjects worst, and for moderate exercise, those with normal build are best.

Exercise: Exercise can increase strength and endurance within limits; these increases frequently are in the range of 30 to 50 percent above beginning levels. There are some indications that at least moderate regular exercise over the years can sometimes stave off some of the typical decline in physical condition with increasing age.

Speed and accuracy.   In the human factors context, speed is described by interpretation of *stimulus*, *response*, and *reaction time*.

Speed.   Most movements are triggered by some external stimulus, such as a visual display indicator or an auditory warning system. The time required to make a response following a stimulus is referred to as reaction time.

There are two types of reaction times: *simple reaction time* and *choice reaction time*. Simple reaction time is the time required to make a specific response when only one particular stimulus can occur, usually when an individual is anticipating the stimulus. In the case of choice reaction time (when there are two or more stimuli and two or more possible responses), reaction time is typically longer, the increase being due to such factors as the time required for identification of the particular stimulus, the need to "record" the stimulus, the time to make a decision, and, of course, the number of stimuli and corresponding responses.

The degree of expectancy influences response and reaction. There is strong evidence to support the contention that the reaction time of a human is longer when the individual is not expecting to have to make a decision to react than when the reaction is in response to an anticipated signal.

Accuracy.   Human performance within the purview of human factors involves consideration of how fast and how well humans accomplish their functions. How well functions are performed directly concerns assessment of human errors. Human errors are categorized as (1) intentional errors, (2) unintentional errors, and (3) omissions. (In this formulation, the term *intentional errors* does not refer to deliberate errors, but rather describes acts that individuals performed intentionally while thinking they were doing the right thing, although in fact they were not.)

Human senses.   The five human senses——sight, hearing, touch, smell, and taste—are basically involved with stimuli external to the human body. To varying degrees, sight, hearing, touch, and smell are among the constituent elements of the human factor in human-machine systems. The sensory receptors (eyes, ears, nasal passages, epidermic nerves) are the sensory receptors which serve as *exteroceptors*. The four senses critical to human factors engineering receive informational signals, discriminate and interpret the data, and transmit instructions to the brain for subsequent muscular reactions which exercise control over the operating device to execute the required function. The predominant senses employed in FIFE for exteroceptor functions are the auditory and visual capabilities. The tactual (touch) and olfactory (smelling) capabilities are used to a limited extent when so dictated in system design.

Human body movements.   The study of human movements as a function of the construction of the musculoskeletal system is referred to as *kinesiology*.

Types of movements.   Certain of the basic types of movements that are performed by the body members are described below, along with their associated jargon in kinesiology:

1. *Flexion:* Bending, or decreasing the angle between the parts of the body
2. *Extension:* Straightening, or increasing the angle between the parts of the body

3. *Adduction*: Moving toward the midline of the body
4. *Abduction:* Moving away from the midline of the body
5. M*edial rotation:* Turning toward the midline of the body
6. *Lateral rotation:* Turning away from the midline of the body
7. *Pronation:* Rotating the forearm so that the palm faces downward
8. *Supination*: Rotating the forearm so that the palm faces upward

Some of these basic movements are illustrated in Fig. 6.1, along with the angular values for each. In this, as in other aspects of biomechanics, there are the ever-present individual differences, including the effects of physical condition and age.

Classification of movements.   Although specific movements of body members can be described in terms of basic types of movements, in describing work activities it usually is preferable to do so in more operational terms. There are ways in which movements can be so classified, as shown below:

1. Positioning movements are those in which the hand or foot moves from one specific position to another, as in reaching for a control knob.
2. Continuous movements are those which require muscular control adjustments of some type during the movement, as in operating the steering wheel of a car or guiding a piece of wood through a band saw.
3. Manipulative movements involve the handling of parts, tools, control mechanisms, etc., typically with the fingers or hands.
4. Repetitive movements are those in which the same movement is repeated, as in hammering, operating a screwdriver, and turning a handwheel.
5. Sequential movements are several relatively separate, independent movements in a sequence.
6. A static adjustment is the absence of a movement, consisting of maintaining a specific position of a body member for a period of time.

Various types of movements may be combined in sequence so that they blend one into another.

1. *Positioning movements.* Positioning movements are made when a person reaches for something or moves something to another location, usually by hand; they are then travel movements of the body member.

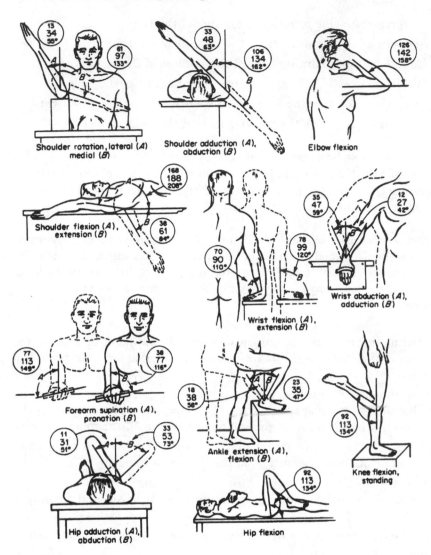

Figure 6.1   Range of certain movements of the upper and lower extremities. The three values (in degrees) given for each angle are the 5th percentile, the mean, and the 95th percentile, respectively, of voluntary movements.

Certain positioning movements can be subdivided into two or three relatively distinct components, namely, reaction time (the time to initiate a response following the stimulus that triggers it), primary or gross travel time (to bring the body member near the terminal), and a secondary or corrective type of motion to bring the body member to the precise position desired.

2. *Continuous movements*. Continuous movements are those that require accurate control over the span of the movement. Deviations from the desired path are produced by tremor of the body member.

3. *Manipulative movements*. Manipulative movements involve the use of the hand, the fingers, or both, as in handling items, assembling parts, or using hand tools or control devices.

4. *Repetitive movements*. Any given type of repetitive movement consists of successive performance of the same action. Such movements may be either self-paced or paced (self-paced tasks being paced by the worker and paced tasks being controlled by some external factor, such as the action of a machine or the movement of a conveyor belt). Probably the critical aspect in the performance of repetitive tasks is the requirement for work pauses from time to time, especially in the case of paced work.

5. *Sequential movements*. Sequential movements are of the same general kind, but vary in some differentiating feature, as in operating a keyboard. In some instances, however, a potpourri of types of movements may occur in sequence, such as those involved in starting a car on a rainy night, which might include turning on the ignition, turning on the lights, and turning on the windshield wipers

6. *Static adjustment*. Maintaining a specific position of a body member over a period of time can produce static reactions, in which certain sets of muscles typically operate in opposition to each other to maintain equilibrium of the body or of certain portions of it. Thus, if a body member, such as a hand, is being held in a fixed position, the various muscles controlling hand movements are in a balance that permits no net movement one way or the other. The tensions set up in the muscles to bring about this balance, however, require continued effort, as most of us who have attempted to maintain an immobile state for any length of time can testify. In fact, it has been stated that maintaining a static position produces more wear and tear on people than some kind of adjustive posture.

Machine considerations in HFE

A system is designed to serve a stipulated purpose. In some instances—especially in the case of complex systems—the performance objectives of the system should be formally set forth before the design process begins.

Given the objectives of a system (especially a complex one), one usually then determines the functions that need to be performed in order to achieve the objectives. This functional analysis initially should be concerned with what functions need to be performed (such as whether they are to be performed by individuals or by machine components).

Functions of machines. The execution of any operational function typically involves a combination of four more basic functions: sensing (information receiving), information storage, information processing and decision, and action functions. The sequential flow and interaction is essentially as follows:

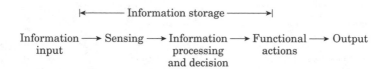

Information storage interacts with the functions of sensing, information processing and decision, and functional actions. It must, therefore, be portrayed in interface.

Sensing (information receiving). One of the basic operative functions is sensing, or information receiving. Some of the information entering a system is from outside the system.

Sensing by a human being would be through the use of the various sense modalities, such as vision, audition, and touch. There are various types of machine sensing devices, electronic, photographic, and mechanical. In some cases sensing by a machine is simply a substitute for the same sensing function performed by a human.

Information storage. For human beings, information storage is synonymous with memory of learned material. Information can be stored in physical machine components in many ways, as on punch cards, magnetic tapes, templates, records, and tables of data. Most of the information that is stored for later use is in coded or symbolic form.

Information processing and decision. Information processing embraces various types of operations performed with information that is received (sensed) and information that is stored. When human beings are involved in information processing, this process, simple or complex, typically results in a decision to act (or, in some instances, a decision not to act). When mechanized or automated machine components are used, their information processing must be programmed in some way in order to cause the component to respond in some predetermined manner to each possible input.

Functional actions. What are called the action functions of a system generally are the operations which occur as a consequence of the decisions that are made. These functions fall roughly into two classes. The first of these is some type of physical control action or process, such as the activation of certain control mechanisms or the handling, movement, modification, or alteration of materials or objects. The other is essentially a communication action, be it by voice (in human beings), signals, records, or other methods. Such functions also involve some physical actions, but these are in a sense incidental to the communication function.

Machine performance criteria.   Machine system performance criteria are those that relate to the performance of the system (or subsystem or component thereof), or, in other words, those that reflect something about the degree to which the system (or subsystem or component) achieves what it is intended to achieve. For example, a computer keyboard might be evaluated in terms of such criteria as number and accuracy of data entries made per unit of time; an earth-moving vehicle might be evaluated in terms of the amount of earth moved per unit of time. Other examples of system criteria are the anticipated life of a system, ease of operation or use, maintainability, reliability, operating cost, and human resources requirements. Some of these criteria are strictly mechanistic, in the sense that they reflect essentially engineering performance (e.g., the maximum rpm of an engine), whereas others reflect more the performance of the system as it is used by the people involved with it (such as errors in cards punched).

It should be obvious that the two classes of criteria are not a definitive dichotomy, but rather tend to form a continuum, with strictly mechanistic system criteria at one end and strictly behavioral criteria at the other end.

Human-machine interaction: input and output

The stimuli that the sense organs receive actually consist of some form of energy, such as light, sound, heat, mechanical pressure, or chemical energy, that is sensed by the sensing devices or converted into coded form for human reception (e.g., certain forms of electromagnetic energy and ultrasonic vibrations).

In these and other types of circumstances, it may be appropriate to transmit relevant information (stimuli) indirectly through some type of display. For our purposes, we shall consider a display to be any method of presenting information indirectly, in either reproduced or coded (symbolic) form. If a decision is made to use a display, there may be some options regarding the sensory modality and the specific type of display to use, since the method of presenting information can influence, for better or worse, the accuracy and speed with which the information can be received.

Types of displays.   Displays can be described as either dynamic or static. Dynamic displays are those that continually change or are subject to time, and include the following types: displays that depict the status or condition of some variable, such as temperature and pressure gauges, speedometers, and altimeters; cathode-ray-tube (CRT) displays, such as radar, sonar, TV, and radio range signal transmitters; displays that present intentionally transmitted information, such as record players, TV, and movies; and displays that are intended to aid the user in the control or setting of some variable, such as the temperature control of an oven. (It might be observed, incidentally, that there are some devices that do double duty as both displays and controls; this is especially the case with devices used for making settings, such as oven controls.) Static displays are those that remain fixed over time, such as signs, charts, graphs, labels, and various forms of printed or written material.

Purposes of displays.   The primary purposes of informational displays are to present the types of information described below.

1. *Quantitative information:* Display presentations which reflect the quantitative value of some variable, such as temperature or speed.
2. *Qualitative information:* Display presentations which reflect the approximate value, trend, rate of change, direction of change, or other aspect of some changeable variable.
3. *Status information:* Display presentations which reflect the condition or status of a system, such as on-off indications; indications of one of a limited number of conditions, such as stop-caution-go lights; and indications of independent conditions of some class, such as a TV channel.
4. *Warning and signal information:* Display presentations used to indicate emergency or unsafe conditions or to indicate the presence or absence of some object or condition.
5. *Representational information:* Pictorial or graphic representations of objects, areas, or other configurations.
6. *Alphanumeric and symbolic information:* Such information usually is static, but in certain circumstances it may be dynamic, as in the case of news bulletins displayed by moving lights on a building.
7. *Time-phased information:* Display presentations of pulsed or time-phased signals, e.g., signals that are controlled in terms of the dura-

tion of the signals and intersignal intervals and their combinations, such as Morse code and blinker lights.

Auditory displays.

Conditions for application of auditory displays.   There are circumstances in which auditory displays would usually be preferable to other types of displays. Some of these circumstances are given below.

1. When the origin of the signal is itself a sound
2. When the message is simple and short
3. When the message will not be referred to later
4. When the message deals with events in time
5. When sending warnings or when the message calls for immediate action
6. When presenting continuously changing information of some type, such as aircraft, radio range, or flightpath information
7. When the visual system is overburdened
8. When speech channels are fully employed (in which case auditory signals such as tones should be clearly distinguishable from the speech)
9. When illumination limits use of vision
10. When the receiver moves from one place to another

In consideration of these guidelines, particular mention should be made of the desirability of restricting auditory messages to those that are short and simple (except in the case of speech), since humans do not do well at short-term storage of complex messages.

Warning and alarm signals.   Because of the unique features of the auditory system, auditory displays lend themselves well to use in signaling warnings and alarms. For this purpose, the various types of available devices have their individual characteristics and corresponding advantages and limitations. A summary of these characteristics and features is given in Table 6.1.

Visual displays.   The visual skills humans have—especially visual acuity, particularly the ability to detect relevant stimuli and to discriminate between and among variations thereof (such as positions of pointers on dials or different letters), and color discrimination—have a direct bearing upon the design of visual displays. The meaningfulness of what humans see in visual displays depends in part upon their perceptual

TABLE 6.1   The Characteristics and Features of Certain Types of Audio Alarms

Alarm	Intensity	Frequency	Attention-getting ability	Noise-penetration ability
Diaphone (foghorn)	Very high	Very low	Good	Poor in low-frequency noise
Horn	High	Low to high	Good	Good
Whistle	High	Low to high	Good if intermittent	Good if frequency is properly chosen
Siren	High	Low to high	Very good if pitch rises and falls	Very good with rising and falling frequency
Bell	Medium	Medium to high	Good	Good in low-frequency noise
Buzzer	Low to medium	Low to medium	Good	Fair if spectrum is suited to background noise
Chimes and gong	Low to medium	Low to medium	Fair	Fair if spectrum is suited to background noise
Oscillator	Low to high	Medium to high	Good if intermittent	Good if frequency is properly chosen

processes and the learning of relevant associations (such as learning the alphabet or the shapes of road signs). Thus, the appropriate design of various types of visual displays must be predicated upon perceptual and learning factors as well as upon the specific visual skills of people.

Conditions affecting visual displays.   The ability of individuals to make visual discriminations is of course dependent upon their visual skills, especially their visual acuity. Aside from individual differences, there are certain variables (conditions) external to the individual that affect visual discrimination. Some of these variables are listed below.

1. *Luminance contrast.* Luminance contrast (frequently called brightness contrast or simply contrast) refers to the differences in luminance of the features of the object being viewed particular, the feature to be discriminated and its background (for example, an arrow on a direction sign against the background area of the sign).
2. *Amount of illumination.* The area or volume made visible.
3. *Time.* Within reasonable limits, the longer the viewing time, the greater the discrimination.
4. *Luminance ratio.* The ratio between the luminance of any two areas in the visual field (usually the area of primary visual attention and the surrounding area).

5. *Glare.* The intensity of the illumination.

6. *Combinations of variables.* Available evidence indicates that various combinations of the above variables, such as a combination of contrast and motion, have interactive effects on visual performance.

7. *Movement.* The movement of a target object or of the observer (or both) decreases the threshold of visual acuity.

Visual acuity.   Visual acuity is the ability of the eyes to differentiate between the detailed features of various objects that we see, whether in reading the fine print in an insurance contract or identifying a person across the street. Acuity depends very largely on the accommodation of the eyes, which is the adjustment of the lens of the eye to bring about proper focusing of the light rays on the retina.

The ability to make visual discriminations regarding certain types of visual stimuli does not necessarily mean that an individual can make discriminations regarding other types of stimuli. Because of this, there are different types of visual acuity. The most commonly used measure of acuity, *minimum separable acuity*, refers to the smallest feature or the smallest space between the parts of a target that the eye can detect. *Vernier acuity* refers to the ability to differentiate the lateral displacement, or slight offset, of one line from another that, if not so offset, would form a single continuous line (such as in lining up the "ends" of lines in certain optical devices). *Minimum perceptible acuity* is the ability to detect a spot (such as a round dot) from its background. In turn, *stereoscopic acuity* refers to the ability to differentiate the different images or pictures of a single object that has depth received by the retinas of the two eyes. (These two images differ most when the object is near the eyes, and differ least when the object is far away.)

Visual display design principles.   In the selection or design of visual displays for certain specific purposes, the basic type of display to use is sometimes virtually dictated by the nature of the information to be presented and its intended use. The following addresses some of the more conventionally used visual displays.

1. *Quantitative scales.*

   a. Digital or open-window displays are preferable if values remain long enough to read.

   b. Fixed-scale, moving-pointer designs are usually preferable to moving-scale, fixedpointer designs.

   c. For long scales, either a moving scale with tape on spools behind a panel or a counter plus a circular scale has practical advantages over a fixed scale.

    d. For values subject to continuous change, display all (or most) of the range used (as with a circular or horizontal scale).

    e. If two or more items of related information are to be presented, consider an integrated display.

    f. The smallest scale unit to be read should be represented on the scale by about 0.05 in or more.

    g. Preferably use a marker for each scale unit, unless the scale has to be very small.

    h. Use conventional progression systems—1, 2, 3, 4, etc.—unless there is reason to do otherwise, with major markers at 0, 10, 20, etc.

2. *Qualitative scales.*

    a. Preferably use a fixed scale with a moving pointer (to show trends).

    b. For groups, use circular scales and arrange null positions systematically for ease of visual scanning, as at 9 o'clock or 12 o'clock positions.

    c. Preferably use extended pointers, and possibly extended lines between scales.

3. *Status indicators.* If the basic data represent discrete, independent categories, or if basically quantitative data are always used in terms of such categories, use a display that represents each.

4. *Signals and warning lights.*

    a. The minimum size used must be consistent with luminance and exposure time.

    b. With low signal-to-background contrast, red light is more visible.

    c. For flashing lights, a flash rate of 1 to 10 per second presumably can be detected by humans.

5. *Representational displays.*

    a. A moving element (such as an aircraft) should be depicted against a fixed background (such as the horizon).

    b. Graphic displays that depict trends are read better if they are formed with lines rather than with bars.

    c. Pursuit displays are usually easier for people to use than compensatory displays.

    d. Cathode-ray-tube (CRT) displays are most effective when there are seven to nine or more scan lines per millimeter (mm).

    e. In the design of displays of complex configurations (such as traffic routes and wiring diagrams), avoid unnecessary detail and use schematic representations if this is consistent with their uses.

6. *Alphanumeric displays.*

    a. The typography of alphanumeric characters (design, size, contrast, etc.) is especially critical under adverse viewing conditions.

b. Alphanumeric characters should be presented in groups of three or four for optimum short-term memory.

c. Capital letters and numerals used in visual displays are read most accurately (a) when the ratio of stroke width to height is about 1:6 to 1:8 for black on white and somewhat higher (up to 1:10) for white on black, and (b) when the width is at least two-thirds the height.

7. *Symbolic displays.* Symbolic displays should be designed on the basis of the following perceptual principles: figure/ground, figure boundaries, closure, simplicity, and unity. In case the symbols do not clearly represent what they are supposed to represent, they should be evaluated experimentally.

Tactual displays.   In everyday life, humans depend upon their cutaneous (or somesthetic or skin) senses much more than they realize. Results of research lead one to consider the skin as housing three more or less separate systems of sensitivity: one for pressure reception, one for pain, and one responsive to temperature changes.

For the most part, tactual (touch) displays have utilized the hand and fingers as the principal receptors of information. For centuries, craft workers have used their sense of touch to detect irregularities and surface roughness in their work.

Not all parts of the hand are equally sensitive to touch. One common measure of touch sensitivity is the two-point threshold, the smallest distance between two pressure points at which the points are perceived as separate. The sensitivity increases (i.e., the two-point threshold becomes smaller) from the palm to the fingertips. Thus, tactual displays that require fine discriminations are best designed for fingertip reception. Tactual sensitivity is also degraded by low temperatures; therefore, tactual displays should be used with extreme caution in low temperatures.

Tactual displays as substitutes for sight.   Tactual displays have been most extensively used as substitutes for seeing—as an extension of our eyes. Some applications of tactual displays as substitutes for seeing, ranging from the relatively mundane to the more sophisticated, are described below.

1. *Identification of controls.* One use of the tactual sense is in the design of control knobs and related devices. The coding of such devices for tactual identification includes their shape, texture, and size.

2. *Reading printed material.* Probably one of the most widely known tactual displays for printed material is braille printing. Braille print for the blind consists of all the possible combinations of six raised "dots," numbered and arranged thus:

1..4
2..5
3..6

A particular combination of these represents each letter, numeral, or common word. The critical features of these dots are position, distance between dots, and dimension (diameter and height), all of which have to be discriminable to the touch.

Tactual displays as substitute for hearing.   Tactual displays have generally found three applications as substitutes for hearing: reception of coded messages, perception of speech, and localization of sound.

1. *Reception of coded messages.* Both mechanical and electrical energy have been used to transmit coded messages. Electrical energy has some advantages over mechanical vibrators. Electrodes are easily mounted on the body surface, and have less bulk and a lower power requirement than mechanical vibrators. Sensitivity to mechanical vibration is dependent upon skin temperature, whereas the amount of electric current required to reach threshold apparently is independent of skin temperature. A disadvantage of electrical energy is that it can elicit pain.

2. *Perception of speech.* Attempts to build successful tactile displays of speech have been disappointing. One reason given for the poor performance of such devices is that the resolving power of the skin, both temporally and spatially, is too limited to deal with the complexities of speech. Researchers believe, however, that improvements in tactual displays in the future will eventually permit tactile comprehension of speech at rapid rates.

3. *Localization of sounds.* Research reports suggest that subjects are able to localize a sound source by use of a simple tactile display with an accuracy level comparable to that with normal audition. The sound intensity is picked up by microphones placed over each ear. The outputs of these microphones are amplified and fed to two vibrators, upon which the subjects rest their index fingers. The intensity of vibration on each finger is then proportional to the intensity of the sound reaching the ear. It is this difference in intensity, especially when one moves one's head, that allows localization of sound.

Olfactory displays.

The olfactory senses. Humans depend on the sense of smell to give them information about things that otherwise is not easily obtainable— for example, the fragrance of a flower, the smell of fresh-brewed coffee, and the odor of sour milk.

The nose is a very sensitive instrument for detecting the presence of odors; this sensitivity depends on the particular substance and the individual doing the sniffing. Isobutyl isobutyrate (a fruity odor), for example, can be detected in concentrations of about three parts per million in water.

Surprisingly, however, the human sense of smell is not outstanding when it comes to making absolute identifications of specific odors (humans are far better at comparing odors on a relative basis). The number of different odors humans can identify depends on a number of factors, including the types of odors to be identified and the amount of training.

Applications of olfactory displays.   Olfactory displays have not found widespread application. Part of the reason for this is that they cannot be depended upon as a reliable source of information because people differ greatly with respect to their sensitivity to various odors; a stuffy nose can reduce sensitivity; people adapt quickly to odors, so that the presence of an odor is not sensed after a short period of exposure; the dispersion of an odor is difficult to control; and some odors nauseate people.

Despite these limitations, olfactory displays have useful applications—primarily as warning devices. Municipal gas companies, for example, add an odorant to natural gas so that people can detect gas leaks in their homes. Another example of an olfactory display uses odor to signal an emergency not associated with the gas itself. Several underground metal mines in the United States have used a "stench" system to signal workers to evacuate the mine in an emergency. The odor is released into the mine's ventilation system and is quickly carried throughout the entire mine.

Olfactory displays are not likely to become widespread in application, but they represent a unique form of information display which in special situations could be creatively integrated to supplement more traditional forms of displays.

Controls.   The primary function of a control is to transmit control information to some device, mechanism, or system. The type of information so transmitted can be divided into two categories: discrete information and continuous information. Discrete information is information that can represent only one of a limited number of conditions, such as on-off, high-medium-low, boiler 1–boiler 2–boiler 3, or alphanumerics such as A, B, and C or 1, 2, and 3. Continuous information, on the other hand, can assume any value on a continuum, such as speed (from 0 to 60 km/h), pressure [from 1 to 100 pounds per square inch (lb/in^2)], position of a valve (from fully closed to fully open), or amount of electric current [from 0 to 10 amperes (A)]. The information transmitted by a control maybe presented in a display, or it may be manifest in the nature of the system response. A secondary function of a control may be to serve, itself, as a display. An example here would be a rotary selector switch where the switch position indicates what information was input to the system.

General types of controls.   There are numerous types of control devices available. Certain types of controls are best suited for certain applications. One simple method of classifying controls is based on the type of information they can most effectively transmit (discrete versus continuous) and the force normally required to manipulate them (large versus small). Some of the more common types of controls and the types of applications for which they tend to be most suited are shown in Table 6.2.

Hand tools.   The basic principles of hand tool design are governed by the biomechanics of the human hand. The key objectives of hand tool design are to minimize strain to the wrist, avoid compression stress on the hand, and avoid repetitive finger action.

The critical wrist configuration.   The flexor tendons of the fingers pass through the carpal tunnel of the wrist. When the wrist is aligned with the forearm, all is well. However, if the wrist is bent, especially in palmar flexion or ulnar deviation (or both), problems occur. The tendons bend and cluster in the carpal tunnel. Continued use in this position will cause tenosynovitis, an inflammation of the tendon sheaths of the wrist. A common type of motion which can lead to tenosynovitis is clothes wringing, in which, for example, the wringing is done by a clockwise movement of the right fist and counterclockwise action of the left.

TABLE 6.2   Type of Information Transmitted

Force Required to Manipulate Control	Discrete	Continuous
Small	Push buttons (including keyboards) Toggle switches Rotary selector switches Detent thumb wheels	Rotary knobs Multirotational knobs Thumbwheels Levers (or joysticks) Small cranks
Large	Detent levers Large hand push buttons Foot push buttons	Handwheels Foot pedals Large levers Large cranks

This type of motion is also involved in inserting screws in holes, manipulating rotating controls such as those found on the steering handles of motorcycles, and looping wire while using pliers.

A key rule is to avoid ulnar deviation, or abnormal bending of the wrist, and allow more natural alignment of the wrist and forearm. By using bent pliers instead of straight pliers, for example, such disorders as tenosynovitis and "tennis elbow" can be avoided.

Tissue compression stress.   Often, in the operation of a hand tool or device, considerable force is applied with the hand, as when squeezing pliers or scraping paint with a paint scraper. Such actions concentrate considerable compressive force on the palm of the hand. Particularly pressure-sensitive areas are those overlying critical blood vessels and nerves, specifically the ulnar and radial arteries. The handle digs into the palm and obstructs blood flow through the ulnar artery. Such obstruction of blood flow, or ischemia, leads to numbness, tingling of the fingers, and thrombosis of the ulnar artery. If possible, handles should be designed to have large contact surfaces to distribute the force over a larger area and to direct it to less-sensitive areas such as the tough tissue between the thumb and index finger.

For a similar reason, the palm of the hand should never be used as a hammer. Not only will such action damage the arteries, nerves, and tendons of the hand, but the shock waves generated may travel to other body regions, such as the elbow or shoulder.

Related to compression stress is the use of finger grooves on tool handles. As anyone who has ever watched a professional basketball game knows, hands come in a wide variety of sizes. A person with thick fingers using a tool with finger grooves often finds that the ridges of the grooves dig into the fingers. A small-handed person may put two fingers into one groove, thereby squeezing the fingers together. It is for this reason that designers recommend not using deep finger grooves or recesses in tool handles if repetitive high finger forces are required.

Repetitive finger action.   Occasionally, if the index finger is used excessively for operating triggers, a condition known as "trigger finger" develops. The condition seems to occur most frequently if the handle of the tool or device is so large that the distal phalanx (segment) of the finger has to be flexed while the middle phalanx must be kept straight.

As a rule, frequent use of the index finger should be avoided, and thumb-operated controls should be used. One must be careful, however, not to hyperextend the thumb. This causes pain and inflammation. Preferable to thumb controls is the incorporation of a finger-strip control, as shown in Fig. 6.2, which allows several fingers to share the load and frees the thumb to grip and guide the tool.

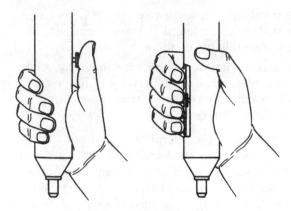

Figure 6.2 *(a)* Thumb-operated and *(b)* finger-strip-
operated pneumatic tool. Thumb operation results
in overex-tension of the thumb. Finger-strip control allows all thefingers to share the load and the
thumb to grip and guidethe tool.

The grip strength of the hand is related to the size of the object being
gripped. Maximum grip strength, for both males and females, occurs
with a grip axis between 2.5 and 3.5 in (66 and 85 mm).

Design for safety.   Designing tools and devices for safe operation includes eliminating pinching hazards and sharp corners and edges. This
can be done by putting guards over pinch points or using stops to prevent handles from fully closing and pinching the palm of the hand.
Sharp corners and edges can be rounded. Power tools such as saws and
drills can be designed with brake devices so that when the trigger is
released, the blade or bit stops quickly. Proper placement of the power
switch for quick operation can also reduce accidents with power tools.
Each type of tool presents its own set of safety considerations. The
designer must consider, in detail, how the tool will be used by the operator, and also how it might be misused.

Gloves.   Gloves are often used in conjunction with hand tools for protection against abrasions, cuts, punctures, and temperature extremes.
Gloves are available in an amazing number of varieties. In general, however, they can be distinguished in terms of the material used for construction (e.g., cotton, leather, vinyl, neoprene, asbestos, and even metal),
the cut (i.e., gunn cut and clute cut, as shown in Fig. 6.3*a*), the design of
the thumb (i.e., straight or wing, as shown in *Fig.* 6.3*b*), and the type
of wristband (e.g., knit, band top, gauntlet, or extended length).

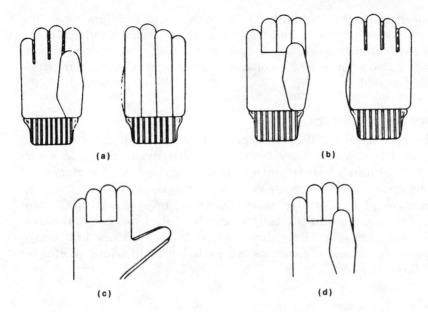

Figure 6.3 Common styles of work gloves distinguished by cut designs—*(a)* Clute cut and *(b)* Gunn cut—and thumb designs—*(c)* wind and *(d)* straight.

On some tasks there will be no differences in performance wearing different gloves, while on other tasks large differences will be found. In general, however, performance on tasks requiring fine motor control and tactile feedback will be adversely affected by wearing gloves, as compared to performance while bare-handed.

Women and left-handers.   Design engineers must consider that women make up approximately 50 percent and left-handers approximately 8 to 10 percent of the world's population. As a consequence, many hand tools and devices are not designed to accommodate everybody. The U.S. Air Force has determined that the average hand length of women is almost 2 cm (0.8 in) shorter than that of men. Less than 1 percent of men have a hand that is as short as the average woman's hand. Further, grip strength of women is on the average only about two-thirds that of men. These differences obviously have implications for tool design.

Anthropometry

Engineering anthropometry concerns the application of scientific physical measurement methods to human subjects for the development of engineering design standards. It includes static and functional (dynamic) measurements of dimensions and physical characteristics of the body as it occupies space, moves, and applies energy to physical objects, as a function of age, sex, occupation, ethnic origin, and other demographic variables.

Human body dimensions

Anthropometric data fall into two general classes, *static* dimensions and *functional* or *dynamic* dimensions. Static (structural) dimensions pertain to the human body in static or fixed positions. Such dimensional measurements would have specific applications, such as in designing helmets, earphones, or eyeglasses. Functional or dynamic body dimensions are taken while the body is involved in some physical movement. Although static body dimensions are useful for certain design purposes, functional dimensions are probably more widely useful for most design problems.

Human body posture

Although anthropometric data have implications for the design of work spaces, seats, equipment, etc., it is also essential to address the matter of posture, since the design of such items can affect the posture of people in their work and other situations. Since posture can influence both the comfort and the physical conditions of humans, it is necessary to consider the effects of human discomfort related to posture. The most important possible physical consequence of improper posture is spinal problems. A major objective in application of anthropometric data, therefore, is the design of systems that humans use to enhance the likelihood of maintaining proper posture.

Anthropometric design principles

In the application of anthropometric data, there are certain principles that may be relevant, each one being appropriate to certain types of design problems.

Design for extreme individuals. In the design of certain aspects of physical facilities, there is some limiting factor that argues for a design that specifically would accommodate individuals at one extreme or the other of some anthropometric characteristic, on the grounds that such a design also would accommodate virtually the entire population. A minimum dimension, or other aspect, of a facility would usually be based on an upper percentile value of the relevant anthropometric feature of the sample used, such as the 95th percentile. Perhaps most typically, a minimum dimension would be used to establish clearances, such as for doors, escape hatches, and passageways. If the physical facility in question accommodates large individuals (say, the 95th percentile), it will also accommodate all those smaller in size. On the other hand, maximum dimensions of some facility would be predicated on lower percentiles (say, the 5th percentile) of the distribution of people on the relevant anthropometric feature. The distance of control devices from an operator is an example; if those with short functional arm reach can reach a control, persons with longer arm reach generally can also do so. In setting such maximums and minimums, it is frequently the practice to use the 95th and 5th percentile values if the accommodation of 100 percent would incur trade-off costs out of proportion to the additional benefits to be derived.

Design for adjustable range.   Certain features of equipment or facilities preferably should be adjustable in order to accommodate people of various sizes. The forward-backward adjustments of automobile seats and the vertical adjustments of typists' chairs are examples. It is fairly common practice to design adjustable items such as these for the range of cases from the 5th to the 95th percentiles. The design of the passenger seats in the coach (tourist class) section of a commercial passenger aircraft is adjustable to accommodate different segments of the population within the standard percentile range according to their sitting height.

The economic rationale is that the amount of seat adjustment required to accommodate the extreme cases (such as below the 5th percentile and above the 95th) is disproportionate to the additional number of passengers which could be accommodated. Providing larger individual seats would reduce the number of seats in an aircraft configuration, which would reduce the potential passenger revenue to the air carrier.

Human Work Space

Work space dimensions

A human work space can consist of many different physical configurations of individuals. This must be considered in work space design, whether it involves facility layout or space requirements attendant to integrating elements of a human-machine system. The critical space parameters are related to human-machine operational relationships and locations of components, visual displays, hand controls, and foot controls. Figure 6.4 illustrates the primary body positions and critical dimensions based on body functions which affect working space design;

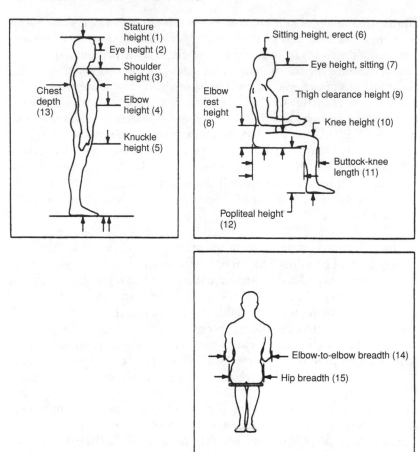

Figure 6.4 Diagrams of structure (static) body features for which data are provided for the U.S. civilian population (20–60 years of age) in Table 6.3. *(From M. S. Saunders and E.J. McCormick, Human Factors in Engineering and Design 7th ed., McGraw-Hill, 1993)*

as the illustration conveys, most work situations involve the human in a standing or sitting configuration.

Table 6.3 provides a summary of data related to the critical dimensions depicted in Fig. 6.4 and details the dimensional data according to gender and specific body features cited in the figure.

Location of components

Operational relationships. The operational relationships among people and between people and physical components can usually be expressed

TABLE 6.3 Selected Body Dimensions and Weights of U.S. Adult Civilians*

Body Dimension	Sex	Dimensions, in			Dimension, cm		
		5th	50th	95th	5th	50th	95th
1. Stature (height)	Male	63.7	68.5	72.6	161.8	173.6	184.4
	Female	58.9	63.2	67.4	149.5	160.5	171.3
2. Eye height	Male	59.5	63.9	68.0	151.1	162.4	172.7
	Female	54.4	58.6	62.7	138.3	148.9	159.3
3. Shoulder height	Male	52.1	56.2	60.0	132.3	142.8	152.4
	Female	47.7	51.6	55.9	121.1	131.1	141.9
4. Elbow height	Male	39.4	43.5	46.9	100.0	109.9	119.0
	Female	36.9	39.8	42.8	93.6	101.2	108.8
5. Knuckle height	Male	27.5	29.7	31.7	69.8	75.4	80.4
	Female	25.3	27.6	29.9	64.3	70.2	75.9
6. Height, sitting	Male	33.1	35.7	38.1	84.2	90.6	96.7
	Female	30.9	33.5	35.7	78.6	85.0	90.7
7. Eye height, sitting	Male	28.6	30.9	33.2	72.7	78.6	84.4
	Female	26.6	28.9	30.9	67.5	73.3	78.5
8. Elbow rest height, sitting	Male	7.5	9.6	11.6	19.0	24.3	29.4
	Female	7.1	9.2	11.1	18.1	23.3	28.1

* Body dimensions are depicted in Figure 6.4.

Source: Kroemer, 1989. (Courtesy of Dr. J.T. McConville, Anthropology Research Project, Yellow Springs, OH 45387, and Dr. K. W. Kennedy, USAF-AMRL-HEG, OH 45433.)

(Continued)

TABLE 6.3 Selected Body Dimensions and Weights of U.S. Adult Civilians* *(Continued)*

Body Dimension	Sex	Dimensions, in			Dimension, cm		
		5th	50th	95th	5th	50th	95th
9. Thigh clearance height	Male	4.5	5.7	7.0	11.4	14.4	17.7
	Female	4.2	5.4	6.9	10.6	13.7	17.5
10. Knee height, sitting	Male	19.4	21.4	23.3	49.3	54.3	59.3
	Female	17.8	19.6	21.5	45.2	49.8	54.5
11. Buttock-knee distance, sitting	Male	21.3	23.4	25.3	54.0	59.4	64.2
	Female	20.4	22.4	24.6	51.8	56.9	62.5
12. Popliteal height sitting	Male	15.4	17.4	19.2	39.2	44.2	48.8
	Female	14.0	15.7	17.4	35.5	39.8	44.3
13. Chest depth	Male	8.4	9.5	10.9	21.4	24.2	27.6
	Female	8.4	9.5	11.7	21.4	24.2	29.7
14. Elbow-elbow breadth	Male	13.8	16.4	19.9	35.0	41.7	50.6
	Female	12.4	15.1	19.3	31.5	38.4	49.1
15. Hip breadth, sitting	Male	12.1	13.9	16.0	30.8	35.4	40.6
	Female	12.3	14.3	17.2	31.2	36.4	43.7
X. Weight (lbs and kg)	Male	123.6	162.8	213.6	56.2	74.0	97.1
	Female	101.6	134.4	197.8	46.2	61.1	89.9

* Body dimensions are depicted in Figure 6.4.

Source: Kroemer, 1989. (Courtesy of Dr. J.T. McConville, Anthropology Research Project, Yellow Springs, OH 45387, and Dr. K. W. Kennedy, USAF-AMRL-HEG, OH 45433.)

in terms of links. Link data can be developed for a wide range of such relationships, although they fall generally into three classes: communication links, control links, and movement links. Communication and control links can be considered as functional. Movement links generally reflect sequential movements from one component to another. Some versions of the three types follow.

1. Communication links
   a. Visual (person to person or equipment to equipment)
   b. Auditory, voice (person to person, person to equipment, or equipment to person)
   c. Auditory, nonvoice (equipment to person)
   d. Touch (person to person or person to equipment)
2. Control links
   a. Control (person to equipment)
3. Movement links (movements from one location to another)
   a. Eye movements
   b. Manual movements, foot movements, or both
   c. Body movements

Human-machine link analysis can be used as an aid in connection with the general location of components or with their relative arrangements. In some circumstances it can be used as the basis for assignment of priorities.

General principles for component location.   There are various components that need to be located within the system or facility. (The term *component* in this context refers to virtually any relevant feature, such as a display, control, material, machine, work area, or room.) It is reasonable to presume that any given component has a generally "optimum" location for serving its purpose. This optimum would be predicated on the human sensory, anthropometric, and biomechanical characteristics that are concerned (reading a visual display, activating a foot push button, etc.) or on the performance of some operational activity (such as reaching for parts, preparing food in a restaurant, or storing material in a warehouse). Along with the concept of optimum location, there are certain principles that can serve as guidelines for component location.

1. *Importance principle.* This principle deals with operational importance, that is, the degree to which the performance of the activity with the component is vital to the achievement of the objectives of the system or some other consideration. The determination of importance is largely a matter of judgment, but, to cite a specific example,

a warning light in an automobile to indicate engine malfunction or low oil supply should be directly in front of the driver.

2. *Frequency-of-use principle.* As implied by the name, this concept applies to the frequency with which some component is used. For example, the activation control of a punch press should be in a convenient location, since it is used very frequently.

3. *Functional principle.* The functional principle of arrangement provides for the grouping of components according to their function, such as the grouping together of displays and controls that are functionally related in the operation of the system. Thus, temperature indicators and temperature controls might well be grouped together, and electric power distribution instruments and controls usually should be in the same general location.

4. *Sequence-of-use principle.* For certain items, there are sequences or patterns of relationship that frequently occur in the operation of the equipment. In applying this principle, the items would be arranged so as to take advantage of such patterns; thus, items used in sequence would be in close physical relationship to one another.

The application of these various principles of arrangement of components must necessarily be predicated on rational, judgmental considerations. In a very general way, however, and in addition to the optimum premise, the concepts of importance and frequency probably are especially applicable to the more basic phase of locating components in a general area in the work space; in turn, the sequence-of-use and functional principles tend to apply more to the arrangement of components within a general area.

Considerations affecting component location.   Although the optimum locations of some specific components probably depend upon situational factors, some generalizations can be made about certain classes of components. Certain of these are noted here.

Visual displays.   The normal line of sight is usually considered to be about 15° below the horizon. Visual sensitivity accompanied by moderate eye and head movements permits fairly convenient visual scanning of an area around the normal line of sight. The area for most convenient visual regard (and therefore generally preferred for visual displays) has generally been considered to be defined by a circle roughly 10 to 15° in radius around the normal line of sight.

Controls.   In minimizing the distances between components, such as the sequential links between controls, there are obvious lower-bound constraints that need to be respected, such as the physical space required when operating individual controls to avoid touching other controls. These lower-bound constraints would be predicated on the combination of anthropometric factors (such as of the fingers and hands) and on the precision of normal psychomotor movements made in the use of control devices.

1. *Hand controls.* The optimum location of hand-control devices is a function of the type of control, the mode of operation, and the appropriate criterion of performance (accuracy, speed, force, etc.).
2. *Foot controls.* Since only the most loose-jointed humans can put their feet behind their head, foot controls generally need to be located in fairly conventional areas, as exemplified by the foot controls on an automobile.

Work space design priorities

In designing workplaces, some compromises are almost inevitable because of competing priorities. In this regard, however, appropriate link values can aid in the trade-off process. Some general guidelines for designing workplaces that involve displays and control are given below.

First priority: Primary visual tasks

Second priority: Primary controls that interact with primary visual tasks

Third priority: Control-display relationships (put controls near associated displays, compatible movement relationships, etc.)

Fourth priority: Arrangement of elements to be used in sequence

Fifth priority: Convenient location of elements that are used frequently

Sixth priority: Consistency with other layout within the system or in other systems.

HFE Environmental Considerations

The physical environments which humans use fit into general categories. The first consists of the physical space and related facilities which people use, ranging from the immediate environment (such as a workstation, a lounge chair, or a typing desk) through the intermediate (such as home, an office, a factory, a school, or a football stadium) to the general (such as a neighborhood, a community, a city, or a highway system). The second category consists of the various aspects of the ambient environment, such as illumination, atmospheric conditions (including pollution), and noise. It should be noted that some aspects of the physical environment in which we live and work are part of the natural environment and may not be amenable to modification. It is therefore necessary to provide protection from certain undesirable environmental conditions, such as heat or cold. Although the nature of people's involvement with their physical environment is essentially passive, the environment tends to impose certain constraints on their behavior (such as limiting the range of their movements or restricting their field of view) or to predetermine certain aspects of behavior (such as stooping down to look into a file cabinet, wandering through a labyrinth in a supermarket to find where the bread is, or trying to see the edge of the road when driving on a rainy night). Since the current model of the human organism is the result of evolutionary processes over millions of years, it has developed substantial ability to adapt to the environmental variables within the world in which we live, including its atmosphere. There are, however, limits to human adaptability. The environmental aspects of science and technology which affect humans have a significant impact on logistics engineering and, consequently, must be incorporated into the system design process.

Heat

The heat exchange process. The human body is continually making adjustments to maintain thermal equilibrium, i.e., a balance between the net heat produced and the net heat loss to or gain from the environment. The primary factors that influence thermal regulation are

1. *Metabolism:* The oxidation of food elements in the body
2. *Convection:* The heat gain or loss by the mixing of air close to the body
3. *Evaporation:* The heat loss by evaporation of body fluids, especially perspiration and exhaled breath
4. *Radiation:* The heat loss to or gain from surrounding environmental sources (the sun, walls, etc.) by direct radiation

Heat can also be exchanged by conduction, the direct contact of the body with some object, but conduction is a very negligible factor and usually is not taken into account in studying heat exchange.

Heat stress.   One of the most direct physiological effects of heat stress is on the temperature of the body. Heat stress usually is accompanied by increases in rectal temperature.

Levels of work in combination with environmental conditions that bring about a rise in core temperature induce other corresponding physiological changes which, if continued, can cause hyperthermia, a condition that renders normal heat loss more difficult. Dehydration, such as from sweat, is another possible consequence of heat stress. A deficit of body liquids brought about during the daily work periods must be replaced during the nonwork intervals.

Cold

Civilization is generally reducing the need for many people to work in cold environments; however, there still are some circumstances in which people must work, and live, in such environments. These situations include outdoor work in winter, Arctic locations (especially in military and exploration activities), cold-storage warehouses, and food lockers. As in the case of heat exposure, there are a number of interrelated factors that affect the tolerance, comfort, and performance of people in cold environments; these include the level of activity, degree of acclimatization, duration of exposure, and insulation.

Performance in cold.   The primary interest in the effects of cold on performance relates to manual tasks of one sort or another. Studies have made it clearly evident that when humans are to perform manual tasks in the cold, provision should be made for maintaining hand and skin temperatures at reasonable levels, such as permitting the individuals to warm their hands indoors or in some other manner.

Physiological effects of cold.   With inadequate protection, exposure to the cold brings about a reduction of both core and skin temperatures. Continued exposure, of course, can bring about frostbite and its related effects, and ultimately death.

Wind chill.   One of the indicators related to cold is the wind chill. It provides a means for making a quantitative comparison of combinations of temperature and wind speed. Table 6.4, derived from wind-chill index data, shows the cooling effects of combinations of certain temperatures

TABLE 6.4   Cooling Effects of Temperature and Wind Speed

Wind speed, mi/h (km/h)		Equivalent temperature when air temperature, °F (°C)					
		40 (4)	20 (−7)	10 (−12)	0 (−18)	−10 (−23)	−20 (−29)
calm		40 (4)	20 (−7)	10 (−12)	0 (−18)	−10 (−23)	−20 (−29)
5	9	37 (3)	16 (−9)	6 (−14)	−5 (−20)	−15 (−26)	−26 (−32)
10	16	28 (−2)	4 (−16)	−9 (−23)	−21 (−29)	−33 (−36)	−46 (−43)
20	32	18 (−8)	−10 (−23)	−25 (−33)	−39 (−37)	−53 (−47)	−67 (−55)
30	49	13 (−11)	−18 (−28)	−33 (−36)	−48 (−43)	−63 (−52)	−79 (−62)
40	64	10 (−12)	−21 (−29)	−37 (−38)	−53 (−47)	−69 (−56)	−85 (−65)

and wind speeds derived from the scale, expressed as equivalent temper-atures. For example, an air temperature of 10° F (−12° C) with a 20-mi/h wind produces the same cooling effect as a temperature of −25° F (−33° C) under calm conditions.

Although acclimatization can increase the tolerance of wind chill, it does not eliminate its effects on manual performance. Cold seems not to affect mental performance or visual performance.

Effects of thermal adjustment

When the body changes from one thermal environment to another, cer-tain physical adjustments are made by the body, especially the following:

Changes from warm to cold environment: (1) The skin becomes cool; (2) the blood is routed away from the skin to the central part of the body, where it is warmed before flowing back to the skin area; (3) rectal temperature first rises slightly and then falls; and (4) shivering and gooseflesh may occur. The body may stabilize with large areas of skin receiving little blood.

Changes from cool to warm environment: (1) More blood is routed to the surface of the body, resulting in higher skin temperature; (2) rectal temperature first falls but rises with continued exposure; and (3) sweating may begin. The body may stabilize with increased sweating and increased blood flow to the surface of the body.

If the change from one situation to another is so extreme as to cause the body temperature to increase appreciably (producing a condition of hyperthermia) or to decrease appreciably (producing hypothermia), there may be serious physiological consequences, and in extreme situations death. If thermal adjustments are dictated by the nature of a functional requirement, it is essential to impose a program of *acclimatization*, which enables the human to make adjustments through gradual, controlled ex-posure to extreme thermal conditions.

Comfort-health index

Because of the complexities of the relationships among the variables that influence the effects of atmospheric conditions, a comfort-health index (CHI) has been developed that summarizes information on tem-perature sensation, comfort, health, and physiological aspects for people engaged in sedentary tasks. This is provided in Table 6.5. It shows the sensory, physiological, and health responses typically associated with prolonged exposures at the equivalent temperature (ET) or CHI values, expressed as a dry-bulb temperature at 50 percent rh.

Management of temperature problems

Where outside heat or cold conditions could be undesirable for people who are to work or live in them, the optimum solution (when possible) is to modify the environments to make them more suitable for people. When this is not possible, certain other actions in the management of personnel may be desirable, such as selection of personnel who can tol-

New T_{off} Scale	Sensation		Physiology	Health
	Temperature	Comfort		
°C ┬ °F	Limited tolerance		Body heating	Circulatory collapse
40 ┤			Failure of regulation	
┤ 100	Very Hot	Very uncomfortable	Increasing stress caused by sweating and blood flow	↑ Increasing danger of heat strokes, cardiovascular embarrassment
35 ┤	Hot			
┤ 90	Warm	Uncomfortable		
30 ┤				
┤	Slightly Warm			
┤ 80			Normal regulation by sweating and vascular change	
25 ┤	Neutral	Comfortable		Normal health
┤			Regulation by vascular change	
┤ 70	Slightly cool			
20 ┤				
┤	Cool	Slightly uncomfortable	Increasing dry heat loss Urge for more clothing or exercise (behavioral regulation)	
┤ 60	Cold			↓ Increasing complaints of dry mucosa and skin
15 ┤			Vasoconstriction in hands and feet	
┤				
10 ┴ 50	Very cold	Uncomfortable	Shivering	↓ Muscular pain, impairment of peripheral circulation

Source: ASHRAE, 1977.

erate the condition (sometimes by tryout for four or five days); permitting people to become gradually acclimatized; modifying the work activities (as by reducing energy requirements); rotating personnel from one job situation to another; providing adequate rest periods; having people wear appropriate clothing; in the case of hot environments, considering conductive cooling; and, also in the case of hot environments, being sure that people drink enough water (or other liquids) to minimize dehydration. In the case of indoor situations, atmospheric control can be effected through heating, air conditioning, circulation of air, humidity control, insulation, and shielding against radiation, as well as by other techniques. Such atmospheric control is, of course, in the domain of the facility engineers, but it is nevertheless critical to the logistics engineer concerned with the well-being and functional efficacy of logistics technicians. Two relevant personnel-related matters are discussed below: clothing and rest periods.

Clothing. Experience has established that in warm and hot environments, one should wear light, loose clothing. And those who live in hot climates have discovered that light-colored clothing (such as white) is cooler than dark clothing; this is because light-colored clothing reflects more heat.

In the case of cold temperatures, the use of warm clothing can extend the tolerance level of people. In connection with the use of apparel in cold conditions, however, researchers have found that adding additional garments does not compensate fully for the discomfort. Although the subjects of research have reported increased feelings of warmth with long-sleeved woolen sweaters, the sweaters did not fully alleviate the feelings of discomfort. Researchers attributed this to the fact that the subjects still reported that their feet felt cold. Thus, for adequate protection in the cold, attention should be given to providing footwear that is as warm as possible.

Rest periods. Persons working in hot environments need to be provided with adequate rest periods in order to reduce the total heat stress to which they are subjected. When rest periods are provided, the rest preferably should be in somewhat neutral situations to permit more complete recovery.

Noise

Defining a possible noise problem essentially consists of two phases. The first of these is the measurement of the noise itself. The overall sound-pressure level (e.g., dBA) will give a gross indication of a potential noise problem. The second is to determine what noise level would be acceptable, in terms of potential hearing loss, annoyance, and impact on verbal communications.

Noise and hearing loss.   There are two primary types of deafness, nerve deafness and conduction deafness. Nerve deafness usually results from damage to or degeneration of the hair cells of the organ of Corti in the cochlea of the ear. Conduction deafness is caused by a condition of the outer or middle ear that affects the transmission of sound waves to the inner ear.

The hearing loss in nerve deafness is typically uneven; usually the hearing loss is greater in the higher frequencies than in the lower frequencies. Once nerve degeneration has occurred, it can rarely be corrected.

Conduction deafness, on the other hand, is only partial, since airborne sound waves strike the skull and may be transmitted to the inner ear by conduction through the bone. It may be caused by a number of different conditions, such as adhesions in the middle ear that prevent the vibration of the ossicles, infection of the middle ear, wax or some other substances in the outer ear, or scars resulting from a perforated eardrum.

This type of deafness can sometimes be arrested, or even improved. Hearing aids are more frequently useful in this type of deafness than they are in cases of nerve deafness.

Noncontinuous noise.   The gamut of noncontinuous noise includes intermittent (but steady) noise (such as machines that operate for short, interrupted periods of time), impact noise (such as that from a drop forge), and impulsive noise (such as from gunfire). In heavy doses, such noise levies its toll in hearing loss, but the combinations and permutations of intensity, noise spectrum, frequency, duration of exposure, and other parameters preclude any simplistic descriptions of its effects.

Physiological effects of noise.   Permanent hearing loss is the consequence of physiological damage to the mechanisms of the ear. Aside from possible damage to the ear itself, there is a probability that continued exposure to noise might induce other temporary or permanent physiological effects.

The physiological reactions to noise usually would not be likely to be of a pathological nature if the noise occurred only a few times. However, there is an accumulating body of evidence that indicates that exposure to high noise levels (such as 96 dB or more) acts as a stressor and over a period of years may produce pathological side effects. European studies have shown that people working where there are high levels of noise have more somatic complaints than workers experiencing low levels of noise. "High-noise" workers complain more of irritability, headaches, tiredness, bad sleep, and heart pains. In such cases the noise levels involved have typically been high, over 95 dBA, and exposure has usually been for more than 10 years.

Noise control.   Controlling a noise problem can be accomplished by attacking the noise at the source, along its path from the source to the receiver, or at the receiver. Control at the source includes proper design and maintenance of the machines, use of vibration-absorbing mountings and mufflers, and use of sound-absorbing materials on the inside and outside surfaces of the machine. Controlling noise along its path includes use of barriers, enclosures, acoustical treatments, and baffles.

When the noise level cannot reasonably be reduced to "safe" limits, some form of ear protection should be considered for those people who are exposed to the noise. The two types of ear-protection devices that are reusable are earplugs and earmuffs.

Relative Capabilities: Humans versus Machines

Humans

Humans are generally better in the abilities to

1. Sense very low levels of certain kinds of stimuli: visual, auditory, tactual, olfactory, and taste.
2. Detect stimuli against high-noise-level backgrounds, such as blips on CRT displays with poor reception.
3. Recognize patterns of complex stimuli which may vary from situation to situation, such as objects in aerial photographs and speech sounds.
4. Sense unusual and unexpected events in the environment.
5. Store (remember) large amounts of information over long periods of time (they are better at remembering principles and strategies than masses of detailed information).
6. Retrieve pertinent information from storage (recall), frequently retrieving many related items of information; however, the reliability of recall is low.

7. Draw upon varied experience in making decisions; adapt decisions to situational requirements; act in emergencies. (They do not require previous programming for all situations.)

8. Select alternative modes of operation if certain modes fail.

9. Reason inductively, generalizing from observations.

10. Apply principles to the solution of varied problems.

11. Make subjective estimates and evaluations.

12. Develop entirely new solutions.

13. Concentrate on the most important activities, when overload conditions require it.

14. Adapt their physical response (within reason) to variations in operational requirements.

Machines

Machines are generally better in their abilities to

1. Sense stimuli that are outside the normal range of human sensitivity, such as x-rays, radar wavelengths, and ultrasonic vibrations.

2. Apply deductive reasoning, such as recognizing stimuli as belonging to a general class (but the characteristics of the class need to be specified).

3. Monitor for prespecified events, especially when infrequent (but machines cannot improvise if unanticipated types of events occur).

4. Store coded information quickly and in substantial quantities (for example, large sets of numerical values can be stored very quickly).

5. Retrieve coded information quickly and accurately when specifically requested (although specific instructions on the type of information that is to be recalled need to be provided).

6. Process quantitative information following specified programs.

7. Make rapid and consistent responses to input signals.

8. Perform repetitive activities reliably.

9. Exert considerable physical force in a highly controlled manner.

10. Maintain performance for extended periods of time (machines typically do not "fatigue" as rapidly as humans).

11. Count or measure physical quantities.

12. Perform a multiplicity of programmed activities simultaneously.

13. Maintain efficient operations under conditions of heavy load (humans have relatively limited channel capacity).

14. Maintain efficient operations under distractions.

Issues of flexibility and consistency

The analysis of the relative advantages of humans and machines often hinges on the critical issues of comparative flexibility and consistency. Human factors experiential data have confirmed that humans are flexible but cannot be depended upon to perform in a consistent manner; machines can be depended upon to perform consistently, but lack the flexibility of humans.

Safety Engineering

General

Overview

It is axiomatic that there is no such thing as an accident-free environment. An important objective of logistics engineering is to take actions that reduce the probability of accidents. Such actions include the design of equipment and facilities in terms of human factor considerations that will contribute to safety, the development of procedures that contribute to safety and the training of personnel to follow such procedures, and the consistent use of appropriate protective devices by personnel.

This chapter will address those aspects of safety which are the predominant concerns of the logistics engineer. These include (1) corporate policies and procedures on safety, (2) safety considerations in product or system design, (3) analysis of potential hazards, (4) safety considerations in system development and production, (5) consumer (product user) safety, and (6) principles of safety with regard to product liability.

Safety terms and definitions

The following terms are frequently cited and pertain to the topics discussed herein.

1. *Damage.* The partial or total loss of hardware caused by component failure; exposure of hardware to heat, fire, or other environments; human errors; or other inadvertent events or conditions
2. *Hazard.* A condition that is prerequisite to a mishap

3. *Hazardous event.* An occurrence that creates a hazard

4. *Hazardous event probability.* The likelihood, expressed in quantitative or qualitative terms, that a hazardous event will occur

5. *Hazard probability.* The aggregate probability of occurrence of the individual events that create a specific hazard

6. *Hazard severity.* An assessment of the worst credible mishap that could be caused by a specific hazard.

7. *Mishap.* An unplanned event or series of events that results in death, injury, occupational illness, or damage to or loss of equipment or property

8. *Risk.* An expression of the possibility of a mishap in terms of hazard severity and hazard probability

9. *Safety.* Freedom from those conditions that can cause death, injury, occupational illness, or damage to or loss of equipment or property

10. *System safety.* The application of engineering and management principles, criteria, and techniques to optimize safety within the constraints of operational effectiveness, time, and cost throughout all phases of the system life cycle

11. *System safety engineer.* An engineer who is qualified by training and/or experience to perform system safety engineering tasks

12. *System safety management.* An element of management that defines the system safety program requirements and ensures the planning, implementation, and accomplishment of system safety tasks and activities consistent with the overall program requirements.

Corporate Policies and Procedures

The success of the corporate safety program depends upon definitive statements of safety objectives by corporate management and their translation into product development, production, and logistics support procedures as well as the design of product hardware and software. A formal safety program that stresses early identification of potential hazards and elimination of the associated risks or their reduction to acceptable levels is the principal contribution of a body of effective corporate safety policies and procedures.

Safety policies

Corporate safety policies should provide goals to ensure that

1. Safety, consistent with consumer market requirements, is designed into the product in a timely, cost-effective manner.

2. Hazards associated with production, consumer use, and logistics support of each product are identified, evaluated, and eliminated, or the associated risks reduced to a level acceptable to management throughout the entire life cycle of the product. Risk should be described in risk assessment terms.

3. Historical safety data, including lessons learned from related product systems, are considered and used.

4. Minimum risk is sought in accepting and using new designs, materials, and production and test techniques.

5. Actions are taken to eliminate hazards or reduce risk to a level acceptable to management, and these actions are documented.

6. Modifications required to improve product safety are minimized through the timely inclusion of safety features during research, development, and production of a system.

7. Changes in design, configuration, or product utilization requirements are accomplished in a manner that maintains a risk level acceptable to management.

8. Consideration is given to safety and ease of disposal of any hazardous materials associated with the product.

9. Significant safety data are documented as "lessons learned" and are submitted to data banks or as proposed changes to the applicable design and specifications.

10. Risks are minimized for actions which must occur under hazardous conditions during the specified periods of performance.

Safety procedures

Procedures implemented by management must consider the safety implications of development, production, testing, distribution, storage, installation, operation, logistics support, and disposal of the products. Hazards to safety and health should be systematically surveyed, identified, and analyzed, with special attention to the following:

1. System, facility, and personnel protective equipment design requirements (e.g., ventilation, noise attenuation, and radiation barriers) to allow safe operation and maintenance. When feasible engineering designs that will reduce hazards to acceptable levels are not available, alternative protective measures must be specified (e.g., protective clothing or specific operation or maintenance practices) to reduce risk to an acceptable level.

2. Changes in functional or design requirements for system hardware! software, facilities, tooling, or support/test equipment that are needed to eliminate hazards or reduce associated risks.

3. Requirements for safety devices and equipment, including personnel safety and life support equipment.

4. Warnings, cautions, and special emergency procedures (e.g., egress, rescue, escape, render-safe, and back-out).

5. Requirements for handling, storage, transportation, maintenance, and disposal of hazardous materials.

6. Requirements for safety training and personnel certification.

7. Hazard potential resulting from
   a. Toxic materials (e.g., carcinogens or suspected carcinogens, systemic poisons, asphyxiants, and respiratory irritants).
   b. Physical agents (e.g., noise, heat or cold stress, ionizing and nonionizing radiation).

8. Potential hazards resulting from unplanned events, including hazards introduced by human errors.

It is management's responsibility to document all existing and potential health and safety hazards and provide for engineering controls, equipment, and protective procedures to reduce the associated risks to acceptable levels.

Management-employee teamwork

Effective safety measures necessarily entail employee awareness, support, participation, and minimum essential safety expertise. This is accomplished through a program based on manager-employee communication, training, and certification.

Communication. The crucial elements of communication are the positive flow of management information to the employees and employee feedback to management. The goal of this interrelationship is

a. To give employees more knowledge of safety factors.

b. To change the attitudes of employees so that they are more inclined to act safely.

c. To ensure that safe behavior takes place. It is obvious that this is the most important aim, and the one at which all safety exhortations must be directed.

In this regard, researchers have noted that in many accidents there are contributory factors that generally are out of the control of the individuals involved—factors such as improper design in terms of human factors, fatigue (which individuals cannot avoid), lack of training, and failure by management to ensure that guards and protective clothing are available. It has been further noted that safety exhortations directed toward workers can be useful only if the accident-causing factors are under the control of the workers. (The responsibility for factors that workers cannot control should, of course, fall in the lap of management.)

Safety exhortations can produce desired employee behavior, but it is essential that they

1. Be specific as to a particular task and situation.
2. Reinforce with a training program.
3. Give a positive instruction.
4. Be placed in close proximity to where the desired action is to take place.
5. Build on existing attitudes and knowledge.
6. Emphasize nonsafety aspects.

Safety exhortations should not

1. Involve horror, because in the present state of our knowledge this appears to bring in the defense mechanisms of those at whom the exhortations are most directed.
2. Be negative, because this can show people the wrong way of acting when what is required is the correct way.
3. Be general in perspective, because most people think they act safely. This type of exhortation is thus seen as relevant only to other people.

Training and certification.   Design engineers need to understand basic system safety principles to design hazard-free systems. An effective training program will include training design engineers as a top priority. Managers need to be educated about the importance of safety in initial design to reduce the need for costly redesign and modifications. Test personnel need to be trained in safe handling, operation, and testing of equipment. Production and logistics personnel need safety training in their functions.

Formal classroom training sessions using a thorough lesson plan containing all the necessary handouts is one of the most effective and efficient training methods. Requiring examinations and final certification helps to ensure that the trainees have understood and hopefully will apply the material presented.

Certification on such subjects as hazard types and their recognition, causes, effects, preventive and control measures, procedures, checklists, human error, safeguards, safety devices, protective equipment, monitoring and warning devices, and contingency procedures should be mandatory for actively involved technical personnel. Specific certification requirements should be established by a program certification board that includes the safety manager as a member.

Design Safety

Design safety is a prelude to consumer safety. The goal is to produce an inherently safe product that will have the minimum operational safety requirements or restrictions. Products must be designed for reasonably foreseeable use, not solely for intended use. This means that an analysis must be made to determine the types of use and misuse a product could be subjected to; this may require a survey of users of similar products or potential users of a new product. Laboratory simulation tests might also be conducted on product prototypes to gain insight into user behavior and product performance. The overall goal of a system safety program is to design products and systems that do not contain hazards. However, the nature of most complex systems makes it impossible or impractical to design them completely hazard-free.

Product safety design requirements

Product safety design requirements should be specified after review of pertinent standards, specifications, regulations, design handbooks, and other sources of design guidance for applicability to the design of the product. Some general safety design requirements are as follows:

1. Delineate the scope of product uses.
2. Identify the environments within which the product will be used.
3. Describe the user population.
4. Postulate all possible hazards, including estimates of probability of occurrence and seriousness of resulting harm.
5. Eliminate identified hazards or reduce the associated risk through design, including material selection or substitution. When potentially hazardous materials must be used, select those with least risk throughout the life cycle of the system.
6. Isolate hazardous substances, components, and operations from other activities, areas, and personnel, and from incompatible materials.
7. Locate equipment to minimize personnel exposure to hazards (e.g., hazardous chemicals, high voltage, electromagnetic radiation, cutting edges, or sharp points) during access for operations, servicing, maintenance, repair, or adjustment.
8. Minimize risk resulting from excessive environmental conditions (e.g., temperature, pressure, noise, toxicity, acceleration, and vibration).
9. Design to minimize risk created by human error in the operation and support of the system.
10. Consider alternative approaches to minimizing risk from hazards that cannot be eliminated. Such approaches include interlocks, re-

dundancy, fail-safe design, system protection, fire suppression, and protective clothing, equipment, devices, and procedures.

11. Evaluate proposed alternatives relative to the expected performance standards of the product, including the following:

   a. Other hazards that may be introduced by the alternatives.

   b. Their effect on the subsequent usefulness of the product.

   c. Their effect on the ultimate cost of the product. d. A comparison to similar products.

12. Protect the power sources, controls, and critical components of redundant subsystems by physical separation or shielding.

13. When alternative design approaches cannot eliminate the hazard, provide warning and caution notes in production, operations, maintenance, and repair instructions, and distinctive markings on hazardous components and materials, equipment, and facilities to ensure personnel and equipment protection.

14. Minimize the severity of injury to personnel or damage to equipment in the event of a mishap.

15. Design software-controlled or monitored functions to minimize initiation of hazardous events or mishaps.

16. Review design criteria for inadequate or overly restrictive requirements regarding safety. Recommend new design criteria supported by study, analyses, or test data.

Warnings and instructions.   Closely allied to the issue of whether a product is unreasonably dangerous are the warnings and instructions that accompany it. A distinction can be made between warnings and instructions, although the line between these is rather vague. Warnings inform the user of the dangers of improper use and tell how to guard against those dangers, if possible, whereas instructions tell the user how to use the product effectively.

Product safety precedence

As hazard analyses are performed, hazards will be identified that will require resolution. Safety precedence defines the order to be followed in satisfying product safety requirements and reducing risks. The order of precedence for satisfying product safety requirements and resolving identified hazards is generally as follows:

1. *Design for minimum risk.* From the first, design to eliminate hazards. If an identified hazard cannot be eliminated, reduce the associated risk to an acceptable level through design selection.

2. *Incorporate safety devices.* If identified hazards cannot be eliminated or their associated risk adequately reduced through design selection,

reduce the risk to an acceptable level through the use of fixed, automatic, or other protective safety features or devices. Provisions should be made for periodic functional checks of safety devices when applicable.

3. *Provide warning devices.* When neither design nor safety devices can effectively eliminate identified hazards or adequately reduce the associated risk, devices should be used to detect the condition and to produce an adequate warning signal to alert personnel of the hazard. Warning signals and their application should be designed to minimize the probability of incorrect personnel reaction to the signals and should be standardized within like types of systems.

4. *Develop procedures and training.* Where it is impractical to eliminate hazards through design selection or adequately reduce the associated risk with safety and warning devices, special procedures and training may be required, both for production personnel and for consumer users of the product.

Hazards Analysis

General

The objective of hazards analysis is to obtain an assessment of the risk attendant on production, support, and consumer operation of the product. Such analysis is based broadly on evaluation of available incident data, including mishap data on similar systems, and evaluation of possible hazards associated with the design and operational function of the product. The scope of product hazard analysis encompasses

1. Hazards associated with product development, production, and logistics support
2. Occupational health assessment
3. Hazards inherent to product design
4. Assessment of risk derived from the potential hazards

Hazards associated with development, production, and logistics support

Analysis of the manufacturing and support environment entails, at a minimum, identification and evaluation of the following:

1. Hazardous components
2. Safety-related interface considerations among various elements of the system
3. Environmental constraints, including the operating environment

4. Operating, test, maintenance, and emergency procedures
5. Facilities, support equipment, and training (e.g., training and certification pertaining to safety operations and maintenance)
6. Safety-related equipment, safeguards, and possible alternative approaches

The purpose of the environmental evaluation is to identify safety-critical areas, hazard severity, hazard probability, and operational constraints, so as to determine safety criteria to be used in the design of the product.

Occupational health hazard assessment

The purpose of occupational health assessment is not to dictate designs based on health protection but to ensure that decision makers are aware of the health hazards involved and their impact so that knowledgeable decisions regarding potential trade-offs can be made.

The first step in the occupational health hazard assessment is to identify and determine the quantities of potentially hazardous materials or physical agents (noise, radiation, heat stress, cold stress) involved with the system and its logistical support. The next step would be to analyze how these materials or physical agents are used in the system and for its logistical support. Based on the use, quantity, and type of each substance or agent, estimate where and how personnel exposure may occur and, if possible, the degree or frequency of exposure involved. The final step would include incorporating in the design of the system and its logistical support equipment/facilities cost-effective controls to reduce exposures to acceptable levels.

The following factors associated with the system and the logistical support required to operate and maintain it should be considered in the occupational health hazard assessment:

1. Toxicity, quantity, and physical state of materials
2. Routine or planned uses and releases of hazardous materials or physical agents
3. Accidental exposure potentials
4. Hazardous waste generated
5. Hazardous material handling, transfer, and transportation requirements
6. Protective clothing/equipment needs
7. Detection and measurement devices required to quantify exposure levels
8. Number of personnel potentially at risk
9. Engineering controls that could be used, such as isolation, enclosure, ventilation, and noise or radiation barriers.

Hazards associated with product design

The purpose of the system design hazard analysis is to identify all components and equipment, including software, whose performance, performance degradation, functional failure, or inadvertent functioning could result in a hazard or whose design does not satisfy specified safety requirements. The analysis shall include a determination of the modes of failure, including reasonable human errors as well as single-point failures. This analysis looks at each component and identifies hazards associated with operating or failure modes and is especially intended to determine how operation or failure of components affects the overall safety of the system. This analysis should identify necessary actions, using the safety precedence criteria to determine how to eliminate or reduce the risk of identified hazards.

Hardware design.   Analysis of hazards associated with the design of hardware components should consider the following:

1. Compliance with safety criteria called out in the applicable system/ subsystem requirements documents.

2. Possible combinations of independent or dependent failures that can cause hazards to the system or personnel. Failures of controls and safety devices should be considered.

3. How normal operations of systems and subsystems can degrade the safety of the system.

4. Design changes to system, subsystem, or interface logic and software that can create new hazards to equipment and personnel.

Software design.   Analysis of potential software hazards is performed to

1. Assure accurate translation of safety specification requirements into computer program requirements.

2. Ensure that the program specifications clearly identify the safety criteria to be used (fail-safe, fail-operational, fail-recovery, etc.).

3. Identify programs, routines, modules, or functions which control or influence safety-critical functions.

4. Analyze those programs, routines, modules, and functions and their system interfaces for events, faults, and environments which could cause or contribute to undesired events affecting safety.

5. Ensure that the actual coded software does not cause identified hazardous functions to occur or inhibit desired functions, thus creating hazardous conditions, and effectively mitigates identified hazardous anomalies in end-item hardware.

Risk assessment

Since the priority for system safety is eliminating hazards by design, a risk assessment procedure that considers only hazard severity will generally suffice during the early design phase to minimize risk. When hazards are not eliminated during the early design phase, a risk assessment procedure based upon hazard probability, as well as hazard severity, shall be used to establish priorities for corrective action and resolution of identified hazards. The process of risk assessment is therefore determined on the basis of categorizing the degrees of severity and evaluating the hazard probabilities. Effective risk analysis and achievement of the objectives of produce safety necessitate that the hazards be differentiated according to hazard severity categories for evaluation in conjunction with the corresponding hazard probabilities (frequencies) of occurrence. A method employed by the Department of Defense (pursuant to MIL-STD-882) and throughout much of private industry portrays individual hazards in terms of hazard description, category-based degree of severity, and definition of the consequences associated with each categorized hazard. Hazard probability is tabulated on basis of projected frequency of occurrence as applicable to individual items and the total population or inventory of the items. The categorization of degrees of severity can be used in conjunction with the corresponding hazard probabilities to construct the hazard risk index, which quantitatively reflects the risk assessment for the system. The following tables illustrate the methodology for this analytical process.

Hazard severity. The following tabulates categorization of hazards by severity, weighted significance, and potential consequences for the system.

Hazard Risk Assessment Severity Definitions/Weight Factor

Description	Category	Mishap Definitions	Weighting Factor
Catastrophic	I	Death to personnel or operators or loss of system	4
Critical	II	Severe injury, severe occupational illness to personnel or major damage to system	3
Marginal	III	Minor injury, minor occupational illness, or minor system damage	2
Negligible	IV	Less than minor injury, less than minor occupational illness to personnel, or less than minor system damage	1

Hazard frequency. The following groups hazards by frequency, weighted significance, and impact on the individual systems and total system population.

Hazard Risk Frequency Definitions/Weight Factors

Description	Level	Individual Systems	Total System Population	Weighting Factor
Frequent	A	Likely to occur frequently	Continuously experienced	5
Probable	B	Will occur several times in the life of a system	Will occur frequently	4
Occasional	C	Likely to occur sometime in the life of a system	Will occur several times	3
Remote	D	Unlikely but possible to occur in the life of a system	Unlikely but can reasonably be expected to occur	2
Improbable	E	So unlikely it can be assumed that occurrence may not be expected	Unlikely to occur, but possible	1

Quantification of hazard severity and frequency. The combinations of the Hazard Severity and Frequency category codes provide a basis for assigning a quantitative value for the individual hazards. The report for each hazard includes an alpha-numeric code denoting its severity and frequency. This is included in the system hazard summary along with all other hazards identified in the system safety analysis. The following scoring matrix portrays the structure of the codes.

Hazard Risk Assessment Scoring Matrix

Hazard Severity Categories

Frequency	I (Catastrophic)	II (Critical)	III (Marginal)	IV (Negligible)
A (Frequent)	I-A	II-A	III-A	IV-A
B (Probable)	I-B	II-B	III-B	IV-B
C (Occasional)	I-C	II-C	III-C	IV-C
D (Remote)	I-D	II-D	III-D	IV-D
E (Improbable)	I-E	II-E	III-E	IV-E

Summarization of the hazard risk analyses. The final scoring of the analysis is determined by multiplying the weighted factors of the individual frequency codes (alpha) and the severity codes (numeric) to produce matrix summary values. The resultant matrix values are then organized to show the management review criteria, based on grouping of hazard matrix scores. The following illustrates this methodology.

Hazard Risk Assessment Summary Matrix

Hazard Severity Categories

Frequency	I (Catastrophic)	II (Critical)	III (Marginal)	IV (Negligible)
A (Frequent)	I-A (4)(5) = 20	II-A (3)(5) = 15	III-A (2)(5) = 10	IV-A (1)(5) = 5
B (Probable)	I-B (4)(4) = 16	II-B (3)(4) = 12	III-B (2)(4) = 8	IV-B (1)(4) = 4
C (Occasional)	I-C (4)(3) = 12	II-C (3)(3) = 9	III-C (2)(3) = 6	IV-C (1)(3) = 3
D (Remote)	I-D (4)(2) = 8	II-D (3)(2) = 6	III-D (2)(2) = 4	IV-D (1)(2) = 2
E (Improbable)	I-E (4)(1) = 4	II-E (3)(1) = 3	III-E (2)(1) = 2	IV-E (1)(1) = 1

Hazard Risk Index		Management Review Criteria
Matrix Score		
15–20	I-A I-B II-A	Risks that are unacceptable—must be reviewed by top management
8–12	I-C I-D II-B II-C III-A III-B	Risks that are undesirable—must be reviewed by the principal for safety
4–6	I-E II-D III-C III-D IV-A IV-B	Risks that are undesirable—must be reviewed by system safety working group
1–3	II-E III-E IV-C IV-D IV-E	Risks that are undesirable—must be reviewed by system safety engineer

This matrix, which is updated as the hazard risks are eliminated or reduced, serves as a tool for management of the system safety program.

Product Development and Production

During the phases of development and production preceding market distribution of a system, it is important from a safety perspective to make sure that the development and production processes are in compliance with military, federal, national, and industry codes imposed contractually or by law to ensure safe design of a system, and to comprehensively evaluate the safety risk being assumed prior to test or delivery of the system to the consumer market.

Safety verification

The following are the functional tasks which should be accomplished in validating the product safety program requirements.

1. Identify critical parts and assemblies, production techniques, assembly procedures, facilities, testing, and inspection requirements which may affect safety.

2. Verify that testing and evaluation are performed on early production hardware so that safety deficiencies can be detected and corrected at the earliest opportunity.

3. Review all test plans and procedures. Evaluate the interfaces between the test system configuration and personnel, support equipment, special test equipment, test facilities, and the test environment during assembly, checkout, operation, foreseeable emergencies, disassembly, and/or teardown of the test configuration. Ensure that hazards identified by analyses and tests are eliminated or their associated risk reduced to an acceptable level.

4. Review technical data for warnings, cautions, and special procedures related to safe operation, maintenance, servicing, storage, packaging, handling, and transportation.

5. Evaluate the interfaces between the system being delivered and personnel and support equipment during transportation, storage, handling, assembly, installation, checkout, and demonstration/test operations. Ensure that hazards identified by analyses are eliminated or their associated risks reduced to an acceptable level.

6. Review procedures for and monitor results of periodic field tests to ensure that acceptable levels of safety are maintained. Identify major or critical characteristics of safety-significant items that deteriorate with age, environmental conditions, or as a result of other factors.

7. Perform or update hazard analyses to identify any new hazards that may result from design changes.

8. Evaluate the results of failure analyses and mishap investigations. Recommend corrective action.

9. Monitor the system throughout its life cycle to determine the adequacy of the design and of operating, maintenance, and emergency procedures.

10. Conduct a safety review of proposed new operating and maintenance procedures or changes in these procedures to ascertain that procedures, warnings, and cautions are adequate and that inherent safety is not degraded.

11. Document hazardous conditions and system deficiencies for development of follow-on requirements for modified or new systems.

12. Update safety documentation, such as design handbooks, product standards, and specifications, to reflect safety lessons learned.

13. Evaluate the adequacy of safety and warning devices, life support equipment, and personnel protective equipment.

14. Identify military, federal, national, and industry safety specifications, standards, and codes applicable to the system and document compliance of the design and procedures with these requirements.

15. Identify and evaluate residual hazards inherent in the system or those that arise from system-unique interfaces, installation, test, operation, maintenance, or support.

16. Identify necessary specialized safety design features, devices, procedures, skills, training, facilities, support requirements, and personnel protective equipment.

17. Identify hazardous materials and the precautions and procedures necessary for safe storage, handling, transport, use, and disposal of these materials.

These guidelines may be further applied to evaluate the need for requirements for certification of safety devices and other special safety features.

Safety review of engineering changes

Correction of a design deficiency or product improvement will often introduce other unanticipated hazards. For this reason, it is important that the safety engineers examine each proposed product modification and investigate all conceivable ways in which the change might result in additional hazards.

Consumer Safety

The critical elements of communication from the producer to the ultimate user of a product are (1) a system of clearly perceived and easily understood warning devices, (2) clearly legible warning and caution statements in user manuals and clearly legible symbols affixed to the product and product packaging, and (3) clearly comprehensible consumer instructions accompanying the product. It is imperative, especially for equipment and appliances, that procedures developed for product test, maintenance, operation, and servicing provide for safe disposal of expendable hazardous materials. In this regard it is also essential to consider constituent materials or manufactured components of the product when access to hazardous material will be required by user personnel during planned servicing, disassembly for inspection, maintenance activities, or reasonably predictable corrective maintenance actions.

Warning devices

In communicating hazards associated with products to the users, the types of warning signals must be commensurate with the dangers inherent in the uses of the product. The common terms used in communicating hazards are differentiated as follows:

1. *Danger* is used where there is an immediate hazard which, if encountered, will result in severe personal injury or death.
2. *Warning* is the signal word for hazards or unsafe practices which could result in severe personal injury or death if encountered.
3. *Caution* is used for hazards or unsafe practices which could usually result in minor personal injury, product damage, or property damage.

Visual and auditory signal systems are the primary forms of warning devices. Accordingly, there are principles which govern the utilization of these systems.

I. *Visual warning systems*
   1. Warning lights should be used to warn of actual or potential dangerous conditions.
   2. Flashing lights should be reserved for extreme emergencies, since they are distracting.
   3. If flashing lights are used, flash rates should be from 3 to 10 per second (4 is best), with equal intervals of light and dark.
   4. A warning light should be at least twice as bright as the immediate background.

5. The warning light should be within 30° of the operator's normal line of sight.
6. Warning lights are normally red because red means danger to most people. (Other signal lights in the area should be of other colors.)

II. *Auditory warning systems*
1. Although "real" signal frequency often cannot be controlled, it is desirable, where possible, to maintain signal frequency at a minimum of 20 signals per hour. If necessary, this should be accomplished by introducing artificial (noncritical) signals to which the operator must respond.
2. Where possible, the operator should be provided with anticipatory information. For example, a buzzer might indicate the subsequent appearance of a critical signal.
3. Whenever possible and however possible, the monitor must be given knowledge of the auditory system.

With respect to both visual and auditory warning signals, it is important to maintain noise, temperature, humidity, illumination, and other related environmental factors at optimum levels to effectively convey danger warnings.

Legal requirements for warnings

The minimum essential legal requirements for an adequate warning as set forth and reinforced by numerous court decisions are as follows:

1. The warning must be in such a form that it could reasonably be expected to catch the attention of a reasonably prudent person in the circumstances of its use.
2. The content of the warning must be of such a nature as to be comprehensible to the average user and to convey a fair indication of the nature and extent of the danger to the mind of a reasonably prudent person.
3. Implicit in the duty to warn is the duty to warn with a degree of intensity that would cause a reasonable person to exercise caution commensurate with the potential danger.
4. The warning must tell the user *how* to act to avoid the hazard. (This point relates to and underscores the importance of clear and unambiguous user instructions.)

Instruction and manuals

An important medium of communication of safety requirements to consumers is product manuals and instructions. A consumer occasionally encounters a set of instructions or a manual relating to the use of a particular appliance that is impossible to understand. (The instructions regarding the use of a videocassette recorder are typically baffling to an adult over 16.) The preparation of clear-cut, simple instructions is an art, and although the details of their preparation are beyond the scope of this text, there are basic guidelines:

1. Less is best—avoid informational overload.
2. Avoid abstract information; use only concrete information.
3. Forget "why"; concentrate on "how."
4. Remember that learning will come from doing.
5. Forget the promotional superlatives. The users already have the product; what they want to do now is set it up and use it with minimum aggravation.
6. Put important material in front of the text or in a prominent position.
7. For *warning* and *caution* statements, use a clearly distinguished typeface preceded the legends "WARNING" and "CAUTION," and use a special paragraph arrangement to distinguish these sections from the rest of the text.

Use of illustrations

Certain types of illustrations are useful either in connection with manuals or instructions or simply by themselves for either giving instructions or serving as warnings. There is a distinction between pictures and symbols. A *picture* is a realistic photograph or drawing of an object about which information is to be conveyed; a *symbol* is a photograph, drawing, or emblem that represents something else, such as the octagonal stop sign. A *pictogram* has a series of associated pictures that are intended to give information about performing a series of actions.

TABLE 7.1　Proposed Methods for Conveying Instructional Information of Various Types

	Concrete	Abstract
Simple	Use pictures/words	Use symbols/words
Complex	Use Pictogram/words	Use words

With respect to the use of illustrations to convey information to users, there is a continuum ranging from simple to complex:

1. *Simple concept:* Instructions are given to perform a single action (why and when are obvious).

2. *Complex concept:* Multiple actions are required, and multiple results can occur (when and why an action is to be performed may not be obvious).

Such concepts can also range from concrete (such as an emergency door) to abstract (with no physical reference, such as the general notion of danger). The various combinations of these distinctions, along with proposed methods of conveying information about them, are shown in Table 7.1.

Figure 7.1 illustrates alternative types of symbols considered for a danger warning symbol for use in providing instructions for actions in an emergency.

It is emphasized that safety instructions alerting users to potential hazards associated with the product must incorporate clear and unambiguous instructions to reinforce the illustrative symbols.

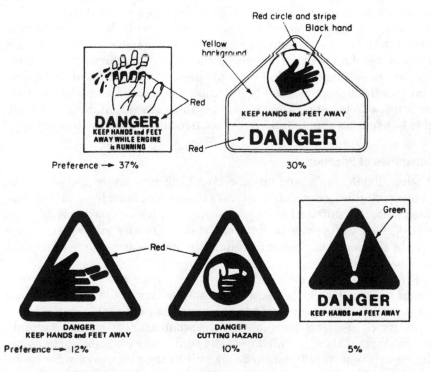

Figure 7.1   Examples of signs used in a simulated emergency. *(From M. S. Sanders and E. J. McCormick, Human Factors in Engineering Design, 7th ed., McGraw-Hill, 1993.)*

Product Liability and Safety

General

The production and sale of products (be they portable drills or nuclear power plants) is not necessarily the end of the producer-consumer relationship. In the case of certain items, the producer or seller maintains relationships with the buyers for service, maintenance, or repair. In recent years in the United States, the matter of product liability has assumed increasing importance, to the point that producers and sellers must give advance attention to such liability in designing, producing, and selling products. The liability issue applies equally to products that are used in industry (machines, tools, etc.) and those purchased by consumers (household appliances, utensils, etc.).

Definition

Product liability is the legal term used to describe an action in which an injured party (the plaintiff) seeks to recover damages for personal injury or loss of property from a manufacturer or seller (the defendant) because the plaintiff believes that the injuries or damages resulted from a defective product. In the current environment, people have increasingly looked to the courts for redress when they suffer injury or damage. This attitude has had a major impact in the area of product liability. Each year an increasing number of cases are brought to court for adjudication. The growth in product liability litigation has created a greater demand for human factor and safety experts, both in the initial design of products to make them safer and in the courtroom as expert witnesses.

Categories of litigation

Product liability is based on case law. Each new court decision adds, changes, clarifies, or sometimes clouds the accumulated legal precedents. Each state has different legal precedents, and hence cases may be tried and decided differently in different states. On any given day a case may be decided that drastically alters the nature and course of product liability cases.

Product liability cases are usually tried under one of the following bodies of law: (1) negligence, which tests the conduct of the defendant; (2) strict liability, which tests the quality of the product; (3) implied warranty, which also tests the quality of the product; or (4) express warranty and misrepresentation, which tests the performance of a product against the explicit representations made about it by the manufacturer or seller.

Product liability cases typically involve three types of defects: manufacturing defects, design defects, and warning defects. The Interagency Task Force on Product Liability has reported that, of all product liability cases, 35 percent involve manufacturing defects, 37 percent involve design defects, and 18 percent involve warning defects. (The remaining 10 percent encompasses all other litigation causes.) In a case for producer's liability, it must be established that a product was defective in manufacture or design. (Suffice it to say, however, that a product can be dangerous without being defective; a knife is a good example.) It must also be established that the product was defective at the time it left the defendant's hands. Products are often abused by users, and through abuse the product may become defective. However, although it might be presumed that in such cases the manufacturer would not be held liable, the manufacturer can in fact be held liable if the *abuse* was *foreseeable*. On the other hand, the mere presence of a defect in a product at the time of injury is not enough to make a case; it must be established that the defect was involved in the injury.

Minimizing product liability

The potential for product liability may be reduced to acceptable and manageable levels through aggressive pursuit of hazard identification and analysis during the product development phases and by designing dangers out of the product. If the nature and utilization of the product so permit, this approach is ultimately more effective than publishing a panoply of warning statements and decorating the product with dazzling danger symbols.

Acquisition and Production

Contracting

Overview

The magnitude and complexities of purchasing and contracting for governmental and private-sector enterprises preclude addressing functional details in this chapter. The text will, therefore, highlight the role of procurement within the organization and the generic principles governing forms and types of contracts and criteria for their application.

Most of the procurement concepts and techniques described in this chapter are derived from those employed by the federal government in general and the Department of Defense in particular. This approach recognizes the range and diversity of available contractual instruments and methodologies which, although applicable to military mission-oriented procurement, would likewise be useful in other governmental activities and private-sector enterprises. Where appropriate, pertinent sections and elements of the Federal Acquisition Regulation (FAR) and the Uniform Commercial Code are referenced.

Terms of Reference

The terms *procurement* and *acquisition* are interchangeable; which term is employed in practice depends upon the individual organization or institution. The terms *purchasing* and *contracting* are likewise interchangeable. The terms *seller*, *contractor*, and *supplier* have similar connotations. The term *contracting activity* is equivalent to *purchaser*. This chapter will focus on the types and forms of contracts and their applications. Because the purchasing activity is critical to logistics efficacy, it is incumbent upon the logistics professional to be familiar with the tools of purchasing and the ways to optimize the utility of the various contractual instruments. Within the context of this chapter, a contract is a written or ratifiable verbal agreement confirming the mutual consent of two parties for delivery of specified services and/or material, within a prescribed time frame, for stipulated financial or equivalent consideration, pursuant to appropriate special clauses detailing ancillary terms of mutual understanding, and subject to legal provisions involving applicable national, state, and local laws and regulations. This general description covers all procurement actions, from minor transactions such as purchases of paper clips to major systems contracts for construction of aircraft carriers.

Role of Procurement within the Organization

The role of procurement within an organization is largely defined by the corporate acquisition strategy and the execution of that strategy by the purchasing organization.

Acquisition strategy

Acquisition strategy within government agencies and private-sector enterprises entails consideration of the following:

1. What to procure
2. From whom to procure it
3. Where to procure it
4. Which prioritized items to procure
5. How much to procure
6. When to procure
7. In what packaging mode to procure
8. Which type and form of contract to use

The detailed elements of acquisition strategy are likely to vary from product to product and commodity to commodity.

What to procure.   The prerequisite to any procurement action is to define what will be procured. The "what" criteria are determined by the purpose, design, and quality of the goods to be purchased.

  Purpose.   Purpose concerns the manner in which the goods will be utilized in support of the end product. In the case of materials and components used in manufacturing marketable items, the purchaser's production personnel are responsible for manufacturing the end product, and are normally in the best position to determine what is to be purchased from vendors and suppliers. Specialized information may be required from engineering, accounting, inventory management, and warehousing.

Design.   Design defines the role of the goods to be procured. In this regard, design is described by drawings, specifications, and related product engineering data. These documents provide information to production (for the planning of production processes), purchasing (for design requirements), and logistics (for inventory definition). Quality assurance assists engineering, production, and purchasing in determining the desired level of quality assurance and quality conformance (inspection and related techniques).

Quality.   Quality characteristics are planned to conform to the product image desired by the enterprise. Within these parameters, numerous cost and performance profiles may need to be evaluated to determine the optimum trade-off between the cost of quality control and the consequences of the risk of not having quality products.

From whom to procure It.   Using the guidelines established by procurement policies concerning alternative sources of supply and reciprocity (buying from customers on a preferential basis), purchasing has the primary responsibility for identifying, evaluating, and establishing contact with potential suppliers for materials and services required by the firm.

Where to procure it.   The location of the procurement source is defined by the designation of a specific supplier. In some cases, suppliers may offer alternative origins from which an order may be filled. Close coordination between the traffic manager and purchasing is required in the selection of transportation mode and routing, the performance of receiving and storage functions upon receipt of the material, and resolution of loss and damage claims against carriers.

Which prioritized items to procure.   Once materials and components have been identified and the product placed into production, inventory replenishment decisions must be made. These are generally based on material component priorities provided by the inventory managers.

How much to procure.   For inventoried purchases, the question of how much to order is subject to continual review. Two aspects are the desired investment (in terms of financial outlay) in inventory and the determination of specific reorder quantities (usually in terms of economic order quantities). The proper reorder quantity is influenced by such factors as customer demand for the end product, production rates, purchase quantity discounts, ordering costs, inventory carrying costs, and the available lead time for acquiring and receiving materials.

When to procure.   When to procure is keyed to delivery requirements and the lead time (order cycle), defined as the total time period between the notification to replenish stock and the actual receipt of materials. The length of the lead time will be influenced by order cycle times at both the firm initiating the order and its supplier and by transportation transit time considerations. "When" is a primary function of inventory control, and is coordinated with production (in terms of production needs) and purchasing (in terms of procurement time requirements).

In what packaging mode to procure.   Purchasing, in coordination with production, traffic management, and warehousing, must determine (1) alternative forms in which items may be procured, (2) constraints that may dictate the forms in which such items must be received, and (3) the best match of the two. In some cases, the proper form in which to order goods may be suggested by material-handling requirements and the pattern (quantities and frequency) in which an item is used in the production process.

Which type and form of contractual instrument to use.   The contract is the legal vehicle which provides for delivery of the goods from the supplier to the users in the most cost-effective and timely manner possible, consistent with the relative degrees of technical and cost risks between the supplier and user. Types and forms of contracts are addressed in further detail later in this chapter.

Functions of purchasing

The major functions of the purchasing organization are (1) providing purchasing services, (2) obtaining the optimum value in goods purchased, (3) controlling financial commitments, (4) conducting negotiations, (5) developing sources of supply so as to assure continuity of availability of goods, and (6) serving as a source of market intelligence.

Purchasing services.   A competent purchasing organization is made up of knowledgeable procurement specialists who issue orders and award contracts on a centralized basis. It is their responsibility to consolidate requirements from all departments and accomplish the acquisitions within the required delivery schedules using the most effective forms of contractual agreement.

Obtaining optimum values.   *Value* is the measure of the worth of goods to the purchaser and is reflected in the price. (This is in contrast to *cost*, which is the measure of the worth of goods to the supplier.) A basic function of purchasing is to coordinate the intracorporate inputs necessary to create the most effective, favorable relationship between value and the price paid by the firm.

Controlling financial commitments.   Inasmuch as purchased materials constitute a significant portion of the cost of goods sold by the corporation, the purchasing manager must operate within limits defined by the company's ability to pay, which is governed by the cash flow posture. This entails close coordination between the financial manager and the purchasing manager.

Conducting negotiations.   A critical function of the purchasing organization is to conduct contract negotiations between the company and its suppliers. In order to bargain effectively over proposal and contract issues, the company must exercise control of the negotiating process and the personnel involved. This is accomplished by designating the purchasing representative to be in charge of negotiations with the suppliers and to coordinate corporate communications with the supplier.

Developing sources of supply.   A company's competitive position maybe enhanced through responsive vendor relationships. Cooperative purchaser/seller relationships enable purchasing managers to better manage the interdependent roles of the purchasers and the suppliers and become conduits of information between the company and its premium vendors.

Market intelligence.   A purchasing manager who is attuned to the dynamics of the marketplace can serve as an effective source of commercial intelligence. Feedback from market community contacts can provide valuable information, such as data on cost/price trends, commodity market fluctuations, indicators of product obsolescence, pending materials shortages, and competitors' marketing strategies.

Procurement Cycle

In general terms, whether the transaction entails an uncomplicated consumer retail purchase or a complex process involving a convolution of statutory, regulatory, and bureaucratic intricacies, the procurement cycle is described by the following sequence:

1. Determination of the user's needs (by the purchaser)
2. Specification or description of these needs (by the purchaser)
3. Solicitation of offers from potential suppliers (by the purchaser)
4. Offers to fulfill the user's needs (by the seller)
5. Contractual agreement (by both purchaser and seller)
6. Delivery of required services and/or material (by the seller)
7. Inspection and acceptance (by the purchaser)
8. Payment for goods (by the purchaser)

The success of the procurement process is determined by the efficacy of the contractual instrument in accommodating the mutual interests of the purchaser and the seller.

Structure of Contractual Instruments

The typical effective contractual instrument covers that which is solicited by the purchaser and that which is offered by the seller in response to the solicitation. Table 8.1, which outlines the basic government procurement instrument as set forth in Section 14.2 of the Federal Acquisition Regulation (FAR 14.2), provides a model for comprehensive contractual coverage.

TABLE 8.1 Outline of Procurement Instrument

Section	Incorporated in solicitation	Incorporated in contract
A. Solicitation/contract form	X	X
B. Supplies or services and prices	X	X
C. Description/specifications	X	X
D. Packaging and marking	X	X
E. Inspection and acceptance	X	X
F. Deliveries or performance	X	X
G. Contract administration data	X	X
H. Special contract requirements	X	X
I. Contract clauses (legal provisions)	X	X
J. Listing of contract contents	X	X
K. Representations and certifications of offerers	X	X
L. Instructions, conditions, and notices to offerers	X	
M. Evaluation factors for award	X	

Types of Contracts and Forms of Application

There are two main types of contractual instruments, *cost*-type contracts and *fixed-price*-type contracts. Within each type of contract, there is an array of forms of contractual instruments; these differ in the phase of the product life cycle to which they are appropriate, the nature of the goods being procured, and the relative risk assumed by the purchaser and the seller.

Cost-type contracts

Cost-type contracts are appropriate for acquisitions involving technical uncertainties which can affect performance, delivery schedule, and costs. A cost-type contract affords flexibility between purchaser and seller and provides opportunities to exploit major technological advances. The most common forms of cost-type contracts are cost, cost-sharing, cost-plus-fixed-fee, as well as cost-plus-incentive-fee contracts.

Cost contracts.   This form is a cost reimbursement contract pursuant to which the contractor (seller) is reimbursed for costs allowable by mutual agreement between the two parties. This form has no provision for fee (profit) and is generally limited to contracts with nonprofit organizations. Under certain conditions, a commercial enterprise would perform under such a "cost only" arrangement if other nonpecuniary features were granted as incentive. Such arrangements would be based on trade-off considerations where potentially lucrative inducements (e.g., patent rights or exclusive technical data rights) would offset financial reward.

Cost-sharing contracts.   The cost-sharing form provides for the sharing of contract costs between the contractor and the purchaser on the basis of a predetermined purchaser-seller cost-sharing ratio. This form would be appropriate where the contractor stands to profit from nonfinancial incentives, such as exclusive rights to technological advances which offer profit potential in future market ventures or new product lines.

Cost-plus-fixed-fee.   The cost-plus-fixed-fee (CPFF) form of contract provides for reimbursement to the contractor for those incurred costs which are allowable and allocable to the contract program pursuant to the terms of the contract *plus* payment of a fixed fee for performance of the contract. Under the terms of a CPFF contract, maximum technical and cost risk is assumed by the purchaser. Financial consideration initially comprises the negotiated estimated cost and a fixed fee based on a negotiated percentage of the originally negotiated estimated cost. The fixed fee amount does not vary with the levels of costs incurred by the contract, even if final contract costs exceed the originally negotiated estimate. An underrun of costs would result in an increased profit percentage; an overrun, or "cost growth," as it may be euphemistically described, would result in a depressed profit percentage. [It should be noted that contracts which provide for profit based solely on a percentage of costs (i.e., *all* incurred costs, irrespective of the ultimate total) are prohibited by federal law for U.S. government contracts.]

Cost-plus-incentive-fee.   The cost-plus-incentive-fee (CPIF) form is similar to other forms of cost-type contracts in that all allowable and allocable costs as defined by the contract terms are reimbursed to the contractor. There are recognized elements of technical and cost risks, but not to the extent of those acknowledged by cost-plus-fixed-fee contracts. Under the CPIF form of contract, the purchaser assumes a major share of the risk. The distinguishing characteristic of CPIF contracts is the incentive fee structure, which stipulates that the contractor assume a certain degree of risk. The incentive arrangement provides for an increased fee for effective cost management and conservation of resources and a reduced fee (possibly a negative fee) for excessive cost growth and ineffective cost control.

The distinctive features of CPIF financial considerations are as follows:

1. *Negotiated target cost,* based on a realistic "most probable" cost outcome.

2. *Target fee,* based on a negotiated percentage of the target cost.

3. *High cost estimate*, based on pessimistic or "worst-case" prospects.

4. *Minimum fee,* based on the assumption of minimally effective cost performance.

5. *Low cost estimate,* based on an optimistic or "best-case" cost scenario.

6. *Maximum fee,* based on optimistic or "best-case" prospects.

7. *Share ratio,* a negotiated ratio for cost and risk sharing between the purchaser and the contractor. The formula for determining the share ratio is

$$\text{Contractor's share } (\%) = \frac{\text{maximum fee} - \text{minimum fee}}{\text{high cost estimate} - \text{low cost estimate}} \times 100$$

$$\text{Purchaser's share } (\%) = (1 - \text{contractor's share}) \times 100$$

In citation of the share ratio, the purchaser's share is noted first in sequence, e.g., if the ratio is 40/60, the purchaser's share is 40 percent and the contractor's share is 60 percent. The sharing ratio is keyed to the target fee and bounded by the low cost estimate and the high cost estimate.

Figure 8.1 illustrates the cost-fee trade-off sensitivity in typical CPIF contracts. It is noted that Fig. 8.1a portrays a CPIF services contract with a 70/30 share ratio and conventional fee parameters. Figure 8.1b describes a research and development contract with a 40/60 share ratio and a "negative fee" swing.

Significance of fee to return on investment. An elusive feature of cost-plus-fee contracts is the significance of the fee when it is described as a return on investment rather than as a percentage of costs. Cost-type contracts usually provide for reimbursement to the contractor of accrued costs on a cyclical billing basis. Given a normal monthly billing cycle coupled with a typical 30-day payment period, the contractor would need to invest 60 days of risk capital prior to starting to receive reimbursements that would sustain the cash flow for the remainder of the performance period. In reality, the return on investment would thus be based on 60 days' operating capital rather than on total costs For example, a $1.272 million, 12-month, CPFF contract might entail $1.2 million in costs ($100,000 per month) plus a 6 percent fee of $72,000. Based on two months' operating capital of $200,000, the fee in this case would constitute a return on investment of 36 percent—a totally different perspective from that implied by expressing profits as 6 percent of costs. For longer performance periods at the same monthly cost rate, the return on investment would be higher, inasmuch as the initial "seed money" would remain unchanged.

Fixed-price-type contracts

Fixed-price-type contracts are appropriate where the user's needs are well defined and firm purchase descriptions and clear and unambiguous specifications of the product requirements are available. Candidates for fixed-price contracts include off-the-shelf, commercially available goods; brand-name items; and systems and equipment for which the technological state of the art has been stabilized and there are established technical data packages which may be used to produce the items. In this situation, the contractor assumes the technical and cost risks and therefore has an incentive to produce the products in compliance with the governing specifications and in the most cost-effective manner possible. A fixed-price contract, once negotiated and signed by

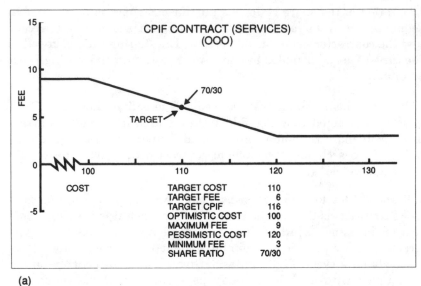

(a)

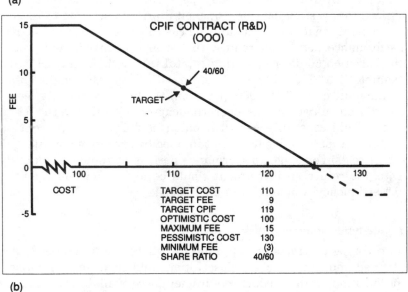

(b)

Figure 8.1 CPIF control cost-profit parameters (services). *(a)* CPIF contract cost-profit parameters (R&D). *(b)* CPIF trade-offs.

both parties, does not allow modification of the contract price, except for possible effects of special performance incentives which may be incorporated in the contract terms and certain engineering changes to enhance the product as proposed by the contractor and approved by the purchaser.

The major forms of fixed-price-type contracts applicable to governmental activities and private-sector enterprises include firm-fixed-price, fixed-price redeterminable, fixed-price-with-economic-adjustment, and fixed-price incentive contracts.

Firm-fixed-price contracts.   The firm-fixed-price (FFP) form of contract is the simplest of all the contract forms. Under the FFP arrangement, a price is negotiated on the basis of terms both the purchaser and the seller consider to be fair and reasonable. More complex contractual programs might entail the supplier's submitting proposals detailing product and management costs plus profit, the proposals being evaluated and analyzed by the purchaser, and the final contract price being established pursuant to negotiation of the constituent cost and profit elements. Once the contract is executed, the contractor is legally bound to provide the items stipulated in the contract according to the prescribed delivery schedules and in compliance with the product specifications. Figure 8.2a portrays the cost-profit interrelationship in a typical firm-fixed-price contract. The figure specifically notes the effects of varying cost levels on the final profit.

Fixed-price redeterminable contracts.   The fixed-price redeterminable (FPR) form incorporates the principle of the firm-fixed-price contract, with the exception of the provision for price redetermination. Price redetermination provides that the contract price is to be renegotiated at prescribed intervals in accordance with established pricing limits. The price may be redeterminable either prospectively (contemplating a prescribed future period of performance) or retrospectively (after a prescribed period of performance). Each negotiated redetermination follows the procedures for a firm-fixed-price contract: prior period empirical cost data and profit history are analyzed, and a price considered by both parties to be fair and reasonable is established accordingly. FPR contracts have features similar to those of cost-plus-fixed-fee contracts; these arrangements recognize a limited degree of technical and cost instability and, therefore, provide for periodic price redetermination within established pricing parameters.

Fixed-price-with-economic-adjustment contracts.   The fixed-price-with-economic adjustment (FPE) form incorporates the principles of a firm-fixed-price contract, plus a provision for recognition of specific economic contingencies. For example, all financial factors relating to a prospective contractual program may be stable, with the exception that certain

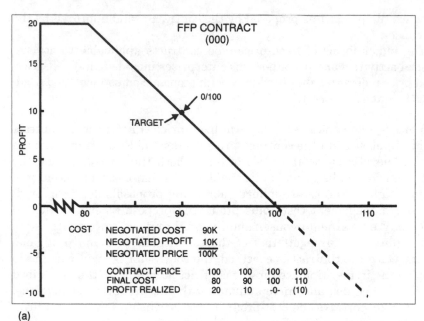

(a)

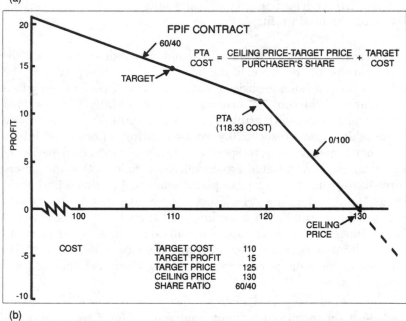

(b)

Figure 8.2 *(a)* FFP contract cost-profit parameters. *(b)* FPIF contract cost-profit parameters.

items of critical material (e.g., petroleum products, gold, manganese, or other precious metals) that are cited in the product specifications may be subject to price volatility as a result of international market turbulence. In such a case, the FPE contract would incorporate a provision stipulating that elements of the contract price affected by the contingencies would be finally determined on the basis of official governmental or recognized, credible market indexes. FPR contract forms are feasible only in situations of unusual market volatility which affects elements of material and labor that are essential to production of the contracted items.

Fixed-price-incentive contracts. The fixed-price-incentive form incorporates the features of firm-fixed-price contracts with a provision that affords an incentive to the contractor by increasing profit commensurate with reduction in costs. Conversely, there is a reduction in profit associated with an increase in program costs. The distinguishing characteristics of the fixed-price-incentive form of contract are

1. *Target cost,* based on the negotiated "most likely" program cost.
2. *Target profit,* based on a negotiated percentage applied to the target cost, giving consideration to factors of technical uncertainty, cost risk, and other related contractual contingencies.
3. *Target price,* determined by the sum of the target cost and target profit.
4. *Sharing ratio,* based on the negotiated purchaser/contractor allocation of cost risk liability.
5. *Ceiling price*, the negotiated maximum price to be paid by the purchaser, irrespective of attendant contingencies.
6. *Point of total assumption (PTA),* based on computation of cost and profit apportionment, after which the contractor assumes a reduction in profit equal to the corresponding increase in cost. The PTA calculation is based on the following:

$$\text{PTA cost} = \frac{\text{ceiling price} - \text{target price}}{\text{purchaser's share}} \times \text{target cost}$$

Figure 8.2b illustrates the cost-profit trade-off sensitivity and the effect of the PTA for a typical fixed-price-incentive contract with a firm target price. It is noted that an FPI form of contract provides considerable latitude for increases in profit commensurate with reductions in cost.

After the PTA, the reduction in profit is equal to the increase in cost up to the point where cost equals the ceiling price, after which the contractor is in a loss situation.

Special Purchase Arrangements

To provide for supplies and services to sustain corporate maintenance and production requirements, conventional forms of contracts must be tailored to accommodate peculiar demand patterns. This also presumes purchaser-vendor relationships that are responsive to the dynamics inherent in corporate operations. Typical of such special arrangements utilized in both governmental and private-sector activities are (1) indefinite delivery contracts, (2) blanket purchase agreements, (3) time and material contracts, and (4) labor-hour contracts. All of these are adaptations of the firm-fixed-price form of contract.

Indefinite delivery contracts

Indefinite delivery contracts are categorized as definite quantity contracts, requirements contracts, and indefinite quantity contracts.

Definite quantity contracts. This special application of the firm-fixed-price form of contract is typified by competitive selection of an individual vendor to provide a stated quantity of specified items of goods and services at negotiated, firm unit prices, to be delivered at any time during a prescribed performance period. The frequency and timing of item deliveries are flexible and may be directed by the purchaser at any point during the contractual performance period.

Requirements contracts. The requirements contract is an adaptation of the fixed-price form which contemplates that all requirements for specific items of goods and services generated by the purchaser will be filled by a competitively selected individual vendor. The period of performance and unit prices are defined by the contract. The quantities are generally not defined, although it is appropriate and will facilitate negotiation if minimum quantities for this prescribed contractual period are specified.

Indefinite quantity contracts. This application of firm-fixed-price arrangement is based on competitive selection of an individual supplier to provide undefined quantities of designated items of goods or services in accordance with negotiated unit price schedules, during a fixed period of time prescribed by the contract. This type of agreement stipulates minimum quantities and stated limits for item deliveries.

Blanket purchase agreements

Blanket purchase agreements (BPAs) are used frequently in procurement strategies involving a multiplicity of vendors. Under BPA procedures, each vendor agrees to deliver specified items of goods and services in accordance with a uniform pricing schedule for a fixed period of time prescribed by the contract. Typically, orders for required goods or services are rotated among the participating vendors.

The BPA arrangement has considerable utility, in that once a basic pricing schedule and terms of contract have been negotiated and contractually executed, corporate requirements during the performance period may be fulfilled by calls to one of the participating vendors without the need to negotiate and prepare individual purchase orders. For this reason, BPA arrangements have frequently been referred to as *call contracts*.

Governmental and commercial organizations frequently use blanket purchase agreements to allocate business among local small-business enterprises to assure broad-based distribution of the associated economic benefits and to foster community goodwill.

Special commercial applications of BPA arrangements.   In the private sector, notable innovations have resulted from the tailoring of blanket purchase agreement arrangements for specific corporate operational policies and objectives. The more prominent of such arrangements include maintenance, repair, and operating supply (MRO) contracts, purchase order draft procedures, and reciprocal purchase agreements with supplier-vendors.

MRO contracts.   The maintenance, repair, and operating supply contracting system involves (1) selecting a qualified source for the widest possible array of supplies, (2) placing the burden of inventory acquisition and availability on the vendor, (3) installing a system by which departments requisition supplies directly from the vendor, with purchasing provided with copies of the transactions, and (4) selecting the sole source on a permanent basis, but subject to periodic audit and review. Among the advantages of MRO contracting are (1) reduced inventory obsolescence, (2) predetermined item pricing, (3) reduction in inventory levels, (4) more responsive deliveries, (5) reduced pilferage, (6) freeing of capital from inventory investment for more profitable opportunities, and (7) reduction of duplicative efforts between vendors and purchasers. An effective MRO system requires preliminary education of both vendor and purchaser personnel and participation of all concerned personnel in the negotiations. This will promote understanding of the governing procedures by the system participants.

Purchase order drafts. Purchase order draft (POD) systems involve (1) the preparation of a purchase order and an accompanying blank check by the purchaser, (2) the completion and deposit (at the vendor's bank) of the check by the vendor at the time the goods are shipped, and (3) auditing by the purchaser (often by computer) of receiving reports for vendor errors. POD arrangements entail an honor system on the part of the participating vendors. Although some degree of dishonesty and fraudulent behavior on the part of vendors cannot be discounted, cumulative corporate experience with POD systems has endorsed the success of the concept, with a negligible incidence of fraudulent vendor activity. The cost savings that can be accrued through a reduction in written orders and the associated documentation overwhelmingly offset the costs associated with potential vendor dishonesty.

Reciprocal purchase agreements. Purchase and sales reciprocity, or "entrepreneurial interrelationship," refers to purchasing from customers in a mariner designed to enhance sales efforts or, conversely, the use of purchasing data to bring buying pressure to bear on a firm's customers.

Reciprocity is probably the most controversial of all practices associated with procurement. While it may assist sales efforts, it can also hamper a company to the extent that it encourages purchasing on bases other than quality, price, or service. The legality of reciprocal purchasing policies has also been questioned.

In recent years the federal government has charged a number of companies with violating various antitrust provisions of the Sherman and Clayton Acts by coercing or attempting to coerce suppliers to purchase their products, or products of subsidiaries, under threat of withdrawing their business from the suppliers. The FTC's (Federal Trade Commission) attitude has been that reciprocity does not necessarily have to be coercive to be considered in violation of Section 5 of the Federal Trade Commission Act; that is, it would be illegal in cases where there was (1) systematic use by a sales department or purchasing department in communicating with suppliers or (2) a discernible pattern of dealing between supplier and purchaser notwithstanding better price, quality, or service available from competitors. In practice, only those reciprocal trading practices that tend to restrain trade and competition are likely to attract unfavorable attention from the Justice Department. Reciprocal practices generally occur independently of logistics considerations. All other factors being equal, the practice of reciprocity often increases the number of suppliers and supply points served by the logistics network in cases where it is practiced with a large number of customers and suppliers. Also, the average size of shipments will be decreased if purchase quantities are apportioned among a larger number of vendors, thus increasing ordering and transportation costs.

Time and materials contracts

The time and materials (T&M) application of the firm-fixed-price form of contract is typically used by governmental and private-sector organizations for projects predominantly involving labor and work materials. T&M contracts are based on fixed hourly rates for labor and specified material charges. Labor hours are categorized by skill levels and corresponding wage rates. Each categorized hourly rate is fixed, based on negotiation and agreement on

1. The direct labor wage rate
2. Overhead and other project-related burden applied as a percentage of direct labor
3. The general and administrative (G&A) rate, a negotiated management charge applied as a percentage of direct labor plus overhead/burden costs
4. Profit, based on a negotiated percentage applied to the sum of direct labor, overhead/burden, and G&A costs

The associated material charges are usually based on actual material cost plus a negotiated handling charge applied as a percentage of material cost. Profit is not allowed on material charges.

Labor-hour contracts.   Labor-hour contracts incorporate the same procedures and methods as time and materials contracts with the exception that there are no provisions for materials charges; only fixed labor-hour charges are allowed.

Basic ordering agreement

The basic ordering agreement (BOA) is a vehicle of convenience established when the purchaser intends to procure certain types of items or technical services from a selected vendor over a period of time. The BOA permits the purchaser to issue individual orders for tasks or goods as the need for them arises during the contract period. The BOA is an instrument employed frequently by the federal government, and would have application to private-sector activities. The classic basic ordering agreement provides for

1. Incorporation of contractual terms and legal provisions for application to future task or supply orders during the contract period
2. Description of the types of goods and services to be provided
3. A prestipulated price schedule and ordering methods

It must be noted that the basic ordering agreement is *not* a contract per se. The BOA is a legal vehicle established to accommodate future individual orders pursuant to the negotiated agreement; the individual orders are considered individual contracts under the purview of the BOA. The financial obligation assumed by the purchaser is detailed in each individual BOA order.

Basic agreement

The basic agreement is a legal instrument frequently used by the Department of Defense which is also applicable to commercial activities. It incorporates legal provisions and special clauses into a family of contracts and orders with a designated single contractor. The basic agreement is an effective legal instrument to use when a purchaser contemplates a multiplicity of major purchases from a selected vendor over a defined period of time and it is determined that the same structure of legal provisions and special clauses would be applicable to all contracts negotiated with the designated contractor. The basic agreement is *not* a funded contract, nor does it provide a statement of work or specifications. It contemplates future awards to a selected contractor and affords the efficiency of not having to repetitively negotiate basic terms and conditions for each individual contractual action. After the original negotiation, all subsequent contracts pursuant to the basic agreement incorporate all the embodied provisions and clauses by citation of and reference to the basic agreement.

TABLE 8.2 Application of Forms of Contracts during Product Life Cycle

Type of effort	Form of contract
Basic research	Cost, CPFF
Applied research	Cost, CPFF
Exploratory research	Cost, CPFF
Advanced development	CPFF
Engineering development	CPFF, CPIF
First production	FPI
Follow-on production	FPI, FFP
Inventory support	FFP

Contractual criteria during the product life cycle

The varying risks inherent in various phases of the product life cycle dictate that the types and forms of contracts utilized reflect the risks attendant to a specific phase. For example, the considerable technical uncertainty and cost risk entailed in basic and advanced research would logically stipulate a cost or cost-plus-fixed-fee contract for such efforts. At the other end of the risk spectrum, a contract for replenishment of inventories after development, production, and market distribution of the product would call for a firm-fixed-price contract. Table 8.2 summarizes the product life cycle phases and the appropriate forms of contracts commensurate with the technical and cost risk for each phase.

Critical-Path
Analysis

General

Critical-path analysis (CPA) is a management control technique which identifies project-sensitive events (or milestones) and provides management visibility for the sequences, interactions, interdependent sensitivities, and schedule status of those events from commencement to completion of the project. Critical-path analysis has also been referred to as critical-path scheduling, which describes a set of graphic networking techniques used in management planning and control of projects. In any given project, there are three factors of concern: time, cost, and resource availability. Critical-path techniques have been developed to deal with each of these, individually and in combination. PERT (program evaluation and review technique) and CPM (critical-path method), the two best-known critical-path analysis techniques, portray a project in network form and relate its component tasks in a way that focuses attention on those that are crucial to the project's completion. For critical-path scheduling techniques to be most applicable, a project must have the following characteristics:

1. It has well-defined jobs or tasks whose completion marks the end of the project.
2. The jobs or tasks are independent; they may be started, stopped, and conducted separately within a given sequence.
3. The jobs or tasks are ordered; they must follow one another in a prescribed sequence.

Construction, aerospace, and shipbuilding projects commonly meet these criteria, and critical-path analysis techniques find wide application within these industries. Project management and critical-path analysis techniques are common among firms in rapidly changing industries.

The basic forms of PERT and CPM focus on finding the longest time-consuming path (the critical path) through a network of tasks as a basis for planning and controlling a project. There are fundamental differences between the two techniques in the way the networks are structured and in their terminologies. The primary difference lies in the fact that PERT permits explicit treatment of probability in its time estimates, whereas CPM does not address probability. This distinction reflects PERT's origin in the scheduling of advanced development projects that are characterized by uncertainty and CPM's origin in the scheduling of the fairly routine activity of plant maintenance.

CPA terms of reference

Prerequisite to a discussion of critical-path analysis is a basic understanding of the most frequently cited technical terms and symbols. The following definitions are used in the remainder of this chapter.

Symbols.

a = optimistic time estimate

b = pessimistic time estimate

m = most likely time estimate

t_e = expected elapsed time

T_E = earliest expected date

T_L = latest allowable date

S = slack time

Terms.

Activity: A work effort in a program, represented by an arrow. An activity cannot be started until the event preceding it has occurred, and it may represent a process, a task, a procurement cycle, waiting time, or simply a connection or interdependency between two events on the network.

Beginning event: An event which signifies the beginning of one or more activities on a network.

Critical path: That particular sequence of events and activities on the network path that has zero slack; i.e., the most time consuming path through the network.

Earliest expected date T_E: The earliest calendar date on which an event can be expected to occur. The T_E value for a given event is equal to the sum of the expected elapsed times t_e for the activities on the longest path from the beginning of the program to the given event.

Ending event (successor): The event which signifies the completion of one or more activities; the ending point in time of an activity.

Event: A specific, definable accomplishment in a program plan, recognizable at a particular instant in time. Events do not consume time or resources.

Expected elapsed time t_e: The elapsed time which an activity is estimated to require. The expected elapsed time can be identical to a single time estimate for the work to be accomplished, or can be derived from the calculation of a statistically weighted average time estimate that incorporates the optimistic *(a)*, most likely *(m)*, and pessimistic *(b)* estimates for the work to be accomplished:

$$\frac{a + 4m + b}{6} = t_e$$

Interface event: An event which signals the transfer of responsibility, end items, or information from one part of the plan to another. Examples of interface events are the receipt of an item (hardware, drawing, specification) or the release of an engineering drawing to manufacturing.

Latest allowable date T_L: The latest calendar date on which an event can occur without delaying the completion of the program. The T_L value for a given event is calculated by subtracting the sum of the expected elapsed times t_e for the activities on the longest path between the *given* event and the *end* event of the program from the latest date allowable for completing the program. T_L for the end event in a program is equal to the directed date T_D of the program. If a directed date is not specified, $T_L = T_E$ for the end event.

Milestone: A milestone is synonymous with an event in a network.

Most likely time estimate m: The most realistic estimate of the time an activity might consume. This time would be expected to occur most often if the activity could be repeated numerous times under similar circumstances.

Network: A flow diagram of the activities and events which must be accomplished to reach the program objectives, showing their logical and planned sequences of accomplishment, interdependencies, and inter-relationships.

Optimistic time estimate a: The time in which the activity can be completed under ideal circumstances. It is estimated that an activity would have no more than one chance in one hundred of being completed within this time.

Pessimistic time estimate b: An estimate of the longest time an activity would require under the most adverse conditions, barring acts of God.

Slack: The difference between the latest allowable date and the earliest expected date $(T_L - T_E)$. Slack is a characteristic, as such, of the network paths. An event with zero slack is on the critical path.

Zero time activity: An activity which constrains the beginning of a following activity or occurrence of the event to which it leads by requiring that the event from which it proceeds occur first.

PERT

The following steps are required in developing a PERT network.

Step 1.   Identify each activity to be done in the project. The output from this step is simply a list of project-sensitive activities.

Step 2.   Determine the sequence of activities and construct a network reflecting the events and the interactive relationships. This is an important step, since it forces the analyst to consider the interrelationships of activities and present them in visual form. PERT networking follows a structure of *activity on arrow, event on node;* that is, arrows denote activities and nodes (or "bubbles") denote events. Activities consume time and resources, and events mark their start or completion. Thus, "bake cake" would be an activity and "cake finished" would be an event.

In the network segment illustrated in Fig. 9.1, three events and two activities are specified. Each event node signifies two events: the end of

Figure 9.1   PERT cake bake.

one activity and the beginning of another. Event 2 marks not only the completion of the preparation of the cake mix but the start of baking the cake. Event 3 signifies the finishing of baking the cake, and perhaps the start of serving the cake, etc.

When constructing a network, it is important to ensure that the activities and events are in the proper order and that the logic of their interactive relationships is maintained. For example, it would be illogical to have a situation in which Event A precedes Event B, B precedes C, and C precedes A. Also, in many projects, problems arise in showing the precise form of dependencies, and the networking device termed a *dummy activity* must be employed. Dummy activities consume no resources and are depicted as having zero activity time. Examples of situations in which dummies are used are portrayed in Fig. 9.2.

Step 3. Ascertain time estimates for each activity. The PERT algorithm requires that three estimates be obtained for each activity.

a = *optimistic time,* the minimum reasonable period of time in which the activity can be completed. (There is only a small probability, typically assumed to be 1 percent, that the activity can be completed in a shorter period of time.)

m = *most likely time,* the best estimate of the time required. (This would be the only time estimate submitted if one were using CPM.) Since m is the time thought most likely to appear, it is also the mode of the distribution discussed in Step 4.

b = *pessimistic time,* the maximum reasonable period of time the activity would take to complete. (There is only a small probability, typically assumed to be 1 percent, that it would take longer.)

Step 4. Calculate the expected elapsed time t_e for each activity. The formula for this calculation is as follows:

$$t_e = \frac{a + 4m + b}{6}$$

On the basis of the calculation of the expected elapsed time t_e for each activity, the PERT factors for the cumulative earliest expected time T_E and the latest allowable time T_L and slack time S for each event can be calculated preliminarily to determine the *critical path* of the PERT network.

Step 5. Determine the critical path, as defined by the longest sequence of connected activities through the network and as the path in which all events have zero slack time S, pursuant to the formula

$$S = T_L - T_E$$

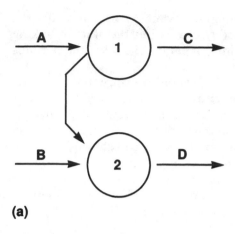

(a)

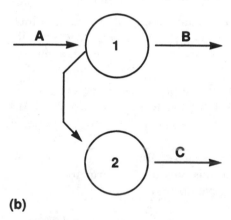

(b)

Figure 9.2 Examples of PERT dummy activities. *(a)* This indicates that D depends on A and B, whereas C depends only on A. *(b)* This indicates that C cannot start until resources are available from A. $t_e \neq 0$.

Example of construction of a PERT network

Tabulating sequential interactions of project-sensitive events. Table 9.1 illustrates the tabulation of eight milestone events structured on the basis of the following attributes for each event:

1. Previous events providing inputs to the event with which the event has interactive relationships

TABLE 9.1 Structure of Interactive Relationships and Estimated Completion-Time Parameters for Significant Events

Event	Previous event	t_a	t_m	t_b
8	7	20	30	40
	6	15	20	35
	5	8	12	15
7	4	30	35	50
	3	3	7	12
6	3	40	45	65
	2	25	35	50
5	2	55	70	95
4	1	10	20	35
3	1	5	15	25
2	1	10	15	30

2. Optimistic completion time estimate t_a

3. Pessimistic completion time estimate t_b

4. Most likely completion time estimate t_m

First step in construction of PERT network. Figure 9.3 represents the PERT network for this example. It is recommended that in the development of the activity arrows and event nodes, the analyst start by plot-

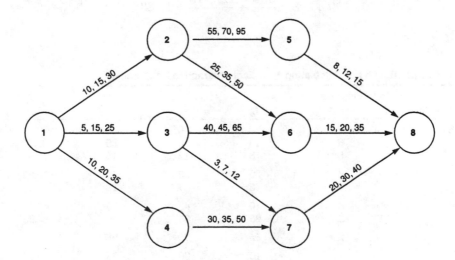

Figure 9.3 Evolution of PERT network depicting activity arrows, event nodes, and t_a, t_m, and t_b estimates.

ting the terminal (completion) event, in this case Event 8, and work from the terminal event to the commencement event, in this case Event 1. The three time estimates a, m, and b are always noted in sequence above the activity arrow; e.g., for the activity arrow from Event 5 to Event 8, the optimistic time estimate of 8, the most likely time estimate of 12, and the pessimistic time estimate of 15 are cited above the activity arrow.

Calculation of expected elapsed activity times. Table 9.2 incorporates the results of the calculation of expected elapsed activity times for each event based on the interactive links between the event and previous input events; i.e., for event 8, the following computations apply:

$$t_e = \frac{a + 4m + b}{6} = \text{expected elapsed time}$$

For Events 7–8:

$$t_e, 7\text{-}8 = \frac{20 + (4)(30) + 40}{6} = 30$$

For Events 6–8:

$$t_e, 6\text{-}8 = \frac{15 + (4)(20) + 35}{6} = 21.67$$

For Events 5–8:

$$t_e, 5\text{-}8 = \frac{8 + (4)(12) + 15}{6} = 11.83$$

TABLE 9.2   Results of Calculating t_e for Each Interactive Relationship

Event	Previous event	t_a	t_m	t_b	t_e
8	7	20	30	40	30
	6	15	20	35	21.67
	5	8	12	15	11.83
7	4	30	35	50	36.67
	3	3	7	12	7.17
6	3	40	45	65	47.5
	2	25	35	50	35.83
5	2	55	70	95	71.67
4	1	10	20	35	20.83
3	1	5	15	25	15
2	1	10	15	30	16.67

Plotting t_e values in PERT network.   After the calculation and tabulation of all t_e values determined by the interactive relationships of the individual events, the interconnecting activity lines are updated, as shown in Fig. 9.4. At this point the activity arrows portray both t_a, t_m, and t_b estimates and the t_e values associated with the individual activities.

By PERT network convention, the t_a, t_m, and t_b estimates are cited above the activity arrow; the calculated t_e values are cited below the arrow.

Principle of event arrow definition.   As indicated in Fig. 9.4, PERT discipline dictates that each event in a PERT network must be preceded by an activity arrow (input) and succeeded by an activity arrow (output), with two exceptions:

1. The commencement event has only *output* activity.
2. The terminal event has only *input* activity.

To illustrate this concept, Event 1 and Event 8 of Fig. 9.4 respectively portray commencement and terminal events.

Calculation of T_E for network events.   The earliest expected time T_E for each event is calculated by adding the expected elapsed time t_e for the preceding input activity to the T_E calculated for the preceding event. Figure 9.5 portrays the currently evolving network with the calculated T_E values for each event. Using the activity between Event 1 and Event 2 as an example, the T_E for Event 2 is 16.67 based solely on

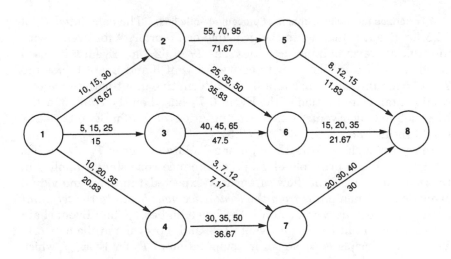

Figure 9.4   Evolution of PERT network incorporating the t_a, t_m, and t_b estimates and the calculated t_e factors.

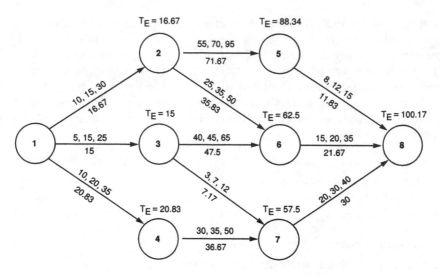

Figure 9.5   Evolution of PERT network incorporating T_E for each event.

the t_e for the input activity; Event 1, being a commencement activity, has no T_E value. The T_E for Event 5 is 88.34, based on addition of the input activity t_e of 71.67 and the T_E of 16.67 accumulated for the preceding Event 2.

Note that Event 1, the commencement event, has no T_E. The T_E for each other event represents the cumulative total of all preceding activity t_e values along the common chain of events.

T_E priorities for multiple chains of preceding activities.   The calculated T_E of 62.5 for Event 6 highlights the principle of T_E priority for events with multiple converging input activities. The T_E of 62.5 for Event 6 is based on the t_e of 47.5 for the activity between Event 3 and Event 6 plus the T_E of 15 accumulated for Event 3. The alternative input path to Event 6 would entail computation of the Event 6 T_E based on the chain of activities defined by Events 1–2–6. Computation of T_E for Events 1–2–6 would result in an Event 6 T_E of 52.5.

In case of such conflicts, the *highest* calculated value of T_E always has priority, and this principle of T_E predominance consistently applies in the computation of the flow of earliest expected times for individual events on common paths *from* the *commencement* event to the *terminal* event. Inasmuch as Event chain 1–3–6 has a higher T_E than Event chain 1–2–6, the cumulative T_E for Events 1, 3, and 6 prevails in the network. The same principle is borne out in computation of T_E for Event 7, which also has more than one converging input activity path. For the same reason as described for Event 6, the higher T_E of Event chain 1–4–7 prevails over the T_E of Event chain 1–3–7 in determining T_E for Event 7.

Calculation of T_L for network events. The latest allowable time T_L for each event is determined by establishing T_L for the terminal event as equal to the terminal event T_E (which principle of terminal time equivalency is stipulated by PERT discipline) and, working *from* the *terminal* event to the *commencement* event, subtracting from each individual event T_L the t_e for the *preceding* activity. On basis of the T_L and T_E time values calculated for each event, the event slack time S can be determined. Figure 9.6 incorporates the calculated T_L values in addition to the previously calculated network times for each event and thereby sets the stage for computation of slack factors and determination of the critical path.

Using the activity from Event 7 to Event 8 (terminal event) for illustration, the T_L for Event 7 is calculated by subtracting the activity time t_e of 30 from the Event 8 T_L of 100.17 to produce the Event 7 T_L of 70.17. The same principle applies throughout the network, except for specific rules applicable to cases in which multiple activity chains emanate from the individual event, as illustrated by Event 2 and Event 3.

T_L priorities for multiple chains of emanating activities. The T_L of 31 for Event 3 illustrates the principle of T_L priority relating to events with

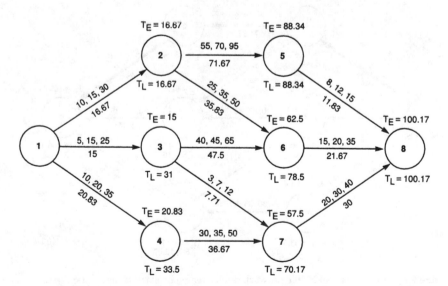

Figure 9.6 Evolution of PERT network incorporating T_L values in addition to the previously calculated PERT time values.

multiple emanating activities. The T_L of 31 for Event 3 is derived by subtracting the t_e of 47.5 established for the activity between Event 3 and Event 6 from the Event 6 T_L of 78.5. The alternative emanating path into Event 3 entails computation of the T_L for Event 3 based on the path defined by Events 8–7–3. Computation of T_L based on Events 8–7–3 would result in an Event 3 T_L of 63.

In case of such conflicts, the *lowest* cumulative value of T_L has priority, and this principle of T_L predominance consistently applies in the computation of latest allowable times from the terminal event to the commencement event. Inasmuch as Event chain 8–6–3 produced a lower T_L than Event chain 8–7–3, the calculated cumulative T_L for Events 8–6–3 prevails in the network. The same principle is borne out by computation of T_L for Event 2, which also has more than one emanating activity path. For the same reason as described for Event 3, the lower T_L of Event chain 8–5–2 prevails over that of Event chain 8–6–2 in determining T_L for Event 2.

Calculation of PERT network slack times. Each individual event slack time S is calculated by subtracting the earliest expected time T_E for that event from the latest allowable time T_L. Figure 9.7 depicts the network showing the slack times for the individual events. For example, the slack time of 16 for Event 6 is determined by subtracting the T_E of 62.5 from

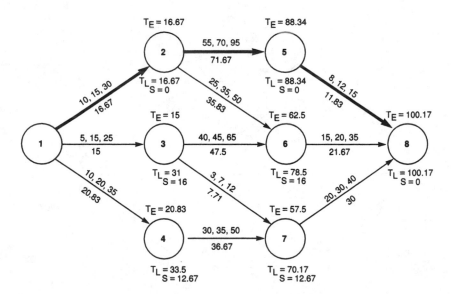

Figure 9.7   Final phases of PERT network depicting event slack times (critical path emphasized).

the T_L of 78.5; for Event 2, the TL of 16.67 less the T_E of 16.67 produces zero slack.

The objective of the calculation of event slack times is to isolate that chain of activities and events for which there is zero slack time for each event.

Determination of critical path.   The chain of events for which there is zero slack time defines the critical path of the PERT network. Inspection of the individual event slack times in the example described herein reveals that the chain of Events 1–2–5–8 (Fig. 9.7) has zero slack, and thereby defines the critical path of the network based on a total activity time of 100.17. (The cumulative critical path activity time is always equal to the T_E and T_L of the terminal event; note that the T_E and T_L of Event 8 are both equal to 100.17.) Slippage in the completion dates of events on the critical path will force slippage in the project completion dates. The critical path denotes the chain of interactive events which requires the longest time to complete. Conversely, the critical path also describes the shortest time in which the project can be completed. The critical path is the focus of management efforts to either reduce or control completion times of the individual events in order to accelerate the project completion time or, at least, assure completion on schedule.

Critical-Path Method

The major distinction between the critical-path method (CPM) and PERT is the use of probability statistics in the latter. Otherwise, despite some differences in network construction and terminology, the approaches are similar.

Development of a CPM network

The following steps are required in developing and solving a CPM network.

1. Identify each activity to be done in the project. In CPM, the term *job* is often used to refer to the task being performed, rather than separating activities and events as in PERT. Since current practice seems to be to treat the terms as synonymous, it is appropriate to refer to the CPM tasks to be done as activities.
2. Determine the sequence of activities and construct a network reflecting precedence relationships. CPM is activity-oriented, with arrows denoting precedence only. A typical segment using the cake-baking example from PERT (Fig. 9.1) would show the activity above the node

rather than on the arrow. To repeat, nodes in CPM represent activities in the PERT sense, not events.

3. Ascertain time estimates for each activity. This is the "engineering estimate" time (or it can be thought of as equivalent to expected time, which is derived statistically in PERT). While the CPM procedure has no provision for probabilistic estimation of this value, the individual who provides the estimate may use a simple statistical model in arriving at a figure. For example, he or she may feel that two times are equally likely and therefore take the average as an estimate.

4. Determine the critical path. As in PERT, this is the path with zero slack. To arrive at slack time requires the calculation of four time values for each activity:

- Early start time (ES), the earliest possible time at which the activity can begin

- Early finish time (EF), the early start time plus the time needed to complete the activity

- Late start time (LS), the latest time at which an activity can begin without delaying the project

- Late finish time (LF), the latest time at which an activity can end without delaying the project

Determination of CPM network slack and critical path.   The procedure for arriving at the four prerequisite time values and for determining slack and the critical path can best be explained by referring to the sample network shown in Fig. 9.8. The letters denote the activities and the numbers the activity times.

1. *Find ES time.* Take 0 as the start of the project and set this equal to ES for Activity A. To find ES for B, add the duration of A (which is 2) to 0 and obtain 2. Likewise, ES for C would be 0 + 2, or 2. To find ES for D, take the larger ES and duration time for the preceding activities; since B = 2 + 5 = 7 and C = 2 + 4 = 6, ES for D = 7. These values are entered on the diagram. The largest value is selected because activity D cannot begin until the longest time-consuming activity preceding it is completed.

2. *Find EF times.* The EF for A is its ES time, 0, plus its duration of 2. B's EF is its ES of 2 plus its duration of 5, or 7. C's is 2 + 4, or 6, and D's is 7 + 3, or 10 (Fig. 9.8c). In practice, one computes ES and EF together while proceeding through the network. Since ES plus activ-

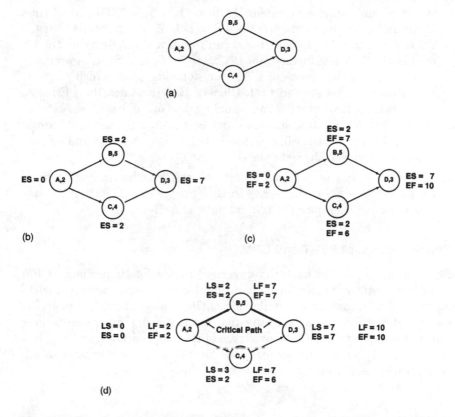

Figure 9.8 Steps to develop and solve a CPM network. *(a)* Basic CPM network. *(b)* Find early start (ES) times. *(c)* Find early finish (EF) times. *(d)* Find late start (LS) and late finish (LF) times.

ity time equals EF, the EF becomes the ES of the following event, and so forth.

3. *Find LS and LF times.* While the procedure for making these calculations can be presented in mathematical form, the concept is much easier to explain and understand if it is presented in an intuitive manner. The basic approach is to start at the end of the project with some desired or assumed completion time. Working back toward the beginning, one activity at a time, it is possible to determine how long the starting of this activity may be delayed without affecting the start of the one that follows it. In reference to the sample network (Fig. 9.8), assume that the late finish time for the project is equal to the early finish time for activity D—that is, 10. If this is the case, the

latest possible starting time for D will be $10 - 3$, or 7. The latest time C can finish without delaying the LS of D is 7, which means that C's LS is $7 - 4$, or 3. The latest time B can finish without delaying the LS of D is also 7, which means that B's LS is $7 - 5$, or 2. Since A precedes two activities, the choice of LF value depends upon which of those activities must be started first. Clearly, B determines the LF for A, since its LS is 2, whereas C can be delayed by one day without extending the project. Finally, since A must be finished by day 2, it cannot start any later than day 0, and hence its LS is 0. These LS and LF values are entered in the network (Fig. 9.8d).

4. *Determine slack time for each activity.* Slack for each activity is defined as either LS – ES or LF – EF. In this example, only Activity C has slack (1 day); therefore the critical path is A–B–D.

Relative Merits of PERT and CPM

PERT is heuristic and basically governed by stochastic methods. CPM is deterministic and based on conventional input-output methods, without regard to probablistic distribution of time values. PERT, in practice, has greater utility as a planning tool for assessing future situations affected by uncertainty. CPM, by virtue of its comparative simplicity, has greater utility as a management tool for oversight and control of ongoing project activities.

10

Work Breakdown Structure

General

The work breakdown structure (WBS) is a product-oriented family tree, composed of hardware, software, services, and data, which completely defines a program or project. The logistics engineering process plays a critical role in identifying the product elements of the VMS. Although of military origin, the WBS can be useful for commercial system programs. The WBS displays and defines the product(s) to be developed and/or produced and relates elements of work to be accomplished to the end product. The WBS is the foundation for

1. Program and technical planning
2. Cost estimation and budget formulation
3. Schedule definition
4. Statements of work and specification of contract line items
5. Progress status reporting and problem analysis

The WBS is essential to providing management with the ability to exercise technical, schedule, and financial control of the program. The WBS is, consequently, the basis for related cost and schedule status reports. As an adjunct to PERT or CPM network programs, the WBS establishes the framework for

1. Defining the work requirements
2. Designating the project significant events
3. Determining the individual event time requirements and project completion date
4. Constructing the network
5. Summarizing the cost and schedule status of the program for progressively higher levels of management

A WBS displays and defines the product(s) to be developed or produced and relates the elements to one another and to the end product. A WBS element is a discrete, identifiable item of hardware, software, data, or service. During the acquisition process, both the purchasing organization and the contractor have opportunities to tailor the WBS. This tailoring should have the goal of adding or deleting elements in order to enhance the effectiveness of the WBS at satisfying both technical management and cost/schedule management objectives. The WBS serves as a framework for the contractor's overall management system. The more frequently utilized WBS formats which will be addressed in the chapter are the project summary WBS and the contract WBS (CWBS). The example used for illustration is a hypothetical U.S. Navy sonar system, which incorporates the governing principles of MIL-HDBK-881.

Project Summary WBS

The project summary WBS consolidates the major structural elements of all WBS related to development, production, and deployment, as defined by three indenture levels of program detail. Figure 10.1 describes the top system indenture (Level 0); development, production, and deployment combined with equipment groups common to these activities (Level 1); and those subsystem elements common to each of the Level 1 items (Level 2).

Contract WBS

The CWBS is the complete WBS applicable to a particular contract or procurement action. It will generally contain the applicable portion of the project summary WBS plus any additional levels of detail necessary for planning and control.

The CWBS outlines program tasks and establishes their relation to program organization, configuration items, and objectives. It establishes a logical, indentured framework for correlating schedule, cost,

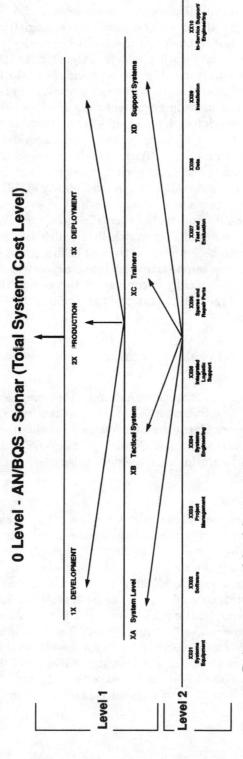

0 Level - AN/BQS - Sonar (Total System Cost Level)

1X DEVELOPMENT   **2X PRODUCTION**   **3X DEPLOYMENT**

Level 1

XA System Level   XB Tactical System   XC Trainers   XD Support Systems

Level 2

| XX01 Systems Equipment | XX02 Software | XX03 Project Management | XX04 Systems Engineering | XX05 Integrated Logistic Support | XX06 Spares and Repair Parts | XX07 Test and Evaluation | XX08 Data | XX09 Installation | XX10 In-Service Support Engineering |

Figure 10.1 Project summary work breakdown structure.

233

performance, and technical objectives and ensures that all derivative plans contribute directly to program objectives. Logistics engineering plays a key role in the extension of the CWBS hardware elements. The dependence of hardware work package extension on the functional analysis, synthesis, and trade-off process provides correlation and traceability of the CWBS to system requirements. As an integrated data system, the logistics engineering documentation also provides a common interface between specialty engineering efforts [such as technical performance measurement, risk management, and integrated logistic support (ILS)] and program-level activities (such as project planning, cost/schedule management, and engineering management). It also plays a key role in ensuring the correlation and traceability of WBS product elements. In addition to providing a framework for cost accumulation and work package schedule and status identification, the CWBS also provides a common structured framework into which the PERT or CPM network methods can be integrated, thus affording graphic visibility of the status of milestone completion relative to expenditures of allocated resources. The CWBS typically encompasses the second through fourth system indentures. Figure 10.2 portrays the scope of the CWBS relative to the project WBS.

Development of Cost Work Breakdown Structure

A work breakdown structure element may represent an identifiable item of hardware or software, a set of data, or a service such as integrated logistic support. Hardware and software breakout is controlled by the system engineering process. Control of the WBS is usually maintained through configuration management (discussed in Chap. 26).

CWBS work packaging

The lowest WBS elements coincide with the most efficient and cost-effective way of controlling the schedule, cost, and technical performance of the program. Where mass production is involved, cost accounts are divided into recurring, nonrecurring, and material costs.

CWBS elements should be selected to permit structuring of budgets and identification/tracking of costs to the level required for control. This is accomplished by assigning job orders or customer orders to the cost-account level for in-house effort, and by structuring line items or work assignments on contracts in accordance with the WBS. Ordinarily, a cost account will be established at the lowest level in the CWBS at which costs are recorded and can be compared with budgeted costs. This cost account (a WBS element) is a natural control point for cost/schedule

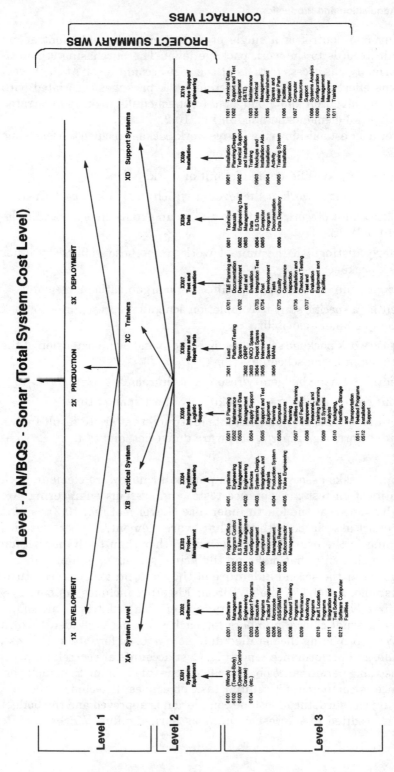

Figure 10.2 Illustration of integrated project and cost work breakdown structure. The project WBS encompasses Levels 0, 1, and 2; the CBWS encompasses Levels 1, 2, and 3.

planning and control of a single organizational element. Contractors maintain records to the work package level. The purchasing organization normally has access to costs at the cost account level. At the lowest level, the effort is broken into discrete work packages associated with both an organization and a budgeted (cost/schedule) task, as illustrated by the Level 3 elements detailed in Fig. 10.2.

Criteria for establishing an effective work package include the following:

1. Represent a specific, definable unit of work.
2. Define a unit of work at the level at which the work is performed.
3. Relate a unit of work directly to and as an extension of a specific element of WBS.
4. Clearly distinguish the unit of work from those defined by other work packages.
5. Assign a unit of work to a specific single organizational element.
6. Identify a specific start-to-completion schedule representative of task accomplishment capability.
7. Relate work package schedules directly to and as an extension of the detailed program schedule.
8. Identify realistic budgetary/resource requirements.
9. Limit each unit of work to a relatively short span of time.
10. Identify specific accomplishments (outputs) that are to result from a unit of work; e.g., reports, hardware deliveries, and tests.

Support tasks associated with a particular hardware element, such as qualification tests, acceptance tests, and systems engineering, are normally shown at the fourth indenture (Level 3 of Fig. 10.2) as part of the effort associated with that hardware element. Support systems pertaining to the overall system effort (rather than to the individual hardware elements that make up the prime system or equipment) are aggregated at the second indenture of the work breakdown structure. For example, Level 1 of the CWBS in Fig. 10.2 includes support systems directly applicable to the total system. The contractor normally assigns internal charge numbers for each work package identified in the CWBS, providing the detailed data source used for integrated cost, schedule, and performance reporting. Cost accounts are correlated with demonstrable performance objectives. At scheduled completion, task performance is compared with initial task objectives. If technical requirements are satisfied, the milestone completion is approved and the budget value is credited to the cost account as earned value. WBS elements that do not achieve required performance levels are identified, using various techniques, including technical performance measurement methods and critical-path analysis, for management evaluation and corrective action.

CWBS elemental alignment and coding

The purpose of the indentured, elemental structure of the CWBS is to facilitate management identification, visibility, and control of the cost of the constituent work packages. The CWBS method is designed to roll up subsidiary indenture-level packages into their governing element at the next higher indenture level. The work breakdown structure is planned and organized from the top-down. WBS data are analyzed and evaluated from the bottom up. Alphanumeric codes are assigned to CWBS work packages to identify their purposes and levels of indenture within the system scheme portrayed by the CWBS.

CWBS coding protocol.   Referring to the sample WBS in Fig. 10.2, the *total system* level is coded Level 0; i.e., the first indenture level is the sonar system itself. The elements of the second indenture level, Level 1, are identified by two-character codes which identify the phase of the system evaluation (from development through deployment). The three system phases are identified by the first character, a number, as follows:

1. 1X = development phase
2. 2X = production phase
3. 3X = deployment phase

The equipment groupings are each designated by an alphabetical second character, as follows:

1. XA = system level
2. XB = tactical system
3. XC = training system
4. XD = support system

For example, in the WBS in Fig. 10.2, the code 1A would connote the Level 1 element describing *system-level* items applicable to the development-phase program; 2D would include *support system* items applicable to *production-phase* programs. The two sets of elemental codes can be interpreted as follows:

$$\text{All of} \begin{Bmatrix} \textit{XA} \text{ (System Level)} \\ \textit{XB} \text{ (Tactical Systems)} \\ \textit{XC} \text{ (Trainers)} \\ \textit{XD} \text{ (Support System)} \end{Bmatrix} \text{ are elements of} \begin{Bmatrix} \textit{1X} \text{ (Development)} \\ \textit{2X} \text{ (Production)} \\ \textit{3X} \text{ (Deployment)} \end{Bmatrix}$$

Each Level 1 identification code, therefore, specifies the equipment group and the phase of the system evolution.

CWBS level 2 coding.   Third indenture, or Level 2, subsystem or elemental functions are described by four-character alpha/numeric codes which denote their relationships to one of the Level 1 elements. The first two characters incorporate the Level 1 elemental code; the latter two characters denote the specific Level 2 functional element. Using the CWBS in Fig. 10.2 for reference, there are 10 Level 2 codes, which are identified as follows:

XX01 System equipment	*XX06* Spares and repair parts
XX02 Software	*XX07* Test and evaluation
XX03 Project management	*XX08* Data
XX04 Systems engineering	*XX09* Installation
XX05 ILS	*XX10* In-service support/engineering

For example, work package code 2C02 signifies *software* related to *trainers* applicable to the *production* program; i.e.,

> Level 1: 2X (Production)
> XC (Trainers)
> Level 2: XX02 (Software)

In keeping with the CWBS roll-up principle, all Level 2 packages are considered constituent subelements of a Level 1 WBS element.

CWBS level 3 coding.   Level 3 CWBS packages describe fourth-indenture subelements grouped according to their applicability to a specified Level 2 package. The CWBS in Fig. 10.2 lists each of the groups of Level 3 subelements under its governing Level 2 work package. Each Level 3 subelement code comprises six alphanumeric characters, the first four of which describe the governing Level 2 element. For example, Level 3 subelement code lA0102 denotes the *towed body* of the *system equipment* element for the *system level* applicable to the *development* program; i.e.,

Level 1: 1X (Development)
         XA (System level)

Level 2: XX01 (System equipment)

Level 3: XXXX02 (Towed body)

The structure which rolls up lower-level indenture detail into consolidated elements at higher-level indentures provides management the option of having visibility at any level of aggregation within the system configuration.

Documentation of CWBS hierarchy

Figure 10.3 is an extract from an expanded CWBS elemental listing encompassing all CWBS levels, elements, and subelements of the CWBS shown in Fig. 10.2. In full, expanded format, the CWBS menu enables management to identify costs incurred at lower component levels and aggregate the cost data at any of the higher levels in order to focus on those areas that need corrective action or greater emphasis on cost control.

CWBS Level 1   1A System Level—Development Phase
  CWBS Level 2   1A01 System Equipment
    CWBS Level 3   1A0101 Winch
                    1A0102 Towed-body
                    1A0103 Operator control console
                    1A0104 etc.
  CWBS Level 2   1A02 Software
    WBS Level 3   1A0201 Software management
                   1A0202 Software engineering
                   1A0203 --------------------
                   1A0204 --------------------
                   1A0205 etc.

WBS Identures (Partials)

0 Level: AN/BQS Sonar (Total system cost level)

1X   Development

  1A   System level

Figure 10.3 AN/BQS-Sonar (illustrative example). Guide to Interpreting CWBS Levels.

Figure 10.3   *(Continued)*

1A01   System equipment
      1A0101   Winch
      1A0102   Towed-body
      1A0103   Operator control console
      1A0104   etc.

1A02   Software
      1A0201   Software management
      1A0202   Software engineering
      1A0203   Support programs
      1A0204   Executive programs
      1A0205   Tactical programs
      1A0206   Microcode
      1A0207   GSS/SIM/STIM programs
      1A0208   Onboard training programs
      1A0209   Performance monitoring (PM) programs
      1A0210   Fault location (FL) programs
      1A0211   Integration and test software
      1A0212   Software computer facilities

1A03   Project management
      1A0301   Program office
      1A0302   Program control
      1A0303   ILS management
      1A0304   Data management
      1A0305   Configuration management
      1A0306   Computer resources management
      1A0307   Design review
      1A0308   Subcontractor management

1A04   Systems engineering
      1A0401   Systems engineering management
      1A0402   Engineering services
      1A0403   System design, integration, and analysis
      1A0404   Production system support
      1A0405   Value engineering

1A05   Integrated logistic support
      1A0501   ILS planning
      1A0502   Maintenance
      1A0503   Technical data management
      1A0504   Supply support planning
      1A0505   Support and test equipment planning
      1A0506   Computer resources planning
      1A0507   Facilities planning and facilities
      1A0508   Manpower, personnel, and training planning
      1A0509   ILS systems analysis

Figure 10.3 *(Continued)*

    1A0510   Packaging, handling, storage, and transportation (PHS&T)

    1A0511   Related programs

    1A0512   Installation support

1A06 Spares and repair parts

    1A0601   Lead platform/testing spares

    1A0602   Onboard repair parts (OBRP)

    1A0603   Installation and checkout (INCO) spares

    1A0604   Depot spares

    1A0605   Intermediate spares

    1A0606   Maintenance assistance modules (MAMs)

1A07 Test and evaluation

    1A0701   T&E planning and documentation

    1A0702   Development test and evaluation (DT&E)

    1A0703   Integration test

    1A0704   Postdevelopment tests

    1A0705   Quality conformance inspection

    1A0706   Installation and checkout testing (I&C)

    1A0707   Test and evaluation equipment facilities

1A08 Data

    1A0801   Technical manuals

    1A0802   Engineering data

    1A0803   Management data

    1A0804   ILS data

    1A0805   Computer program documentation/data

    1A0806   Data depository

1A09 Installation

    1A0901   Installation planning/design

    1A0902   Technical support and installation training

    1A0903   Mockups and installation aids

    1A0904   Installation activity

    1A0905   Training system installation

1A10 In-service support/in-service engineering

    1A1001   Technical data

    1A1002   Support and test equipment (S&TE)

    1A1003   Maintenance

    1A1004   Technical management

    1A1005   Replenishment spares and repair parts

    1A1006   Facilities operation

    1A1007   Computer resources support

    1A1008   Systems analysis

    1A1009   Configuration management

PROJECT/PROGRAM						CONTRACT WORK BREAKDOWN	DATE	
CONTRACT NO.						STRUCTURE DICTIONARY	SHEET	OF
WBS LEVEL								
1	2	3	4	5	6	ELEMENT TITLE		

ELEMENT DESCRIPTION

WBS LEVEL						ASSOCIATED LOWER LEVEL ELEMENTS
1	2	3	4	5	6	TITLE

Figure 10.4   Work breakdown structure dictionary format.

WBS worksheet

All work packages in a CWBS plan are based on analysis of the inherent, fundamental attributes at all indenture levels. The WBS dictionary format in Fig. 10.4, which is based on the system data elements of MIL-HDBK-881, provides a sample worksheet for construction of an effective cost work breakdown structure which can be tailored to specified organizational requirements.

Learning Curve

General

The learning curve, frequently referred to as the manufacturing progress function, is a technique for projecting the estimated future costs (in money or worker-hours) of an item or group of items in a continuing production sequence. The learning curve factor is an indicator of what future items in a production sequence would cost relative to preceding items in that sequence. The higher the learning curve factor, the higher the ratio of the cost of a future item to that of a previous item; conversely, the lower the learning curve factor, the lower the cost ratio.

The learning curve presumes continuous production of a basic item using the same design or specified process. It precludes engineering changes or modifications to the original item which would alter the basic design requirements and begin a new learning curve cycle for the new configuration. The learning curve principle applies to service processes which involve sequences of standard, repetitive tasks in the same manner as it applies to standard production items.

Types of Learning Curve Functions

There are two types of functional applications of the learning curve technique: (1) the *unit-cost* learning curve and (2) the *cumulative-average-unit-cost* learning curve.

Unit-cost learning curve

The unit-cost learning curve is based on three operative principles:

1. The first principle is that of the learning curve factor, commonly cited as a percentage (e.g., 90 percent), but mathematically applied as a decimal equivalent (e.g., 0.90).

2. The second principle relates to the use of sequential "double-octave" intervals to denote the individual production units subject to the learning curve factor. This principle dictates that succeeding intervals between affected production units, after the first unit, are always double the preceding interval: if the *first* unit subject to learning curve impact after the initial production unit is the *second* unit, succeeding intervals will be defined by the *fourth* unit, the *eighth* unit, the *16th* unit, the *32d* unit, etc.

3. The third principle stipulates that, according to the double-octave discipline, the projected cost of each succeeding interval unit will be determined by multiplying the cost of the *preceding* interval unit by the learning curve factor. For example, if one were to apply a 90 percent learning curve factor to the aforementioned double-octave pattern, the cost of the *second* unit would be 90 percent of that of the *first (initial)* production unit, the cost of the *fourth* production unit would be 90 percent of that of the *second* unit, the cost of the *eighth* production unit would be 90 percent of that of the *fourth* unit, the cost of the *16th* production unit would be 90 percent of that of the *eighth* unit, etc. A later section of this chapter gives details of the algorithmic methodology governing unit-cost learning curve computation.

Cumulative-average-unit-cost learning curve

The cumulative-average-unit-cost learning curve function incorporates mathematical techniques and operative factors similar to those of the unit-cost function. The cumulative-average-unit-cost approach is used to project the average unit production cost for the total production program as of the end of a defined time frame or extended production run, relative to the average unit cost calculated for an initial (or pilot) group of units. For example, one would predict the average unit cost for a cumulative total of X_2 units to be completed at a future point in time on the basis of the average unit cost determined from an initial lot of X_1 units. A later section of this chapter describes the algorithmic methodology applicable to the computation of the cumulative-average-unit-cost learning curve.

Distinction between unit cost and cumulative average unit cost

The conceptual difference between *unit cost* and *cumulative average unit cost* can be explained by a simplistic analogy. Suppose a grocer were to sell a head of lettuce at a different price on each of five days during a week; i.e.,

Monday: $5.00/head
Tuesday: $4.00/head
Wednesday: $3.00/head
Thursday: $2.00/head
Friday: $1.00/head

The *unit cost* as of Friday would be $1.00. The *cumulative average unit cost* as of Friday would be $3.00:

$$\frac{\$5 + \$4 + \$3 + \$2 + \$1}{5 \text{ (days)}} = \$3.00$$

Hence, under the learning curve concept, at any point during an extended production sequence, the unit cost is always lower than the cumulative average unit cost, as the computation of the cumulative average unit cost must necessarily consider the preceding higher unit costs.

Learning Curve from the Mathematical Perspective

Unit-cost formula

The cost of a specific sequential unit in a production lot is

$$C_X = C_1\,(X)^{\log S/\log 2}$$

where C_X = cost of Xth sequential unit assessed in production lot (see the note in Fig. 11.1).
  C_1 = cost of base (initial) unit, or first unit in sequence
  X = sequence number of assessed Xth unit
  S = learning curve factor, in decimal form (e.g., if the learning curve factor is 90 percent, $S = 0.90$)
 $\log S$ = common logarithm of S
 $\log 2$ = common logarithm of 2

Learning curve
double-octave values $C_x = C_1(X)^{\log S/\log 2}$

Applied value of X

1	2	4	8	16	32	64	128	etc.

Base unit			Governed by sequential double-octave intervals					
1	5	10	20	40	80	160	320	etc.
1	10	20	40	80	160	320	640	etc.
1	20	40	80	160	320	640	1280	etc.
1	etc.	etc.	etc.	etc.	etc.	etc.	etc.	etc.

Example: If the *first* assessed unit is the 10th unit, so that $X = 2$, then for the 40th unit in this production sequence, $X = 8$. For the 160th unit in the sequence, $X = 32$.

Note: The learning curve factor is sequentially imposed on succeeding units at double-octave intervals. For example, if the assessment sequence is 1, 2, 4, 8, etc., and the learning curve factor is 90 percent:

  1st unit is basic cost
  2d unit is 90 percent of 1st unit
  4th unit is 90 percent of 2d unit
  8th unit is 90 percent of 4th unit
  etc.

Figure 11.1 How to determine the Xth assessable unit.

Figure 11.1 illustrates the procedure for determining the cost of X (the assessable unit) based on production sequence and double-octave intervals.

Relative to the matrix provided in Fig. 11.1, experience has indicated that for pilot programs involving prototype or first-article production groups, the sequence 1, 2, 4, 8, 16, etc., would be applied to determine a base group average on which to project future production lots. For established products, the assessed unit for which $X = 2$ in the learning curve algorithm is more likely to be later in the production sequence.

Example of Unit-Cost Learning Curve Calculation The learning curve factor of 90 percent is initially applied on the 10th unit. The first article produced costs $10.00. What is the projected cost of the 320th unit?

If

$$C_X = C_1 X^{\log S/\log 2}$$

where $C_1 = \$10.00$
       $S = 90\% = 0.9$
       $X = 64$ (from Fig. 11.1)
       $\log S = \text{LOG } 0.9 = -0.04576$
       $\log 2 = 0.301$

then

$$C_x = (10,000)[(64)^{-0.04576/0.301} = (10)(64)^{-0.152}$$
$$= (10)(0.5314)$$
$$= \$5.31$$

The unit cost of the 320th unit based on the first assessment and initial interval defined at the 10th unit is $5.31.

Cumulative-average-unit-cost formula

The cumulative average unit cost of an extended production group, based on the average cost of the initial lot, is

$$C_{\bar{x}_2} = C_{\bar{x}_1}\left(\frac{\bar{x}_2}{\bar{x}_1}\right)^{\log S/\log 2}$$

where $C_{\bar{x}_2}$ = average unit cost of extended production group
$C_{\bar{x}_1}$ = computed average unit cost of initial lot, or the first lot point
x_2 = number of units in extended production lot
x_1 = number of units in initial lot
S = learning curve factor in decimal form (e.g., if the learning curve factor is 85 percent, $S = 0.85$)
$\log S$ = common logarithm of S
$\log 2$ = common logarithm of 2

Learning curve discipline permits the use of computed averages or the following:

1. If the initial lot is 10 or less, the first lot point, $C_{\bar{x}_1}$, can normally be the median (midpoint) cost value between the first and tenth units.
2. If the initial lot is more than 10 units, $C_{\bar{x}_1}$ will be the cost of the sequential unit at one-third the lot size (e.g., for an initial lot of 30 units, $C_{\bar{x}_1}$ will be the unit cost of the tenth unit).

Example of Cumulative Average-Unit-Cost Learning Curve Calculation   It is necessary to negotiate a final unit price for a contract for production of 1000 skyhooks. Based on an initial pilot program of 30 units, the computed average unit cost was $10.00 and the learning curve factor was calculated to be 90 percent. What should be the final unit cost negotiated for the production contract for 1000 skyhooks?

If

$$C_{\bar{x}_2} = C_{\bar{x}_1}\left(\frac{x_2}{x_1}\right)^{\log S/\log 2}$$

and

$$C_{\bar{x}_1} = \$10.00$$
$$x_1 = 30$$
$$x_2 = 1000$$
$$S = 90\% = 0.9$$
$$\log S = \log 0.9 = -0.04576$$
$$\log 2 = 0.301$$

then

$$C_{\bar{x}_2} = (10.00)\left(\frac{1000}{30}\right)^{-0.04576/0.301}$$
$$= (10)(33.33)^{-0.152}$$
$$= (10)(0.5868)$$
$$= \$5.87$$

The cumulative average unit cost as of the 1000th unit would be $5.87 based on the calculated average of $10.00 per unit derived from the initial 30-unit pilot group.

Cumulative Effects of Learning Curve Factors

A critical point of negotiation relative to production contracts is the learning curve factor to be applied in determining the price for the total production lot. The production contractor will aggressively pursue the highest possible factor, which would have the effect of ensuring that the negotiated sequential unit costs used as the basis for determining the price for the total production run would not decrease significantly relative to the preceding units. This scenario tends to promote higher negotiated prices. The purchaser would endeavor to negotiate the lowest possible learning curve factor, which would mean that sequential unit costs relative to the costs of the preceding units would tend to decrease at a *faster* rate. This scenario would promote a lower negotiated price for the production run.

Comparative effects of differing unit-cost learning curve factors

Figure 11.2 depicts an example comparing the effects of a 90 percent learning curve factor and an 85 percent learning curve factor. The difference is relatively minor at the outset but becomes increasingly significant as the forecast is projected further into the future production sequence. Also, note that computation of the unit costs at the double-octave

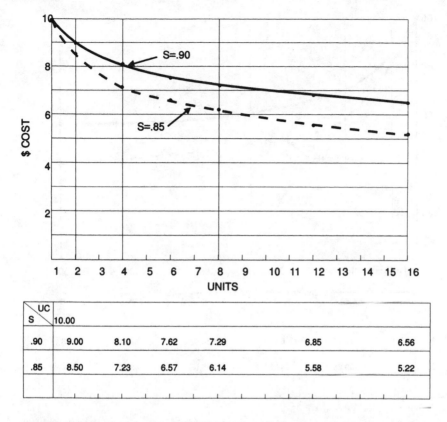

Figure 11.2   Unit-cost learning curve. Comparison of 90 percent and 85 percent unit-cost learning curves when base unit cost is $10.00 and assessment sequence is 1, 2, 4, 8, 16, etc.

intervals (1, 2, 4, 8, and 16) is relatively simple (see Fig. 11.1). Computation of the unit cost at intrainterval points, such as the cost of the sixth unit in Fig. 11.2, requires the use of the unit-cost learning curve algorithm, as described earlier in the chapter.

Comparative effects of differing cumulative-average-unit-cost learning curve factors

The increasingly significant differences in sequential cost estimates evident in prediction of the unit costs are equally evident in comparative analysis of cumulative average unit costs. Figure 11.3 compares the effects of 90 percent and 85 percent learning curve factors when applied to a five-year production run of 100 units per year. The first 10 units of the first year's 100-unit output are used as a pilot group to determine

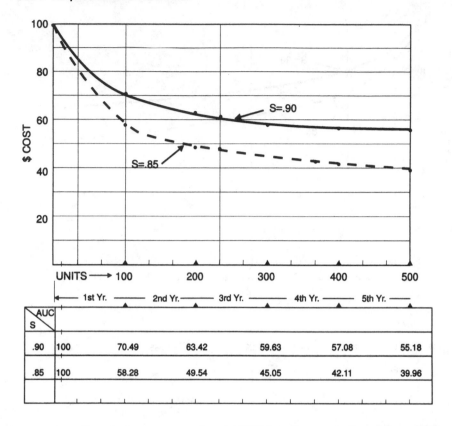

Figure 11.3 Cumulative-average-unit-cost (AUC) learning curve. Comparison of 90 percent and 85 percent cumulative-average-unit-cost learning curves when average unit cost of pilot production lot of 10 units is $100.00 and subsequent annual production lots over a five-year program are 100, 200, 300, 400, and 500 units.

the base average unit cost for estimating the cumulative average unit cost as of the end of each 100-unit annual output throughout the five-year program. Calculation of the cumulative average unit cost for each of the annual production lots is based on the cumulative-average-unit-cost learning curve algorithm explained earlier in the chapter.

Inspection of the trend in the 90 percent learning curve in Fig. 11.3, provides the projections in Table 11.1.

The cumulative-average-unit-cost learning curve highlights the cost advantage of multiyear procurements. By contracting for the 500 units over a five-year period, a 22 percent reduction in unit cost is achieved compared to the unit cost for the first year. The same learning curve concept also provides a quantitative basis for determining cancellation

TABLE 11.1   Cumulative Average Unit Cost—Five-Year Projection

Year	Annual unit output	Cumulative average unit cost	Cumulative units produced	Cumulative costs ($)
1	100*	70.49	100	7,049
2	100	63.42	200	12,684
3	100	59.63	300	17,889
4	100	57.08	400	22,832
5	100	55.18	500	27,590

*The pilot group is incorporated in the first year's computation.

charges, in the event it should be necessary to terminate the production program prior to the end of the five-year contract schedule. For example, for termination at the end of year 3 with 300 units, the termination charges to which the production contractor would justifiably be entitled would be

$$300 \text{ units} \times (\underset{\text{3d yr AUC}}{59.63} - \underset{\text{5th yr AUC}}{55.18}) = \$1335$$

The cancellation charge is the conventional means by which a production contractor who has negotiated in good faith on the basis of a projected longer-term production requirement may be compensated for the cost impact of a cancellation.

Comparison of Unit-Cost Learning Curve and Cumulative-Average-Unit-Cost Learning Curve Trends

Earlier in this chapter, it was emphasized that for any designated unit in a production sequence, the cumulative average unit cost will always be higher than the cost of that specific unit. Figure 11.4 portrays this comparison. In the figure, the inherent algorithmic factors are common to both learning curves:

1. The learning factor is 90 percent.
2. The cost of the first unit is $10.00.

   Based on these common elements, the double-octave pattern (applicable to unit cost) is 1, 10, 20, 40, 80, etc. The computed base group average cost (applicable to cumulative average unit cost) is $9.50, the median between $10.00 (first unit cost) and $9.00 (unit cost for first assessed unit). The computations for the respective values of cumulative

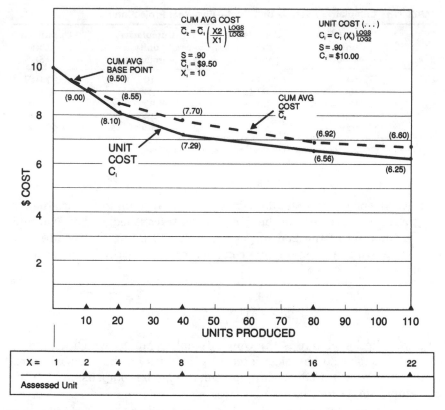

Figure 11.4   Comparison of cumulative average unit cost and unit cost learning curve trends using common learning curve and base cost factors.

average unit cost and unit cost are made at common points in the production sequence.

It is also characteristic that although both learning curves tend to flatten as the production run is extended, there is always a persistent gap between cumulative average unit cost and unit cost for any specific unit in the sequence.

Derivation of Unit Cost Learning Curve Factor (S) from Scatter Plot Data

Although familiarity with the previously discussed mathematical applications to learning curve scenarios is an essential prerequisite for the logistics engineer, it is also important to understand how to derive the learning curve factor (S). This can be accomplished by logarithmic methodologies described in Chapter 1. Determination of the learning curve factor (S) is based on the series of steps described in the following.

The unit cost learning curve formula is:

$$C_{x_2} = C_{x_1} (X_2^{\log s/\log 2})$$

where X_2 = specifically designated unit in production sequence after initial unit

C_{x_1} = cost of first unit

C_{x_2} = cost of specifically designated unit in projected production sequence

S = learning curve factor

Apply process of algorithmic derivation based on manipulation of common logarithm (Log) methodology:

$$\text{Log } C_{X_2} = \text{Log } C_{X_1} + (\text{Log } S/\text{Log } 2)(\text{Log } X_2)$$
$$(\text{Log } 2)(\text{Log } C_{X_2}) = (\text{Log } 2)(\text{Log } C_{X_1}) + (\text{Log } S)(\text{Log } X_2)$$
$$(\text{Log } S)(\text{Log } X_2) = (\text{Log } 2)(\text{Log } C_{X_2}) - (\text{Log } 2)(\text{Log } C_{X_1})$$
$$\text{Log } S = \frac{(\text{Log } 2)(\text{Log } C_{X_2} - \text{Log } C_{X_1})}{(\text{Log } X_2)}$$

Therefore:

$$S = \text{Antilog} \left| \frac{(\text{Log } 2)\left(\text{Log } C_{X_2} - \text{Log } C_{X_1}\right)}{\text{Log } X_2} \right.$$

The starting point is a sample from the initial pilot group of production items. In the case of DoD production programs this sample may be drawn from the low rate initial production (LRIP) lot. The example portrayed in Figure 11.5 illustrates a hypothetical situation involving a sample of 31 observations where the key resource is labor hours required (e.g., as reflected in the job routing records) for processing individual items through the production process. The basic approach is to calculate the value of S for each observation by exercising the algorithm described above. The projected value of S for the total production run is based on the mean value of the pilot sample of 31 observations. Examination of the results from analysis of the details in Figure 11.5 indicates that the projected learning curve factor should be 94.5%.

$$S = \text{Antilog} \left[\frac{(\text{Log } 2)\left(\text{Log } C_{X_2} - \text{Log } C_{X_1}\right)}{\text{Log } X_2} \right]$$

Sample Task Nr.	X_2	C_{X_1}(h)	C_{X_2}(h)	S (Learning Curve Factor)
1	2	10	9.75 (9h 45min)	0.975
2	3	10	9.60 (9h 36min)	0.975
3	4	10	9.25 (9h 15min)	0.962
4	5	10	9.33 (9h 20min)	0.970
5	6	10	9.25 (9h 15min)	0.978
6	7	10	8.33 (8h 20min)	0.837
7	8	10	8.50 (8h 30min)	0.947
8	9	10	8.40 (8h 24min)	0.946
9	10	10	8.25 (8h 15min)	0.944
10	11	10	8.10 (8h 6min)	0.941
11	12	10	8.20 (8h 12min)	0.946
12	13	10	8.15 (8h 9min)	0.946
13	14	10	8.30 (8h 18min)	0.952
14	15	10	8.00 (8h)	0.944
15	16	10	8.10 (8h 6min)	0.949
16	17	10	7.90 (7h 54min)	0.944
17	18	10	7.80 (7h 48min)	0.942
18	19	10	7.85 (7h 51min)	0.945
19	20	10	8.35 (8h 21min)	0.959
20	21	10	7.90 (7h 54min)	0.948
21	22	10	7.70 (7h 42min)	0.943
22	23	10	7.65 (7h 39min)	0.943
23	24	10	7.60 (7h 36min)	0.942
24	25	10	7.55 (7h 33min)	0.941
25	26	10	7.50 (7h 30min)	0.941
26	27	10	7.70 (7h 42min)	0.947
27	28	10	7.45 (7h 27min)	0.941
28	29	10	7.40 (7h 24min)	0.940
29	30	10	7.35 (7h 21min)	0.939
30	31	10	7.30 (7h 18min)	0.938
31	32	10	7.25 (7h 15min)	0.938
			251.76	29.303

Sample Size = 31 Task Observations

$X_2 = 32$

$\sum S = 29.303$ Mean Pilot Lot Value of $S = \dfrac{29.303}{31} = 0.945258$

$X_2 = 2$ (@ 94.5%)

Figure 11.5   Example of Determination of Unit Cost Learning Curve Factor (S) from LRIP/Pilot Production Lot-Analyzing Unit Production Task Time.

12

Financial Analysis

General

Financial analysis concerns the prediction and analysis of monetary dynamics in terms of defined time parameters and interest or discount rates. This chapter addresses nine methods of financial analysis which are commonly used in engineering economics:

1. Future value
2. Present value
3. Uniform series compound amount
4. Annuity value
5. Uniform series present worth
6. Capital recovery
7. Effective annual interest
8. Cost gradient: equivalent uniform periodic cost
9. Cost gradient: present-value cost

Each method of financial analysis is appropriate in particular situations, depending upon the type of question posed by the analyst. The student of logistics will find an electronic calculator to be a valuable aid in achieving understanding and proficiency in application of the various financial formulas. This chapter employs the engineering economics approach to financial analysis. The nine formulas are divided into three groups:

1. Time-value analysis
2. Effective interest analysis
3. Cost gradient analysis

The following sections describe the three algorithmic groups.

Time-Value Analysis

The time-value group encompasses future value, present value, uniform series compound amount, annuity value, uniform series present worth, and capital recovery. To develop expertise in applying these methods, one should become familiar with a memorization technique that has been employed successfully by practitioners of engineering economics.

Memorization approach

Each of the six time-value formulas uses one or more of the three basic building blocks. The first three formulas can be constructed in a sequence of four steps.

Step 1.   Use FP, FA, and PA to build a frame for the building blocks as follows:

$$F = P$$
$$F = A$$
$$P = A$$

The frame is actually an outline for the formulas.

Step 2.   Incorporate building block 1—$(1 + i)^n$—into all of the formulas

$$F = P[(1 + i)^n]$$
$$F = A[(1 + i)^n]$$
$$P = A[(1 + i)^n]$$

Step 3.   Additionally, incorporate building block 2— $-\dfrac{1}{i}$ —into FA and PA.

$$F = P[(1+i)^n]$$
$$F = A\left[\frac{(1+i)^n - 1}{i}\right]$$

$$F = A\left[\frac{(1+i)^n - 1}{i}\right]$$

Step 4. Additionally, incorporate building block 3—$(1 + i)^n$ as a denominator—into PA.

$$F = P[(1+i)^n]$$

$$F = A\left[\frac{(1+i)^n - 1}{i}\right]$$

$$P = A\left[\frac{(1+i)^n - 1}{i(1+i)^n}\right]$$

These steps have produced the following set of time-value financial formulas.

Formula	Constituent factors	Financial term for formula
$F = P[(1 + i)^n]$	F = future value P = present amount n = number of periods i = interest rate per period (decimal equivalent to %)	Future value

Typical question posed by analyst: If I invest a lump sum P now, at i annual interest, how much F will I accumulate at the end of n years?

Formula	Constituent factors	Financial term for formula
$F = A\left[\frac{(1+i)^n - 1}{i}\right]$	F = future value A = periodic input amount n = number of periods i = interest rate per period (decimal equivalent to %)	Uniform series compound amount

Typical question posed by analyst: If I invest A amount at the end of each year at i annual interest, how much F will accumulate by the end of n years?

Formula	Constituent factors	Financial term for formula
$P = A\left[\dfrac{(1+i)^n - 1}{i(1+i)^n}\right]$	P = present value A = periodic input or output n = number of periods i = interest rate per period (decimal equivalent to %)	Uniform series present worth

Typical questions posed by analyst:

1. If I need to receive A amount of annual revenue at the end of each year for n years, what lump sum P must I invest at i annual interest at the present time?
2. If I can afford to pay A amount at the end of each year for n years at i annual interest, how much P will I be able to borrow now (e.g., for capital equipment, automobile, house)?

Completing the time-value series

Once one has mastered the memorization technique for future value, uniform series compound amount, and uniform series present worth, the remaining three time-value formulas are within easy grasp. For this purpose, the factors FP, FA, and PA are individually reversed to become PF, AF, and AP, and their corresponding operative factors reciprocated. This reciprocative transformation has produced the remaining series of time-value formulas, as follows:

$$F = P\left[(1+i)^n\right] \quad \text{becomes} \quad P = F\left[\frac{1}{(1+i)^n}\right]$$

$$F = A\left[\frac{(1+i)^n - 1}{i}\right] \quad \text{becomes} \quad A = F\left[\frac{i}{(1+i)^n - 1}\right]$$

$$P = A\left[\frac{(1+i)^n - 1}{i(1+i)^n}\right] \quad \text{becomes} \quad A = P\left[\frac{i(1+i)^n}{(1+i)^n - 1}\right]$$

Formula	Constituent factors	Financial term for formula
$P = F\left[\dfrac{1}{(1+i)^n}\right]$	P = present value F = future value n = number of periods i = interest rate per period (decimal equivalent of %)	Present value

Typical question posed by analyst: If I need to accumulate F amount in n years, assuming i annual interest, what lump sum P must I invest at the present time?

Formula	Constituent factors	Financial term for formula
$A = F\left[\dfrac{i}{(1+i)^n - 1}\right]$	A = periodic input amount F = future amount n = number of periods i = interest rate per period (decimal equivalent of %)	Annuity value

Typical question posed by analyst: If I need to accumulate F amount in n years, assuming i annual interest, how much A do I need to invest at the end of each year?

Formula	Constituent factors	Financial term for formula
$A = P\left[\dfrac{i(1+i)^n}{(1+i)^n - 1}\right]$	A = periodic input or output amount P = present value n = number of periods i = interest rate per period	Capital recovery

Typical questions posed by analyst:

1. If I am able to invest P amount in a lump sum at the present time at i annual interest for n years, how much annual revenue A will I receive at the end of each year until the funds are depleted?

2. If I borrow P amount at i annual interest to be repaid by the end of n years, what annual payment A must be made at the end of each year?

It is important to note that the periodic factors and interest factors used in financial formulas must be consistent; for example, when n is expressed in months or quarters, interest i must correspondingly be expressed in monthly interest or quarterly interest.

Examples of time-value computations

The following cases illustrate the application of the six previously described formulas used in time-value financial calculations.

Future value.

$$F = P(1 + i)^n$$

where P = present one-time investment
$\quad i$ = interest rate per period
$\quad n$ = number of periods
$\quad F$ = future value received after n periods

Example   If P = \$1000 present investment lump sum, i = 10 percent (or 0.10) interest per year, and n = 5 years, how much will be accumulated after 5 years?

$$\begin{aligned} F &= (1000)(1 + 0.10)^5 \\ &= (1000)(1.61051) \\ &= \$1610.51 \end{aligned}$$

Present value.

$$P = F\left[\frac{1}{(1+i)^n}\right]$$

where F = future amount desired
$\quad i$ = interest rate per period
$\quad n$ = number of periods
$\quad P$ = present investment required

Example   If F = \$1610.51 future sum desired, i = 10 percent (or 0.10) interest per year, and n = 5 years, how much must be invested in a lump sum at the current point in time to achieve \$1610.51 in 5 years?

$$\begin{aligned} P &= (1610.51)\left[\frac{1}{(1+0.10)^5}\right] \\ &= (1610.51)\left[\frac{1}{1.61051}\right] \\ &= (1610.51)(0.620921323) \\ &= \$1000 \end{aligned}$$

Note: Refer to the previous example for confirmation.

Uniform series compound amount.

$$F = A\left[\frac{(1+i)^n - 1}{i}\right]$$

where A = amount invested each period
    i = interest rate per period
    n = number of years
    F = future sum accumulated after n periods of investing A dollars
      per period

Example If A = $1000 invested at the end of each year, i = 10 percent (or 0.10) interest per year, and n = 5 years, how much will be accumulated at the end of 5 years?

$$F = (1000)\left[\frac{(1+0.10)^5 - 1}{10}\right]$$

$$= (1000)\left(\frac{1.61051 - 1}{0.10}\right)$$

$$= (1000)\left(\frac{0.61051}{0.10}\right)$$

$$= (1000)(6.1051)$$

$$= \$6105.10$$

Annuity value.

$$A = F\left[\frac{i}{(1+i)^n - 1}\right]$$

where F = future amount desired
    i = interest rate per period
    n = number of periods
    A = sum invested at the end of each period

Example If F = $6105.10 desired to be received at the end of 5 years, i = 10 percent (or 0.10) interest per year, and n = 5 years, how much must be invested at the end of each year to accumulate $6105.10 by the end of 5 years?

$$A = (6105.10)\left[\frac{0.10}{(1+0.10)^5 - 1}\right]$$

$$= (6105.10)\left(\frac{0.10}{1.61051 - 1}\right)$$

$$= (6105.10)\left(\frac{0.10}{0.61051}\right)$$

$$= (6105.10)(0.163797)$$

$$= \$1000$$

Note: Refer to the previous example for confirmation.

Uniform series present worth.

$$P = A\left[\frac{(1+i)^n - 1}{i(1+i)^n}\right]$$

where A = amount required per period
i = interest rate per period
n = number of periods
P = present sum required to generate A dollars per period at i interest rate for n periods

Example If A = $1000 required as income per year, i = interest rate of 10 percent (or .10) per year, and n = 5 years, how much must be invested at the present time to receive $1000 at the end of each year for 5 years?

$$P = 1000\left[\frac{(1+0.10)^5 - 1}{0.10(1+0.10)^5}\right]$$

$$= 1000\left[\frac{1.61051 - 1}{(0.10)(1.61051)}\right]$$

$$= (1000)\left(\frac{0.61051}{0.161051}\right)$$

$$= (1000)(3.79079)$$

$$= \$3790.79$$

Capital recovery.

$$A = P\left[\frac{i(1+i)^n}{(1+i)^n - 1}\right]$$

where P = present value or present sum
i = interest rate per period
n = number of periods
A = periodic investment or repayment required

Example If P = $3790.79, lump sum borrowed from bank, i = interest rate of 10 percent (or 0.10) per year, and n = 5 years, how much must be paid to the bank at the end of each year to repay the $3790.79 loan?

$$A = (3790.79)\left[\frac{0.10(1+0.10)^5}{(1+0.10)^5 - 1}\right]$$

$$= (3790.79)\left[\frac{(0.10)(1.61051)}{1.61051 - 1}\right]$$

$$= (3790.79)\left(\frac{0.161051}{0.61051}\right)$$

$$= (3790.79)(0.26379748)$$
$$= \$1000$$

Note: Refer to the previous example for confirmation.

Effective Interest Analysis

Effective interest analysis focuses on the *effective annual interest rate*. This rate is based on the interest accumulated per year when the interest is compounded on a daily basis. This method of computing interest is illustrated by a financial institution's quoting the annual return on a certificate of deposit in terms of *annual interest* and *annual yield*; the *annual yield* is actually the *annual interest* compounded on a daily basis.

Effective annual interest formula

There are complicated methods of calculating daily compounded interest; however, the analyst equipped with an electronic calculator will find the following method used in engineering economics to be the most convenient:

$$\text{Effective annual interest rate} = e^i - 1$$

where i = true annual interest rate
$\quad\quad e$ = base number for the natural logarithm

Example The stated annual interest rate for a financial credit program is 18 percent. The actual interest on charges to the program is compounded on a daily basis, commencing on the dates of the individual charges. Calculate the effective annual interest rate on the charges.

$$i = 0.18$$
$$\text{EAI} = \text{Effective annual interest rate} = e^{0.18} - 1$$
$$= 1.1972 - 1$$
$$= 0.1972 \text{ or } 19.72\%$$

It is obvious from the above example that daily compounding has the effect of significantly inflating the true annual interest factor, especially at higher rates. (This revelation could also prove painfully enlightening to some who might use credit plans.)

Cost Gradient

Cost gradient techniques address the effects of adding predetermined cost increments to periodic amounts subject to prescribed interest or inflation rates over a projected number of periods. There are two types of cost gradients:

1. Equivalent uniform periodic cost
2. Present-value cost

The cost gradient techniques are applicable when future periodic supplemental costs must be considered along with normal discount factors in evaluating the ramifications of future cost situations. The two basic cost gradient formulas are detailed in the following sections.

Equivalent uniform periodic cost

$$A = g\left[\frac{1}{i} - \frac{n}{(1+i)^n - 1}\right]$$

where g = increment in cost per period
i = interest rate per period
n = number of periods
A = equivalent uniform periodic cost (periodic average over n periods)

Example Maintenance on a system will increase an estimated $1000 each year for 10 years. The first year, maintenance is covered by warranty. The second year, maintenance should be $1000; the third year, maintenance would be $2000, etc. Calculate the equivalent uniform annual cost for maintenance. Interest is 10 percent per annum.

$$g = \$1000 \qquad i = 10 \text{ percent or } 0.10 \qquad n = 10$$

$$A = 1000\left[\frac{1}{0.10} - \frac{10}{(1+0.10)^{10} - 1}\right]$$

$$= 1000\left(10 - \frac{10}{2.5937 - 1}\right)$$

$$= 1000(10 - 6.2747)$$

$$= 1000(3.72529)$$

$$= \$3725.99$$

Present-value cost

$$P = g\left[\frac{(1+i)^n - 1}{(i)^2(1+i)^n} - \frac{n}{1(1+i)^n}\right]$$

where g = increment in cost per period
$\quad$ i = interest rate per period
$\quad$ n = number of periods
$\quad$ P = present-value cost

Example Maintenance on a system will increase an estimated $1000 each year for 10 years. The first year is covered by warranty; the second year, maintenance should be $1000; the third year, maintenance should be $2000, etc. Calculate the present value (or present worth).

$$g = \$1000 \qquad i = 10 \text{ percent or } 0.10 \qquad n = 10$$

$$P = 1000\left[\frac{(1+0.10)^{10} - 1}{(0.10)^2(1+0.10)^{10}} - \frac{10}{(0.10)(1+0.10)^{10}}\right]$$

$$= 1000\left[\frac{2.5937 - 1}{(0.01)(2.5937)} - \frac{10}{(.10)(0.25937)}\right]$$

$$= 1000\left[\frac{1.5937}{(0.01)(2.5937)} - \frac{10}{0.25937}\right]$$

$$= 1000(61.445 - 38.555)$$
$$- 1000(22.89)$$
$$= \$22,890$$

Life-cycle cost applications

The cost gradient methods are effective companions to the time-value techniques in performing life-cycle cost analyses.

13

Depreciation of Assets

General

Depreciation is the financial management discipline whereby investments in capital assets are amortized over a defined fiscal period. The predominant factors constituent to the depreciation process are

1. Acquisition value
2. Salvage value
3. Service life
4. Periodic book value
5. Periodic depreciation allowance
6. Reserve for depreciation

The rules for asset depreciation permit allocation of corporate revenue or other resource apportionment to periodically replenish the original capital investment (less residual salvage value) at a periodic rate governed by an asset's recognized service life. Depreciation methods are determined primarily in accordance with procedures currently approved by tax and accounting authorities. Tax rules concerning capital depreciation generally provide for allocation of funds to reserves for depreciation established by the organization to repay the original investments. The legal allocation of corporate revenues to allowances for depreciation has the effect of depressing corporate income and reducing income tax liability, while providing for reconstituting the resources that enable future acquisition of replacement assets.

Depreciation Methods

The basic methods of depreciation with which the logistics analyst should be familiar are

1. Straight-line method
2. Declining-balance method
3. Sum-of-the-years'-digits method

Most financial regulations dictate that the method established upon acquisition and declaration of the asset must remain in effect until the asset is completely amortized. The authorities who stipulate the criteria for fiscal asset depreciation allow for few exceptions.

Computation of Depreciation Schedules

The following gives the algorithmic procedures, along with examples, for three commonly used depreciation methods.

Straight-line method of depreciation

The straight-line method is the simplest to calculate and administer. Table 13.1 illustrates.

Under this method, the remaining basis can be shown as a descending straight line with a negative slope determined by the calculated straight-line rate. The example is shown graphically in Fig. 13.1.

Declining-balance method of depreciation

The declining-balance rate (DBR) is established on the basis of a multiple of the straight-line rate. For example, a 150 percent declining-balance rate indicates a rate 1.5 times the straight-line rate; a 175 percent

TABLE 13.1 Example of Straight-Line Rate*
A new machine costs $10,000
Estimated salvage value is $1,500
Expected life of 5 years

End of year	Cost or remaining basis	Rate	Depreciation allowance	Reserve
Start	$8500	—	$ —	$ 1,500
First	6800	20%	1700	3,200
Second	5100	20%	1700	4,900
Third	3400	20%	1700	6,600
Fourth	1700	20%	1700	8,300
Fifth	0	20%	1700	10,000

*Straight-line rate = 100% estimated salvage percentage/estimated service life (years).

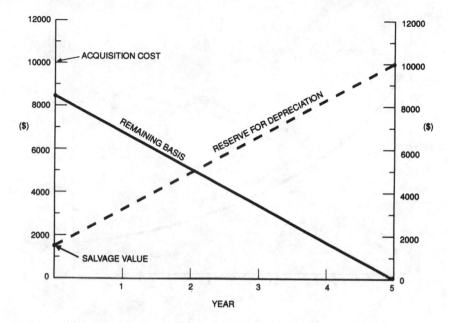

Figure 13.1   Depreciation trend effects of Table 13.1.

DBR indicates a rate 1.75 times the straight-line rate, etc. Table 13.2 depicts an example of a *double-declining-balance* rate (or 200 percent declining-balance rate). For this example, the straight-line annual depreciation rate would be 20 percent; because of the effect of the double-

TABLE 13.2   Example of Double-Declining-Balance Method

A new machine costs $10,000
Estimated salvage value is $1500
Expected life of 5 years
DBR = 2

End of year	Cost or remaining basis	Rate	Depreciation allowance	Reserve
Start	$10,000	—	$ —	$ 0
First	6,000	40%	4000	4000
Second	3,600	40%	2400	6400
Third	2,160	40%	1440	7840
Fourth	1,500	40%	660*	8500
Fifth	1,500	40%	0	8500

*Using DDB method, salvage value is not subtracted from cost first to determine total allowable depreciation. Instead, the remaining basis is tracked so that allowable depreciation is only that amount that will reduce the remaining basis to the salvage value, $1500. In this case, at the end of year 4, only 660 of the 864 ($2160 × 40%) is allowable because any depreciation amount greater than 660 would reduce the asset basis to below $1500. Note there is no depreciation in year 5.

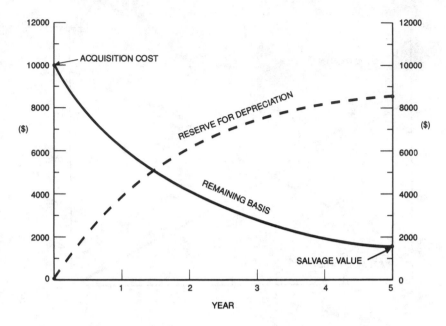

Figure 13.2   Depreciation trend effects of Table 13.2.

declining-balance method, a 40 percent annual depreciation allowance is applied.

Figure 13.2 graphically depicts the trend effects of the double-declining-balance (DDB) rate in the example.

Sum-of-the-years'-digits for the *Y*th year method of depreciation

The sum-of-the-years'-digits (SOD) method reflects a compromise in the pace of depreciation between the conservative straight-line method and the more liberal declining-balance method. Table 13.3 illustrates the SOD method.

$$\text{SOD (}Y\text{th year)} = \frac{(n - y)(100\% - \text{estimated salvage }\%)}{\sum_{x=1}^{x=n} X}$$

where n = expected life in years

y = number of years in service before Yth year, and

$$\sum_{x=1}^{x=n} X = \frac{(n)(n+1)}{2}$$

Figure 13.3 graphically portrays the SOD example.

TABLE 13.3 Example of Sum-of-the-Years'-Digits Method

A new machine costs $10,000
Estimated salvage value is $1500
Expected life of 5 years

End of year	Cost or remaining basis	Rate	Depreciation allowance	Reserve
Start	$8500	—	$ —	$ 1,500
First	5667	5/15	2833	4,333
Second	3400	4/15	2267	6,600
Third	1700	3/15	1700	8,300
Fourth	567	2/15	1133	9,433
Fifth	0	1/15	567	10,000

Interrelationship of acquisition cost, remaining basis, and reserve for depreciation

Tables 13.1, 13.2, and 13.3 emphasize a basic tenet governing depreciation schedules. Irrespective of the method used, at any given point during the service life of a capital asset, the sum of the remaining basis and the reserve for depreciation must equal the acquisition cost. Inspection of the figures depicting the depreciation methods addressed in this chapter will confirm this principle.

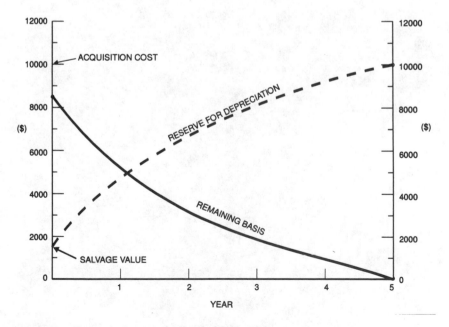

Figure 13.3 Depreciation trend effects of Table 13.3.

Applicability of Depreciation to Life-Cycle Costing

To compare multiple capital acquisition alternatives, it would probably be necessary to incorporate asset depreciation factors into the life-cycle cost (LCC) comparative analysis. Depreciation costs would be used along with the financial analysis formulas described in Chap. 12. The straight-line method is considered the most practical depreciation method for LCC analysis.

14

Life-Cycle Cost Analysis

General

Because of the uniqueness of each life-cycle cost (LCC) case and the fact that the scientific methodology used for case evaluation necessarily reflects this uniqueness, many professionals regard LCC analysis as more of an art than a science. The analyst employs artistic creativity to define the scope of the analysis, delimit the problem, identify the critical parameters, determine the pertinent factors, and design the mathematical tools necessary to accomplish the analysis. Accordingly, this chapter does not attempt to provide a set of standard LCC analysis recipes; rather, it is predicated on the assumption that better tutorial value would be achieved by addressing basic principles and discussing typical basic LCC scenarios to impart the LCC concept and sense of mental discipline.

Approach to LCC analysis

The life-cycle cost (LCC) of a system or equipment includes all the costs associated with activities broadly pertaining to research and development; design, test, and evaluation; production; construction; product distribution; system/equipment operation; retirement; and disposal. Simply stated, LCC is the total cost of ownership of a product. The concept of life-cycle ownership has a multiple perspective. This chapter, therefore, focuses on (1) the life cycle from the perspective of an organization that has conceived, developed, engineered, and produced a system and retained it for use until retirement and disposal; (2) the life cycle from the perspective of the producer of a marketable item for which are after-sale considerations—product distribution, customer support, area service centers, etc.; and (3) the life cycle from the perspective of a purchaser of a system "ready for service" which will be operated and logistically supported by the purchaser until retirement and disposal.

In the first case, LCC is based on total "cradle-to-grave" system costs. For the second case, LCC includes the costs of the earlier phases through production, plus whatever involvement in product distribution through retirement is necessary to accommodate customer pressures. In the third case, LCC encompasses the acquisition price plus the subsequent costs of operation, logistical support, retirement, and disposal of the system.

Principles of Life-Cycle Cost Analysis

From the disciplinary perspective of logistics engineering, the life-cycle milestone phases are

1. Conceptual definition (milestone 0)
2. Demonstration and validation (milestone I)
3. Full-scale engineering development (milestone II)
4. Production and distribution (milestone III)
5. Operation and customer support (milestone IV)
6. Retirement and disposal (milestone V)

The producer-user will be concerned with all phases. The market item producer has technical and financial responsibilities with respect to the earlier phases plus whatever is involved in product distribution, customer support, and product service. The purchaser-user, after purchase of the system, has technical and financial responsibilities for operational logistics support and retirement and disposal.

The production or acquisition of systems that meet performance and readiness requirements, on schedule, and at an affordable life-cycle cost presumes that purchasers and production managers have comprehensive and accurate LCC estimates for all product milestone phase decisions. LCC is considered part of the engineering and logistics trade-off studies used in selecting the most cost-effective system design, which will respond to the market expectation or users' needs at the lowest ultimate overall cost. The objective of LCC analysis is to provide quantified and qualified time-phased cost information to be used for cost measurements that support resource allocation planning, management, and control.

Figure 14.1 illustrates that, based on experience with Department of Defense systems (which may plausibly be extended to commercial systems) approximately 85 percent of life-cycle costs are determined by

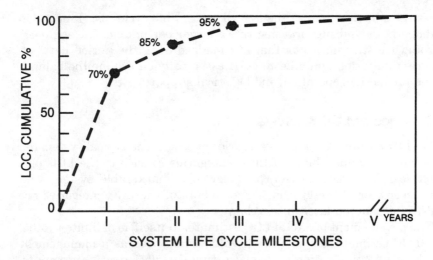

Figure 14.1 System life-cycle milestones.

milestone II. The system characteristics incorporated in design constraints, engineering drawings, and product specifications formulated during the earlier phases tend to significantly affect the performance parameters and logistical support requirements throughout the system service life. For this reason, substantial management attention must be paid to the impacts of decisions made during the conceptual definition and demonstration and validation phases of the evolution of the system.

Figure 14.2 portrays the sequence and distribution of actual resource consumption during the six life-cycle phases. Research and development

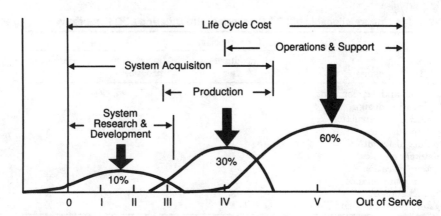

Figure 14.2 Distribution of life-cycle costs.

constitutes a small part of the overall system life-cycle cost. Production and operation and support constitute a larger percentage. The engineering and logistics decisions that are made during the period between program initiation and milestone III essentially determine the balance of the production, operation, and logistical support costs.

Categories of LCC Analysis

Table 14.1 outlines the major categories of cost in the typical system life cycle and the applicability of those categories to each of the LCC perspectives, i.e., producer-user, producer of a marketable system, and purchaser-user of the system. The elements of the LCC categories are addressed in detail later in this chapter.

The development of a total LCC estimate, in itself, is of limited value. The LCC estimate has significant value when used as a management tool for conducting design trade-offs, estimating design-cost constraints, and evaluating changes in logistic support policy.

Approach to LCC planning

The goals of LCC analysis are (1) to identify the total cost of alternative means of meeting users' needs, achieving production schedules, and attaining product performance objectives, and (2) to estimate the cost impact of the various design and support options. To achieve these goals, cost is established as a parameter of equal importance with technical, supportability, and schedule requirements. The acquisition management technique which supports this policy is called the design-to-cost (DTC) concept. The relationship between LCC and DTC is governed by

TABLE 14.1   Interrelationship of LCC Categories with Differing LCC Perspectives

	LCC perspective		
LCC category	Producer-user	Market-item producer	Purchaser-user
Research and development	X	X	
Engineering design	X	X	
Production/construction	X	X	
System purchase			X
Product distribution		X	
Customer support		X	
System operation	X		X
System maintenance	X		X
Product support	X		X
System retirement and disposal	X		X

the LCC analysis, which forms the foundation for the selection and allocation of DTC goals and thresholds. Once these are established, the production manager must manage the program development and production while making decisions that control the life-cycle cost without exceeding the DTC constraints. Effective and efficient management of LCC requires a plan for such management early in the system life cycle. This plan defines the roles and responsibilities of the participating corporate elements and provides a milestone schedule to be implemented by the production manager in order to monitor and manage the program to satisfy LCC objectives and to permit early corrective actions in high-risk or problem areas of design. In addition, the plan should

1. Describe the LCC evaluation and tracking methodology.
2. Specify key LCC-driving parameters of the system.
3. Identify the requirements for cost estimates relating to a baseline (e.g., a prior system or an existing system that is already on the market).
4. Require LCC analyses for significant design and program trade-off studies.
5. Establish internal documentation and technical data requirements for production management, tracking, and verification.
6. As appropriate, provide for an effective incentive award/profit plan to motivate all significant vendors and suppliers to consider and implement alternatives that reduce system LCC.

Elements of Life-Cycle Cost

The major LCC categories depicted in Table 14.1 correlate with the sequential life-cycle phases of a typical system. Each LCC category represents an aggregation of contributory cost elements, each realistically defined in sufficient detail to facilitate subsidiary system-level data collection and subcategorical cost analysis. This methodology provides summary cost data for each LCC category through summation of the subcategorical cost elements.

Table 14.2 relates the LCC categories and the individual groupings of contributory, subcategorical cost elements to the individual life-cycle perspectives. It is emphasized that Table 14.2 cites the more frequently utilized subsidiary cost elements with which the student of logistics should be familiar; it should not be construed as being all-inclusive. It is further noted that the individual subcategorical cost elements are developed on the basis of cost-estimating relationships (CERs) defined by algorithms needed to manipulate the detailed input data. The determination of applicable LCC categories, relevant cost elements, and CER

**TABLE 14.2 Reference Matrix,
LCC Category-Element-Life Cycle Applicability**

LCC applicability: (1) producer-user life cycle, (2) market-item producer life cycle, (3) purchaser-user life cycle

LCC elements					LCC Category					
	Research and development	Engineering design	Production/ construction	System purchase	Product distribution	Customer support	System operation	System maintenance	Product support	Retirement and disposal
Human resources	1, 2	1, 2	1, 2		2	2	1, 3	1, 3	1, 3	1, 3
Nonspecific burden cost	1, 2	1, 2	1, 2		2	2	1, 3	1, 3	1, 3	1, 3
Direct facility cost	1, 2	1, 2	1, 2		2	2	1, 3	1, 3	1, 3	
Support and test equipment	1, 2	1, 2	1, 2			2	1, 3	1, 3	1, 3	
Energy	1, 2	1, 2	1, 2		2	2	1, 3	1, 3	1, 3	
Quality control	1, 2	1, 2	1, 2			2	1, 3	1, 3		
Test and evaluation	1, 2	1, 2	1, 2							
EDM/ADM prototype	1, 2	1, 2	1, 2							
Technical data	1, 2	1, 2	1, 2			2	1, 3		1, 3	
Materials handling				3		2			1, 3	
Packaging		1, 2	1, 2		2					
System purchase price				3						
Warranty costs				3	2					
Initial investment spares, repair parts				3					1, 3	
Training			1, 2			2			1, 3	

TABLE 14.2 Reference Matrix (*Continued*)

LCC elements	Research and development	Engineering design	Production/ construction	System purchase	Product distribution	Customer support	System operation	System maintenance	Product support	Retirement and disposal
					LCC Category					
Transportation			1, 2	3	2	2			1, 3	1, 3
Inventory support			1, 2		2	2			1, 3	1, 3
Warehousing/ storage			1, 2		2	2			1, 3	1, 3
Repairs						2		1, 3		
Scheduled maintenance						2		1, 3		
Salvage (resale) value										1, 3 Negative cost
Trade-in value										1, 3 Negative cost
Scrap value										1, 3 Negative cost
Technician support								1, 3	1, 3	
Replenishment						2	1, 3	1, 3	1, 3	1, 3
Other costs	1, 2	1, 2	1, 2	3	2	2	1, 3	1, 3	1, 3	1, 3

algorithms will be influenced by the LCC perspective governing the LCC analysis. The primary objective of LCC analysis is to provide visibility for the cost impact of a management decision. The format used by the cost analyst is, therefore, critical to satisfying this objective.

The LCC structure based on LCC categories, cost elements, and CER algorithms must be reasonably consistent, yet capable of reporting explicitly the cost drivers affected by the specific decision alternatives. The analytical structure for any LCC application must satisfy the criteria of analytical relevance and significance to avoid the unnecessary use of resources and masking or confusing the major issues under review.

Structure of LCC analysis

The aggregation of resource data to define system life-cycle cost involves a hierarchy of computations:

1. *Total system LCC*, a function of the number of systems in the population subject to analysis multiplied by the calculated LCC per system based on the projected system service life (also referred to as the *operational horizon*).

2. *System LCC*, a summation of the LCC of all pertinent LCC categories, as applicable to an individual system unit.

3. *LCC categories* (see Table 14.1), each of which is a summation of the LCC of cost elements constituent and contributory to the governing LCC category.

4. *LCC elements* (see Table 14.2), each of which is assigned an LCC estimate as derived from massage and summation of the results of those CERs which compose the LCC element.

5. *CERs*, each of which is a mathematical amalgamation of the *requirements factor*, *usage factor*, and *cost factor* developed by the analyst to construct the CER. The next section elaborates on the development and application of CER methodology.

Use of U.S. and SI measurement units.   In order to effectively convey the following details and examples of LCC to the international perspective of the professional logistics community, both U.S. and SI (metric) measurement units are incorporated, as specifically appropriate to the context of the guidance.

Construction of cost-estimating relationships.   (CERs). CER composites must have factorial integrity; e.g., for the energy cost element contributory to the LCC of a ground vehicle, the requirements factor, expressed as miles (kilometres) per year; the usage factor, expressed as miles per gallon (kms/litre); and the cost factor, expressed as cost per gallon (litre), extended over the number of years of the vehicle's service life, would be integrated as follows:

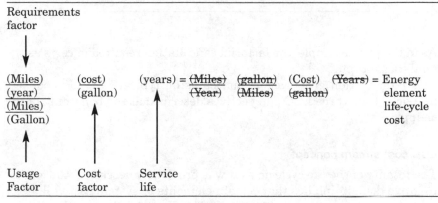

U.S. Measurement Units Equivalent

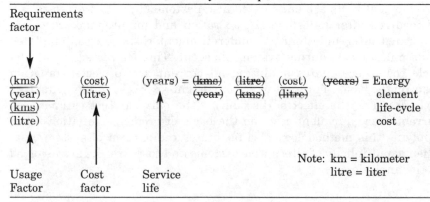

SI (metric system) Equivalent

As illustrated by the above, the amalgamation of an individual CER composite should result in factorial cancellation that produces the desired quantitative expressions; in the above case, the objective was cost over the system life cycle. Another case emphasizing factorial integrity is the LCC of the human resources element (labor) for corrective maintenance of the system attributable to predicted failure frequency, noted as follows:

OH = operating hours
Failures/OH is same as failure rate (λ)
MLH = maintenance labor hours
MLH/failure is based on the number of technicians and mean corrective maintenance time ($\bar{M}$)

Requirements factor	Usage factor	Cost factor		Service life	
~~(OH)~~ ~~(Year)~~	~~(Failures)~~ ~~(OH)~~	~~(MLH)~~ ~~(Failure)~~	(Cost) ~~(MLH)~~	~~(Years)~~	= Human resources element (labor) life-cycle cost

As in the prior example, the factorial cancellation resulted in costs as the analytical output.

A later section gives case scenarios using typical LCC methodology for each of the three LCC perspectives described at the beginning of the chapter.

LCC cost stream concept

The totality of the life cycle or, as it was previously described, the cradle-to-grave domain implies that certain elements of system cost will affect the program at varying points in time. The flow of such costs throughout the life cycle is referred to as the *cost stream*. For example, research and development costs are anticipated at milestone 0, production costs will be incurred after milestone III, operation and maintenance costs are generated after milestone IV, and retirement costs are contemplated after milestone V, during system phaseout. This characteristic of life-cycle costs necessitates analyzing all current and projected program costs on the basis of present value, or current-year cost equivalents. This technique has the effect of discounting the out-year cost estimates to current-year equivalent costs on the basis of predicted inflation rates. Applying this methodology also facilitates comparison of system costs when a multiplicity of alternative systems and cost streams are involved in the analysis.

Representative LCC Scenarios

The following LCC scenarios are representative of approaches to life-cycle cost analysis and reflect the three individual life-cycle perspectives described earlier in the chapter. Since LCC analysis resembles an art form more than a highly disciplined scientific procedure, it must be recognized that applicable LCC categories, the composite of cost elements, and the design of cost-estimating relationships (CERs) will vary among differing LCC case situations.

LCC case: producer-user

In this hypothetical scenario, a multinational corporation is contemplating the installation of a specially designed materials handling system at each of 26 distribution centers at United States and foreign country locations. The new system employs current state-of-the-art technology but will be internally developed, produced, logistically supported, and phased out using company resources. The projected service life is five years after installation based on continual operation, 24 hours a day, except for downtime for necessary repairs and scheduled maintenance. The total projected life cycle from commencement of research and development (R&D) through retirement and phaseout is 10 years. Life-cycle cost estimates developed by company planners encompass requirements for the categories of R&D, engineering design, production, operations, maintenance, product support, and retirement. The LCC categories include cost elements for human resources, test and evaluation (T&E), advanced development model, production prototype, technical data, quality control, packaging and transportation, facility installation, repairs, scheduled maintenance, inventory support, storage, and salvage. The matrix in Table 14.3 organizes the LCC categories and identifies the pertinent cost elements of each category. The sequence and schedule of resource expenditures throughout the planned 10-year life cycle are provided by the cost stream. The corporate management has imposed a $12 million present value design-to-cost (DTC) ceiling and directed that a 6 percent annual inflation (discount) rate be incorporated into future cost-stream annual projections. The LCC analysis will determine whether the program complies with DTC constraints.

Basis of estimates. It may be presumed that all cost values are based on valid engineering estimates. Implicit in the estimates provided in the summary in Table 14.3 are certain principles and stipulations inherent in many LCC analyses:

1. The human resources requirements for the year 3 and year 4 production program assume a 95 percent learning curve which incorporates a 6 percent inflation rate and productivity improvement.

2. The production program plans for 10 units to be produced in year 3 and 16 units in year 4.

3. The facility cost estimate (year 4) under operations is based on initial costs for installation at the 26 distribution centers.

4. Maintenance repairs are based on an MTBF of 2200 hours (approximately three-month intervals) with $\overline{M}_{ct}$ of three hours, involving one technician at $50 per maintenance person-hour at each location. Each corrective maintenance action is assumed to require $500 in materials.

5. Scheduled maintenance is based on an MTBM of 4400 hours (approximately six-month intervals) with $\overline{M}_{pt}$ of 24 hours, involving two technicians

TABLE 14.3 Producer-User LCC Summary Cost Stream ($000)

Cost element	Year 1	Year 2	Year 3	Year 4	Year 5	Year 6	Year 7	Year 8	Year 9	Year 10
Research and development										
Human resources	200.0									
T&E	50.0									
Subtotal	250.0									
Engineer design										
Human resources		500.0								
T&E		100.0								
ADM		50.0								
Subtotal		650.0								
Production										
Human resources			3000.0	4480.0						
Prototype			150.0							
Technical data			50.0							
Quality control			150.0	240.0						
Packaging & transportation			10.0	16.0						
Subtotal			3360.0	4736.0						
Operations										
Facility (installation)				520.0						
Energy					26.0	28.0	29.0	31.0	33.0	
Subtotal				520.0	26.0	28.0	29.0	31.0	33.0	
Maintenance										
Repairs					55.0	58.0	62.0	66.0	69.0	
Scheduled maintenance					130.0	138.0	146.0	155.0	164.0	
Subtotal					185.0	196.6	208.0	221.0	233.0	
Product support										
Inventory support				650.0	689.0	730.0	774.0	821.0		
Storage					130.0	138.0	146.0	155.0	164.0	
Subtotal				650.0	819.0	868.0	920.0	976.0	164.0	
Retirement										
Salvage										(2000.0)
Subtotal										(2000.0)

at \$50 per hour each at each location. Each scheduled maintenance action requires \$100 in materials.

6. Inventory support costs (year 4) are based on ordering annual requirements at one-year lead time intervals prior to anticipated needs for spare parts. This explains the lack of investment in inventory during the last year of service life (year 9).

7. The retirement salvage costs are actually *negative costs*, based on estimates that *salvage value* proceeds for the 26-unit inventory would be \$2 million at year 10. This sum would, therefore, be subtracted from the other cost stream totals.

Summary costs for present-value computation.   The categorized subtotals in Table 14.3 are transferred to the present-value computational summary in Table 14.4 and aggregated to show undiscounted annual totals for each year in the cost stream. These annual totals are then discounted, using the discount factor for each year; i.e.,

$$\text{discount factor} = \frac{1}{(1 + i)^n}$$

where i = inflation rate of 0.06
n = year in cost stream

This results in the present-value equivalent costs for the life-cycle period. (This is also referred to as current-year costs relative to then-year costs.)

Aggregate present-value computation.   The total present value of all life-cycle annual cost inputs is determined by simple summation, as shown in Table 14.5.

The total projected life-cycle cost of approximately \$10.531 million produced by this analysis is within the \$12 million design-to-cost ceiling.

Life-cycle value case: market-item producer

Manufacturers of goods for the competitive consumer market are concerned primarily with life-cycle cost projections for product development, production, distribution, and customer support. Another critical consideration is life-cycle profit, which must be incorporated into the LCC analysis process to provide the analyst with a portrait of the *life-cycle value*. Life-cycle value, in this type of scenario, focuses on profitability as a synthesis of life-cycle cost and life-cycle profit analyses. The hypothetical case of a computer manufacturer contemplating development, production, and marketing of a state-of-the-art desktop personal computer (PC)

TABLE 14.4 Producer-User LCC Present-Value Computations Cost Stream ($000)

LCC Category	Year 1	Year 2	Year 3	Year 4	Year 5	Year 6	Year 7	Year 8	Year 9	Year 10
Research and development	250.0									
Engineering design		650.0								
Production			3360.0	4736.0						
Operations				520.0	26.0	28.0	29.0	31.0	33.0	
Maintenance					185.0	196.0	208.0	221.0	233.0	
Product support				650.0	819.0	868.0	920.0	976.0	164.0	
Retirement										(2000.0)
Undiscounted matrix total	250.0	650.0	3360.0	5906.0	1030.0	1092.0	1157.0	1228.0	430.0	Salvage (2000.0)
6% = .06 Discount factor	.9434	.8900	.8396	.7921	.7473	.7050	.6651	.6274	.5919	.5584
Present value	235.85	578.50	2821.06	4678.14	769.72	769.86	769.52	770.45	254.52	Salvage (1116.80)
Equivalent cost										

TABLE 14.5

Year	Present-value cost input ($000)
1	235.85
2	578.50
3	2821.06
4	4678.14
5	769.72
6	769.52
7	769.52
8	770.45
9	254.52
10	(1116.80)*
Total	10,530.82

*Salvage value (year 10) is a *negative cost,* or cost recovery, based on positive residual value of assets.

illustrates life-cycle considerations pertinent to a market- item producer. For this case, the following assumptions describe the LCC scenario:

1. The company market analysts estimate a potential market of 2000 PC units annually, with 6 to 10 years market viability. For the purpose of this life-cycle value analysis, presume 8 years of market viability and 16,000 units total production.

2. The proposed suggested retail price of $4995 is a life-cycle year 3 market entry baseline which anticipates 3 percent annual price increase and is considered competitive. The wholesale price to the distributors and retailers would initially be $3000, subject to the same rate of price augmentation as the retail policy.

3. The forecast inflation (discount) rate is 5 percent.

4. Projected production labor (human resources) cost is based on a 95 percent learning curve, which incorporates the inflation rate.

5. The total program life cycle is 10 years, commencing at year 0 and extending through the eighth year of the marketing schedule.

The life-cycle cost parameters developed by company planners include the LCC categories of research and development (R&D), engineering design, production, product distribution, and customer support. The detail cost elements encompass requirements for market research and analysis (MR&A), human resources, test facility usage, test and evaluation (T&E), energy, quality control, technical data, packaging, transportation, inventory investment, storage, warranty impact allowance, training, and service technician. The summary matrix in Table 14.6 organizes the applicable LCC categories,

TABLE 14.6 Market-Item Producer LCC Summary Cost Stream ($000)

LCC Category/ Cost Element	Year 1	Year 2	Year 3	Year 4	Year 5	Year 6	Year 7	Year 8	Year 9	Year 10
R & D M, R, & A	50.0									
TOT	50.0									
Engineering design										
Human resources		150.0								
Facility		50.0								
T&E		20.0								
Subtotal		220.0								
Production										
Human resources			2400.0	2280.0	2220.0	2170.0	2130.0	2100.0	2080.0	2060.0
Energy			100.0	105.0	110.0	116.0	122.0	128.0	134.0	141.0
Quality control			200.0	210.0	225.0	232.0	243.0	255.0	268.0	281.0
Technical data			25.0	26.0	28.0	29.0	30.0	32.0	34.0	35.0
Subtotal			2725.0	2621.0	2583.0	2547.0	2525.0	2515.0	2516.0	2517.0
Product distribution										
Human resources			200.0	210.0	220.0	232.0	243.0	255.0	268.0	281.0
Packaging			100.0	105.0	110.0	116.0	122.0	128.0	134.0	141.0
Transportation			150.0	157.0	165.0	173.0	182.0	191.0	201.0	211.0
Inventory	(1) (2)		1500.0	—	—	—	—	—	—	—
Storage			80.0	84.0	88.0	93.0	103.0	103.0	108.0	113.0
Subtotal			2030.0	556.0	583.0	614.0	677.0	677.0	711.0	746.0
Customer Support	(1) (2)									
Human resources			300.0	315.0	331.0	347.0	365.0	383.0	402.0	422.0
Warranty			200.0	210.0	220.0	232.0	243.0	255.0	268.0	281.0
Training			200.0	210.0	220.0	232.0	243.0	255.0	268.0	281.0
Inventory			600.0	—	—	—	—	—	—	—
Storage			100.0	105.0	110.0	116.0	122.0	128.0	134.0	141.0
Transportation			300.0	315.0	331.0	347.0	365.0	383.0	402.0	422.0
Technician			300.0	315.0	331.0	347.0	365.0	383.0	402.0	422.0
Subtotal			2000.0	1470.0	1543.0	1621.0	1703.0	1787.0	1876.0	1969.0

with each category broken down into its constituent cost elements. As with the previous case, the sequence and schedule of resource expenditures are portrayed by the cost stream.

Consideration of alternative opportunities.   For tutorial purposes, let it be assumed that an alternative to the ten-year product venture is to invest capital in U.S. Treasury notes with an annual return of 7 percent. The profitability of the product venture must, therefore, be assessed in light of the return offered by U.S. Treasury notes. Given the two alternatives, the analyst must couple the life-cycle cost analysis with an assessment of the product profit potential in order to comparatively evaluate the return on investment (ROI) of each alternative. The utility of such a comparative assessment is a major strength of the life-cycle value analysis methodology.

Summarizing costs for present-value computation.   The categorized subtotals emphasized in Table 14.6 are transferred to the present-value computational summary in Table 14.7 and, in the same manner as described for the previous example, are aggregated to show undiscounted annual totals throughout the life-cycle cost stream. The discount-rate algorithm for present-value computation is as described in the previous example, except that in this case the discount rate i is 0.05, based on the 5 percent annual inflation rate. Application of the discount factors to the undiscounted values results in the present-value equivalent costs for each year throughout the life cycle.

Based on the data in Table 14.7, the total present-value equivalent cost for the lifecycle cost stream is calculated as shown in Table 14.8.

Life-cycle profitability.   The market-item producer is concerned with the implications of the life-cycle cost stream relative to the flow of revenues expected during the eight-year market schedule. The goal of such comparison is to develop a potential profitability analysis. This assessment is accomplished by comparing projected annual costs with corresponding revenues to be generated during the market period of the life cycle. For this case, gross revenues are based on sales of 2000 units annually at $3000 (wholesale) each starting in year 3, with annual 3 percent price increments throughout the market period. Annual profits are determined by comparing annual revenues with annual expenditures throughout the life cycle on an undiscounted value basis. Table 14.9 portrays the annual profit generation based on projected annual sales.

The data in Table 14.9 confirm that the total undiscounted profit value over the ten-year life cycle is $12.003 million. The breakeven point occurs during year 4, when *cumulative profits* start reflecting a positive value.

TABLE 14.7 Market-Item Producer LCC Present-Value Computations ($000)

LCC category/ cost element	Year 1	Year 2	Year 3	Year 4	Year 5	Year 6	Year 7	Year 8	Year 9	Year 10
Research and development	50.0									
Engineering design		220.0								
Production			2725.0	2621.0	2583.0	2547.0	2525.0	2515.0	2516.0	2517.0
Product distribution			2030.0	556.0	583.0	614.0	645.0	677.0	711.0	746.0
Customer support			2000.0	1470.0	1543.0	1621.0	1703.0	1787.0	1876.0	1969.0
Undiscounted matrix total	50.0	220.0	6755.0	4647.0	4709.0	4782.0	4873.0	4979.0	5103.0	5232.0
5% = 0.05 PV/discount factor	0.9524	0.9070	0.8638	0.8227	0.7835	0.7462	0.7107	0.6768	0.6446	0.6139
Present value	47.62	199.55	5834.97	3823.09	3689.50	3568.33	3463.24	3369.79	3289.39	3211.92
Equivalent cost										

TABLE 14.8

Life-cycle year	Present value ($000) from Table 14.7
1	47.62
2	199.55
3	5,834.97
4	3,823.09
5	3,689.50
6	3,568.33
7	3,463.24
8	3,369.79
9	3,289.39
10	3,211.92
	Total 30,496.98, or approximately $30.497 million

Developing the comparative analysis baseline.   The cumulative profit of $12.003 million as of the end of year 10 (see Table 14.9) represents the projected life-cycle profit. This value must be compared to what the equivalent risk capital would carn if it were alternatively invested in 7 percent U.S. Treasury notes. This enables comparison of the return on investment (ROI) values offered by the two alternatives.

Determining the investment at risk.   From the profit flow depicted by Table 14.9, it is evident that the product program emerged from a loss position to profitability during year 4, which period, consequently, includes the breakeven point. This is a critical point in the life cycle, inasmuch as it indicates the maximum, cumulative undiscounted risk capital committed by the company to subsidize the program; after the breakeven point, the program is self-sufficient and is replenished by revenues from sales.

TABLE 14.9   Market-Item Producer Summary of Annual Profit Generation Based on Undiscounted Values ($000)

Year	Annual costs (from Table 14.7)	Annual revenues	Annual profit generated	Cumulative profit/loss
1	50.0	0	(50.0)	(50.0)
2	220.0	0	(220.0)	(270.0)
3	6,755.0	6,000.0	(755.0)	(1025.0)
4	4,647.0	6,180.0	1,533.0	508.0
5	4,709.0	6,365.0	1,656.0	2,164.0
6	4,782.0	6,556.0	1,774.0	3,938.0
7	4,873.0	6,753.0	1,880.0	5,818.0
8	4,979.0	6,956.0	1,977.0	7,795.0
9	5,103.0	7,164.0	2,061.0	9,856.0
10	5,232.0	7,379.0	2,147.0	12,003.0
Total	41,350.0	53,353.0	12,003.0	

To construct a realistic basis of assessment, it is necessary to convert the cumulative program costs at the breakeven point to the present-value equivalent sum as of year 0. This is accomplished by the following steps.

Step 1.   Estimate the production unit u in year 4 which approximates the breakeven point. This is calculated using an equation that balances accumulated costs and revenues, using the data in Table 14.9:

$$Costs = Revenues$$

$$\text{\$ Year 1} + \text{\$ Year 2} + \text{\$ Year 3} + \quad 0 + 0 + 6000 +$$

$$u\left(\frac{\text{\$Year4}}{\text{Year 4 units produced}}\right) = u\left(\frac{\text{\$Year4}}{\text{Year 4 units sold}}\right)$$

$$50 + 220 + 6755 + u\left(\frac{3823.90}{2000}\right) = 6000 + u\left(\frac{6180.0}{2000}\right)$$

$$6082.14 + u(1.912) = 6000 + u(3.09)$$

$$6082.14 - 6000 = u(3.09) - u(1.912)$$

$$82.14 = 1.178u$$

$$u = \frac{82.14}{1.178}$$

$$u = 69.7538, \text{ or unit } 70$$

Step 2.   Estimate the present value of the accumulated costs at the breakeven point, as denoted by the 70th unit in the year 4 production run. This will determine the risk capital committed by the company before achieving profitability. The following equation based on the *present value* data in Table 14.8 applies:

Year 0 risk
capital equivalent = \$ year 1 + \$ year 2+ \$ year 3

$$+ \text{\$ year 4}\left(\frac{\text{breakeven unit}}{\text{year 4 production}}\right)$$

$$= 47.62 + 199.55 + 5834.97 + 3823.09\left(\frac{70}{2000}\right)$$

$$= 6082.14 + (3823.09)(0.035)$$

$$= 6082.14 + 133.81$$

$$= 6215.95, \text{ or \$6.216 million}$$

The sum of approximately \$6.216 million represents the risk capital committed to the product venture, which could have alternatively been invested in U.S. Treasury notes.

Comparison of alternative ROI prospects.

First alternative. Invest risk capital in U.S. Treasury notes.

$$F = P(1 + i)^n$$

where F = future value of risk capital of $6.216 million
$i = 0.07$ (7 percent annual return)
$n = 10$ years (product life cycle)

$$F = 6.216(1.07)^{10}$$
$$= (6.216)(1.96715)$$
$$= 12.228(\$12.228 \text{ million})$$

$$\text{Projected total profit} = F - P$$
$$= 12.228 - 6.216$$
$$= 6.012$$

Second alternative. Invest in product venture. Future estimated profit (from Table 14.9) is $12.003 million.

Conclusion. Product venture offers significantly higher profit potential than U.S. Treasury notes.

Methodology for determining return on investment (ROI). It is obvious that the product venture is more profitable than the 7 percent return offered by U.S. Treasury notes. It is further appropriate to more precisely estimate the return on investment based on the life-cycle data. This is accomplished by applying logarithmic techniques to the future value formula:

$$F = P(1+i)^n$$

$$(1+i)^n = \frac{F}{P}$$

$$n \log(1+i) = \log\left(\frac{F}{P}\right)$$

$$\log(1+i) = \frac{\log F - \log P}{n}$$

$$\text{antilog}\,[\log(1+i)] = \text{antilog}\left(\frac{\log F - \log P}{n}\right)$$

$$1 + i = \text{antilog}\left(\frac{\log F - \log P}{n}\right)$$

$$i = \text{antilog}\left(\frac{\log F - \log P}{n}\right) - 1$$

Extending this principle to the product venture data,

$P = 6.016$

$F = 6.016 + 12.003 = 18.019$ (The product venture would recover the original risk capital *in addition to* achieving $12.003 million profit.)

$n = 10$ years

i = annual ROI

$$i = \text{antilog} \left(\frac{\log 18.019 - \log 6.016}{10} \right) - 1$$

$$= \text{antilog} \left(\frac{1.25574 - 0.77931}{10} \right) - 1$$

$$= \text{antilog} \, (0.047642) - 1$$

$$= 1.1159 - 1$$

$$= 0.1159, \text{ or } 11.59\%$$

Comparing the 11.59 percent annual ROI projected for the product venture with the 7 percent offered by the U.S. Treasury note reinforces the previous conclusion supporting the product venture.

Life-cycle cost case: purchaser-user.   The life cycle of a system acquired for use by a purchaser is defined by the planned service life of the system. The purchaser-user of a system is primarily concerned with costs for operation, maintenance, product support, and retirement and disposal. The typical purchaser focuses on the *lowest ultimate overall cost* of ownership. Costs for research and development, engineering, development, and production, which the purchaser has no direct control over or cognizance of, are incorporated in the system purchase price. The informed purchaser will, therefore, actively seek and evaluate relevant empirical data on performance, operational efficiency, maintenance, taxes, insurance, resale value, and other costs of ownership which have life-cycle cost implications when evaluating alternative acquisitions.

The new automobile.   One typical purchaser-user scenario that is of interest to many consumers and that involves life-cycle cost analysis is the purchase of a new automobile. This type of acquisition includes the categories of system purchase, operation, support, maintenance, and disposal, with the determinant life-cycle cost elements typically governed by purchase price, energy (gasoline), scheduled maintenance, repairs, taxes, insurance, and salvage values. For this scenario, the automobile will average 15,000 mi (24,140 kms) annual operation and have a planned service life of five years. Let it be assumed that the following alternatives are under consideration:

Model	Characteristics	Acquisition cost
A	Medium size, average performance, U.S., four-door	$ 9,900
B	Small size, high performance, import, two-door	14,900
C	Medium size, high performance, U.S., four-door	11,900

The following assumptions and CERs pertain:

1. The estimated real market value depreciates by 20 percent per year. The resale value at the end of each year will be 80 percent of the market value at the close of the prior year, which is identical to the value at the start of the current year.
2. Depreciation is not included, inasmuch as the purchase price is a designated cost element and the estimated actual market value is used for computation of salvage value at the end of service life.
3. The discount factor is 0.10, based on a 10 percent annual inflation rate.
4. Fuel performance:

   Model A: 15 mi/gal   (6.377 kms/litre)
   Model B: 30 mi/gal   (12.754 kms/litre)
   Model C: 25 mi/gal   (10.628 kms/litre)

5. Projected fuel costs are

   year 1: $1.10/gal   ($0.29/litre)
   year 2: $1.20/gal   ($0.32/litre)
   year 3: $1.30/gal   ($0.34/litre)
   year 4: $1.40/gal   ($0.37/litre)
   year 5: $1.50/gal   ($0.40/litre)

U.S. Measurement Units

$$\text{CER: Cost per year} = \left(\frac{15,000 \text{ mi/yr}}{\text{mi/gal}}\right)(\$/\text{gal})$$

SI (metric system) Equivalent

$$\text{CER: Cost per year} = \left(\frac{24,135 \text{ km/yr}}{\text{kms/litre}}\right)(\$/\text{litre})$$

For example, for model A, year 1,

U.S. Measurement Units

$$\text{Cost} = \left(\frac{15,000}{15}\right)(\$1.10) = \$1100$$

SI (metric system) Equivalent

$$\text{Cost} = \frac{(24,135)}{(6,377)}(\$0.29) = \$1,097.56\ (@\ \$1100)$$

6. Basic liability insurance for each model is \$600/year; costs for collision and other insurance coverage are based on 7.5 percent of market value at the start of the year. For example, for model A, year 1,

$$\text{Cost} = \$600 + (0.075)(9900) = 1342.5 \text{ or } \$1343$$

7. Taxes are based on 5 percent of market value at the start of the year. For example, for model A, year 1,

$$\text{Cost} = (0.05)(9900) = \$495$$

8. Predicted unscheduled maintenance (M_{unsch})
Mean operational intervals between unscheduled maintenance:

year 1: 5000 mi   (8,045 kms)
year 2: 3750 ml   (6,034 kms)
year 3: 3000 mi   (4,825 kms)
year 4: 2500 mi   (4,023 kms)
year 5: 2500 mi   (4,023 kms)

M_{unsch} factors:

Model	$\overline{M}_{\text{unsch}}$ per action	\$/MMH
A	1 h	\$60
B	1 h	\$55
C	1 h	\$50

CER:

U.S. Measurement Units

$$\overline{M}_{\text{unsch}}\ \text{cost} = \left(\frac{15,000\ \text{mi/yr}}{\text{mean miles between } M_{\text{unsch}}}\right)(\overline{M}_{\text{unsch}}\ \text{h})(\$/\text{MMH})$$

SI (metric system) Equivalent

$$M_{\text{unsch}} \text{ cost} = \left(\frac{24{,}135 \text{ km/yr}}{\text{mean kms between } M_{\text{unsch}}}\right) (\overline{M}_{\text{unsch}} \text{ h})(\$/\text{MMH})$$

Note: MMH = maintenance man-hours (human resources).
For example, for model A, year 1,

U.S. Measurement Units

$$\text{Cost} = \left(\frac{15{,}000}{5000}\right)(1)(60) = \$180$$

SI (metric system) Equivalent

$$\text{Cost} = \frac{(24{,}135)}{(8{,}045)}(1)(60) = \$180.30 \ (@ \ \$180)$$

9. Scheduled maintenance (M_{sch}) factors.

Model	Mean operational interval between M_{sch}		$\overline{M}_{\text{sch}}$ per action	\$/MMH
A	5,000 mi	(8,045 kms)	2 h	\$50
B	15,000 mi	(24,135 kms)	2 h	\$60
C	2,500 mi	(4,023 kms)	2 h	\$40

CER:

U.S. Measurement Units

$$M_{\text{sch}} \text{ cost} = \left(\frac{15{,}000 \text{ mi/yr}}{\text{mean miles between } M_{\text{sch}}}\right) (\overline{M}_{\text{sch}} \text{ h})(\$/\text{MMH})$$

SI (metric system) Equivalent

$$M_{\text{sch}} \text{ cost} = \left(\frac{24{,}135 \text{ km/yr}}{\text{mean kms between } M_{\text{sch}}}\right) (\overline{M}_{\text{sch}} \text{ h})(\$/\text{MMH})$$

Note: MMH = maintenance man-hours (human resources).
For example, for model A, year 1,

$$\text{U.S. Measurement Units}$$

$$\text{Cost} = \left(\frac{15,000}{5000}\right)(2)(50) = \$300$$

$$\text{SI (metric system) Equivalent}$$

$$\text{Cost} = \frac{(24,135)}{(8,045)}(2)(50) = \$300$$

Organizing and displaying LCC data. The matrix in Table 14.10 (U.S. Measurement) and Table 14.11 (SI Measurement) organizes the LCC categories of operating, unscheduled maintenance, scheduled maintenance, tax, and insurance costs to display undiscounted (actual) costs from year 0 through the five-year service life. The LCC data incorporate the assumptions, CERs, and mathematical methodologies just described. The summary in Table 14.12 rearranges the aggregate data by model. For each model, it shows the actual (undiscounted) cost data and the present value equivalents of the same cost values, produced by applying the appropriate discount factors. For example, for model A, the discounted purchase price of $9900 is the same as the actual (undiscounted) purchase price because it is a year 0 input. The model A, year 1 undiscounted cost of $3418 represents the total of the constituent model A, year 1 cost elements detailed in Table 14.10 and Table 14.11, i.e.,

Operating cost	$1,100
Unscheduled maintenance	180
Scheduled maintenance	300
Tax	495
Insurance	1,343
	$3,418

The same procedures for aggregation apply to all models for all years of the life cycle. Each year's actual cost total is converted to present value by applying the appropriate discount factor.

The residual market value of $3244 is considered a *negative cost*, and is so indicated in the data in Table 14.12. The cumulative five-year costs are therefore decremented by the proceeds anticipated from recovery of the terminal resale value. The resale value of model A would be computed as shown in Table 14.13, considering annual 20 percent market value depreciation.

TABLE 14.10 Undiscounted Cost ($) per Year of Ownership (U.S. Measurement Units)

Criteria	Auto	Year 1	Year 2	Year 3	Year 4	Year 5	Description
1. Operation	A	1100	1200	1300	1400	1500	Gas costs
	B	550	300	650	700	750	$1.10/gal in year 1
	C	660	720	780	840	900	$1.20/gal in year 2
							$1.30/gal in year 3
				$(15,C20 \div mpg)(\$/gal)$			$1.40/gal in year 4
				$= \text{Cost}$			$1.50/gal in year 5
2. Unscheduled maintenance (labor and material)	A	180	264	363	479	527	Repairs
	B	165	242	333	439	483	
	C	150	220	303	399	439	
3. Scheduled maintenance (labor and material)	A	300	330	360	399	439	Oil changes, lubrication, periodic checks
	B	120	132	145	160	176	
	C	480	528	639	703	773	
4. Property tax (5% begin year market value)	A	495	396	317	254	203	Tax based on end of preceding year market value, computed at 80% of value at start of year
	B	745	596	479	381	305	
	C	595	476	381	306	244	
5. Insurance (collision and liability)	A	1343	1194	1075	980	904	Liability insurance standard at $600/year; collision coverage based on current market value
	B	1718	1494	1315	1172	1058	
	C	1493	1314	1171	1057	966	
	Year 0 = 1.000						
10% discount rate factor		0.909	0.826	0.751	0.683	0.621	

Operation (formula box):

$$\left(\frac{15,000}{\text{mean miles between } M_{unsch}}\right)(\overline{M}_{unsch}\ hr)\left(\frac{\$}{MMH}\right) = \text{Cost}$$

Scheduled maintenance (formula box):

$$\left(\frac{15,000}{\text{mean miles between } M_{sch}}\right)(\overline{M}_{sch}\ hr)\left(\frac{\$}{MMH}\right) = \text{Cost}$$

TABLE 14.11 Undiscounted Cost ($) per Year of Ownership: SI (metric) Measurement Units

Criteria	Auto	Year 1	Year 2	Year 3	Year 4	Year 5	Description
1. Operation	A	1100	1200	1300	1400	1500	Gas costs
	B	550	600	650	700	750	$0.29/litre in year 1
	C	660	720	780	840	900	$0.32/litre in year 2
							$0.34/litre in year 3
							$0.37/litre in year 4
							$0.40/litre in year 5
			(24135÷kms/litre) ($/litre) = Cost				
2. Unscheduled maintenance (labor and material)	A	180	264	363	479	527	Repairs
	B	165	242	333	439	483	
	C	150	220	303	399	439	
3. Scheduled maintenance (labor and material)	A	300	330	360	399	439	Oil changes, lubrication, periodic checks
	B	120	132	145	160	176	
	C	480	528	639	703	773	
4. Property tax (5% begin year market value)	A	495	396	317	254	203	Tax based on end of preceding year market value, computed at 80% of value at start of year
	B	745	596	479	381	305	
	C	595	476	381	306	244	
5. Insurance (collision and liability)	A	1343	1194	1075	980	904	Liability insurance standard at $600/year; collision coverage based on current market value
	B	1718	1494	1315	1172	1058	
	C	1493	1314	1171	1057	966	
	Year 0 = 1.000						
10% discount rate factor		0.909	0.826	0.751	0.683	0.621	

$$\left(\frac{24135}{\text{mean kms between } M_{munch}}\right) (\overline{M}_{munch}\,\text{hr}) \left(\frac{\$}{MMH}\right) = \text{Cost}$$

$$\left(\frac{24135}{\text{mean kms between } M_{sch}}\right) (\overline{M}_{sch}\,\text{hr}) \left(\frac{\$}{MMH}\right) = \text{Cost}$$

TABLE 14.12 Purchaser-User LCC Summary

Automobile	Purchase price (Year 0)	Consumer operation and support cost					5-Year LCC (present value) total	Residual market value*	Net ownership cost
		Year 1	Year 2	Year 3	Year 4	Year 5			
A									
Actual	9,900	3418	3384	3415	3512	3573			
Discounted	9,900	3107	2795	2565	2399	2219	22,985	(3244)	19,741
B									
Actual	14,900	3298	3064	2922	2852	2772			
Discounted	14,900	2998	2531	2194	1948	1721	26,292	(4882)	21,410
C									
Actual	11,900	3378	3258	3216	3241	3252			
Discounted	11,900	3071	2691	2415	2214	2019	24,310	(3899)	20,411
10% discount factor	1.000	0.909	0.826	0.751	0.683	0.621			

*Residual market value as of beginning of sixth year. Annual market decline presumes 20 percent reduction from value as of end of preceding year. Baseline value is year 0 purchase price.

TABLE 14.13

Year	Computation	Start of year base value
0	N/A	9900
1	(0.8)(9900)	7920
2	(0.8)(7920)	6336
3	(0.8)(6336)	5069
4	(0.8)(5069)	4055
5	(0.8)(4055)	3244*

*Terminal market value.

For each model, the total present value costs are determined by summing the annut1 present value costs and decrementing the total by the terminal salvage value. For example, for Model A,

Purchase (year 0)	9,900
Year 1 costs	3,107
Year 2 costs	2,795
Year 3 costs	2,565
Year 4 costs	2,399
Year 5 costs	2,219
Cost total	22,985 (present value)
Less terminal resale value	(3,244)
Net ownership cost	$19,741

Conclusions. The net ownership costs shown in Table 14.12 suggest that model A is the preferred acquisition alternative, given the assumptions and criteria applied to the analysis. The marginal superiority of model A over model C would, in a realistic scenario, require more scrutiny of these two options. For example, if it were feasible to change the scheduled maintenance interval of model C from 2500 mi (4023 kms) to 5000 mi (8045 kms), the results of the analysis would favor model C.

15

Performance-Based Logistics Engineering

General

Performance-Based Systems Engineering is an incentive-based process which embodies the phases of (1) development of the statement of work and performance specifications, (2) execution of the most effective contractual instrument with realistic metrics, and (3) evaluation of the contractor's performance for the purpose of awarding incentives. The total process demands a technical team bound by the synergistic coalescence of system design engineering, logistics analysis, and contracting. Performance-Based Systems Engineering provides the cornerstone for long-term logistics performance of the system (or equipment). This is accomplished by optimizing the key performance parameters (KPPs) during the earlier phases of development. The systems engineering parameters developed during the system development and demonstration process prior to production define the baselines for logistics performance after production and deployment or market distribution of the system. Department of Defense experience has confirmed that 75 to 85 percent of the system life cycle operating and support costs are governed by the engineering design and support configuration established prior to commencement of production. The technical and economic opportunities during the preliminary phases should, therefore, be exploited to the maximum extent practicable. The key to optimizing the key performance parameters is effective formulation of financial incentives for the system development contractor or equivalent organizational entity. The major considerations are:

1. Focus on logistics impact at the organization (retail) levels.
2. Emphasize engineering design and development objectives prior to production.
3. Define the key performance parameters which are measurable and which offer the greatest potential to enhance logistics supportability.
4. Establish *threshold and objective* values of the key performance parameters. The *Defense Acquisition Guidebook* describes the performance threshold as the minimum acceptable value that, in the user's judgment, is necessary to satisfy the need. The performance objective is defined as the value desired by the user that, if at all possible, could be achieved by the developer.
5. Structure incentives to be commensurate with the range of performance values within the thresholds and objectives.
6. Evaluate the performance of the development contractor (or equivalent organizational entity) based on achieved performance parameters within the range of incentives.
7. Use the most effective and practicable form of incentive contract applicable to development efforts prior to production.

Within the context of the scenario for a hypothetical system, this chapter highlights Reliability, Maintainability, and Availability to illustrate the methodology and potential values of Performance-Based Systems Engineering. The key performance parameters are *Mean Time between Failure; Mean Corrective Maintenance Time* (also referred to as *Mean Time to Repair*); *Mean Time between Preventive Maintenance; Mean Preventive Maintenance Time;* and *Inherent Availability* (also referred to as *Design Availability*). A Cost Plus Incentive Fee (CPIF) contract is assumed as the appropriate acquisition vehicle for execution of the program. The program cost is presumed to be $10 million. An additional $1.5 million is designated for financial considerations which include (1) the technical incentives, based on the key performance parameters, and (2) the base fixed fee, which can be adjusted if the contract program exceeds the originally negotiated cost. The Federal Acquisition Regulation (FAR) stipulates that when technical financial incentives are involved in a CPIF program there must be a mechanism to constrain cost growth (overrun); this is most effectively accomplished by invoking a sharing ratio on the base fixed fee that is triggered when program costs exceed the originally negotiated total.

Designation of Key Performance Parameters (KPPs)

From a system perspective, the following KPPs are considered major indicators of logistics resource requirements and impact on the logistics infrastructure:

1. *Mean Time between Failure* (MTBF). This KPP metric denotes the average interval between corrective maintenance (repair).

2. *Mean Corrective Maintenance Time* ($\overline{M}_{ct}$). This is a measure, also known as *Mean Time to Repair (MTTR)*, that enables estimation of maintenance hours required to restore an individual system. This factor multiplied by the number of technicians required translates to man-hours per repair action.

3. *Mean Time between Preventive Maintenance* (MTBM$_{pt}$). This factor projects the average interval between scheduled (or planned) maintenance actions.

4. *Mean Preventive Maintenance Time* ($\overline{M}_{pt}$). This provides for the number of maintenance hours per scheduled maintenance action required to retain a system in operational status under a planned maintenance program. This factor multiplied by the number of technicians required translates to man-hours per scheduled maintenance action.

5. *Inherent Availability* (A_i). This is a mathematical amalgamation of the parameters of *MTBF* and *MTTR*, as described in Chapter 4, herein.

Specification of the Individual KPP Threshold and Objective Values

The technical team must develop a performance which sets forth (1) the *threshold* Key Performance Parameters, which are considered acceptable and which would have a reasonable likelihood of achievement, and (2) the *objective* KPP values, which represent the desired performance levels and which would define the maximum technical incentive that could be earned by the developer. For purpose of this case, Figure 15.1 portrays the threshold and objective parameters for the designated KPPs.

Performance Parameters

KPP	Threshold (Acceptable)	Objective
MTBF	226	350
MTTR/$\overline{M}_{ct}$	2.83	2.25
MTBM$_{pt}$	237	400
$\overline{M}_{ct}$	2.09	1.5
A_i	0.98763	0.99361

Figure 15.1 Threshold and objective performance parameters.

Prioritization of the KPPs Based on Relative Essentiality

The technical team should establish the relative order of importance of the designated key performance parameters in order to assign allocation of the incentive pool to individual KPPs and incorporate order of importance in the contractual solicitation. For the purpose of this case Figure 15.2 notes the essentiality values assigned to the individual performance parameters. It is noted that the total of the individual essentiality values must equal 1.00.

Key Performance Parameter (KPP)	Essentiality Value
MTBF (Mean Time Between Failure)	0.35
MTTR (Mean Time to Repair)	0.25
$MTBM_{pt}$ (Mean Time between Preventive Maintenance)	0.20
$\overline{M}_{pt}$ (Mean Preventive Maintenance Time)	0.15
A_i (Inherent Availability)	0.05
	1.00

Figure 15.2   Essentiality values for KPPs.

For the sake of consolidation of the parametric analyses previously accomplished, Figure 15.3 provides a tabular recapitulation of the individual KPP threshold values, along with their relative essentiality weighted values expressed in percentage factors, as determined by conversion from the decimal equivalents. In similarity to the methodology used for Figure 15.2, the sum of the relative essentiality weighted values must equal 100 percent.

KPP	Threshold	Objective	Relative Essentiality Weighted Value
MTBF	226	350	35% (0.35)
$MTTR/\overline{M}_{ct}$	2.83	2.25	25% (0.25)
$MTBM_{pt}$	237	400	20% (0.25)
$\overline{M}_{pt}$	2.09	1.5	15% (0.15)
A_i	0.98763	0.99361	5% (0.05)
			100%

Figure 15.3   KPP thresholds, objectives, and relative essentiality.

Definition of the Individual KPP Allocated Cost Factors

An important element in the final computation of the technical performance incentive schedule is the allocation of the estimated costs required for commitment to efforts dedicated to the individual performance parameters within the total program cost of $10 million (excluding fees). This process provides the basis for the allocated cost factors which, in this case, are delineated by Figure 15.4.

KPP	Allocated Costs	Allocated Cost Factor (CF)
$\overline{\text{MTBF}}$	$ 3,000,000	30% (0.30)
$\text{MTTR/}\overline{\text{M}}_{ct}$	3,500,000	35% (0.35)
$\overline{\text{MTBM}}_{pt}$	1,000,000	10% (0.10)
$\overline{\text{M}}_{pt}$	2,000,000	20% (0.20)
A_i	500,000	5% (0.05)
	$10,000,000	100% (1.00)

Figure 15.4 Allocated individual KPP cost factors based on $10 million contract.

Construction of KPP-Incentive Tableau

The incentive fee structure for the individual key performance parameters is formulated to identify the maximum percentage of the incentive pool allocated to each KPP. Table 15.1 displays the incentive tableau for all the performance parameters. The maximum allocated incentive values are mathematically derived from of the following factors:

- Weighted Value (WVi), which is drawn from *relative essentiality weighted value* for each KPP shown in Figure 15.3
- Cost Factor (CFi), which is drawn for the *allocated cost factor* for each KPP shown in Figure 15.4
- (WVi)(CFi), which is the multiplicative product of the *relative essentiality weighted value* and *allocated cost factor*. This calculation is accomplished for each KPP, after which the individual mathematical values are totaled and identified as Σ(WVi)(CFi).

The *Incentive Ratio* (R) for each KPP is determined by the ratio of (WVi)(CFi) to the total of all factors, i.e.,

$$R = \frac{(WVi)(CFi)}{\sum (WVi)(CFi)}$$

As an example and with reference to Table 15.1, the incentive ratio for MTBF is obtained by dividing 0.105 by 0.245, which results in 0.429. It

TABLE 15.1 Key Performance Parameter (KPP)—Incentive Tableau

KPP	Weighted Value (WVi)	Cost Factor (CFi)	(WVi)(CFi)	Incentive* Ratio(R)	Maximum Incentive Pool Allocation(R × 100%)
MTBF	0.35	0.30	0.105	0.429	43%
$\overline{\text{MTTR/M}}_{ct}$	0.25	0.35	0.0875	0.357	36%
$\underline{\text{MTBM}}_{pt}$	0.20	0.10	0.02	0.082	8%
$\overline{M}_{pt}$	0.15	0.20	0.03	0.122	12%
A_i	0.05	0.05	0.0025	0.010	1%
	1.00	1.00	Σ(WVi)(CFi) 0.245	1.000	100%

$$*R = \frac{(WVi)(CFi)}{\sum (WVi)(CFi)}$$

Note: Incentive Pool Normally Based on Designated % of Contract Cost

is noted that the sum of all the individual KPP incentive ratios must equal 1.000. The *Maximum Incentive Pool Allocation* for each KPP is identified by converting the decimal value of the incentive ratio to a percentage equivalent, i.e., the allocation of MTBF (0.429) is converted to 43 percent. The sum of all KPP maximum incentive pool allocations must equal 100 percent.

Development of Incentive Fee Ranges for Individual KPPs

For each key performance parameter, the incentive structure is framed by the threshold (acceptable) performance level and objective (desired) performance determined by the technical team in preparation of the contractual statement of work. Pursuant to judgment and determination by the technical team for this case, the incentive range for each KPP will be incrementally scaled from zero as a minimum to the maximum percentage allocation, as identified in Table 15.1. The authorized range of incentive fees will be scaled so as to correlate with the incremental intervals between the threshold performance level and objective performance parameter. Figure 15.5, which provides the incentive structure for the MTBF performance parameter, illustrates this approach. Of the supplemental financial considerations totaling $1.5 million, $1.0 million is designated for the KPP technical incentives; $500,000 is allowed for the base fixed fee under the contract. The based fixed fee is subject to a cost sharing ratio, which would decrement the base fee in the event of cost "overrun." In addition to the MTBF technical incentive structure portrayed in Figure 15.5, the other KPP incentive structures are displayed as follows:

$$\text{MTTR}/\overline{M}_{ct}: \text{Figure 15.6}$$

$$\text{MTBM}_{pt}: \text{Figure 15.7}$$

$$\overline{M}_{pt}: \text{Figure 15.8}$$

$$A_i: \text{Figure 15.9}$$

The percentage allocation for each KPP would be applied to the total $1.0 million

Incentive pool, e.g., the maximum MTBF allocation of 43 percent equates to $430,000.

System Program: *XYZ*
Contract: *XXXXXXXX*

Key Performance Parameter: MTBF
Threshold: 226 h
Objective: 350 h
Maximum Incentive: $430,000 (43%)

Performance Achieved	Authorized Incentive (%)	Authorized Incentive ($)
≥ 350 h	43%	$430,000
≥ 340 h ≤ 349 h	37%	$370,000
≥ 330 h ≤ 339 h	34%	$340,000
≥ 320 h ≤ 329 h	31%	$310,000
≥ 310 h ≤ 319 h	28%	$280,000
≥ 300 h ≤ 309 h	25%	$250,000
≥ 290 h ≤ 299 h	22%	$220,000
≥ 280 h ≤ 289 h	19%	$190,000
≥ 270 h ≤ 279 h	16%	$160,000
≥ 260 h ≤ 269 h	13%	$130,000
≥ 250 h ≤ 259 hr	10%	$100,000
≥ 240 h ≤ 249 h	7%	$ 70,000
≥ 230 h ≤ 239 h	4%	$ 40,000
≥ 226 h ≤ 229 hr	1%	$ 10,000
≤ 226 h	0	0

Figure 15.5 Incentive structure for MTBF.

System Program: *XYZ*
Contract: *XXXXXXXX*

Key Performance Parameter: MTTR/$\overline{M}_{ct}$
Threshold: 2.83 h
Objective: 2.25 h
Maximum Incentive: $360,000 (36%)

Performance Achieved	Authorized Incentive (%)	Authorized Incentive ($)
≥ 2.83 h	0	0
≥ 2.78 h ≤ 2.82 h	3%	$ 30,000
≥ 2.73 h ≤ 2.77 h	6%	$ 60,000
≥ 2.68 h ≤ 2.72 h	9%	$ 90,000
≥ 2.63 h ≤ 2.67 h	12%	$120,000
≥ 2.58 h ≤ 2.62 h	15%	$150,000
≥ 2.53 h ≤ 2.57 h	18%	$180,000
≥ 2.48 h ≤ 2.52 h	21%	$210,000
≥ 2.43 h ≤ 2.47 h	24%	$240,000
≥ 2.38 h ≤ 2.42 h	27%	$270,000
≥ 2.33 h ≤ 2.37 h	30%	$300,000
≥ 2.28 h ≤ 2.32 h	33%	$330,000
≥ 2.25 h ≤ 2.27 h	35%	$350,000
≤ 2.25 h	36%	$360,000

Figure 15.6 Incentive structure for MTTR/$\overline{M}_{ct}$.

System Program: *XYZ*
Contract: *XXXXXXXX*

Key Performance Parameter: MTBM$_{pt}$
Threshold: 237 h
Objective: 400 h
Maximum Incentive: $80,000 (8%)

Performance Achieved	Authorized Incentive (%)	Authorized Incentive ($)
≥ 400 h	8%	$80,000
≥ 378 h ≤ 399 h	7%	$70,000
≥ 356 h ≤ 377 h	6%	$60,000
≥ 334 h ≤ 355 h	5%	$50,000
≥ 312 h ≤ 333 h	4%	$40,000
≥ 290 h ≤ 311 h	3%	$30,000
≥ 268 h ≤ 289 h	2%	$20,000
≥ 246 h ≤ 267 h	1%	$10,000
≥ 237 h ≤ 245 h	0.5%	$ 5,000
≤ 237 h	0	0

Figure 15.7   Incentive structure for MTBM$_{pt}$.

System Program: *XYZ*
Contract: *XXXXXXXX*

Key Performance Parameter: M$_{pt}$
Threshold: 2.09 h
Objective: 1.5 h
Maximum Incentive: $120,000 (12%)

Performance Achieved	Authorized Incentive (%)	Authorized Incentive ($)
≥ 2.09 h	0	0
≥ 2.00 h ≤ 2.08 h	1%	$ 10,000
≥ 1.95 h ≤ 1.99 h	2%	$ 20,000
≥ 1.90 h ≤ 1.94 h	3%	$ 30,000
≥ 1.85 h ≤ 1.89 h	4%	$ 40,000
≥ 1.80 h ≤ 1.84 h	5%	$ 50,000
≥ 1.75 h ≤ 1.79 h	6%	$ 60,000
≥ 1.70 h ≤ 1.74 h	7%	$ 70,000
≥ 1.65 h ≤ 1.69 h	8%	$ 80,000
≥ 1.60 h ≤ 1.64 h	9%	$ 90,000
> 1.55 h ≤ 1.59 h	10%	$100,000
> 1.50 h ≤ 1.54 h	11%	$110,000
≤ 1.50 h	12%	$120,000

Figure 15.8   Incentive structure for M$_{pt}$.

Inherent (or Design) Availability (A$_i$) is an algorithmic reflection of the interplay between MTBF and MTTR, i.e.,

$$A_i = \frac{MTBF}{MTBF + MTTR}$$

In this regard, using Ai as a performance metric is subject to the judgment and discretion of the technical team regarding the uniqueness and importance of this factor as a design parameter.

System Program: *XYZ* Contract: *XXXXXXXX*	Key Performance Parameter: A_i Threshold: 0.98763 Objective: 0.99361 Maximum Incentive: $10,000 (1%)	
Performance Achieved	Authorized Incentive (%)	Authorized Incentive ($)
≥ 0.99361	1.0%	$10,000
$\geq 0.99276 \leq 0.99360$	0.8%	$ 8,000
$\geq 0.99191 \leq 0.99275$	0.6%	$ 6,000
$\geq 0.99106 \leq 0.99190$	0.4%	$ 4,000
$\geq 0.98763 \leq 0.99105$	0.2%	$ 2,000
≤ 0.98763	0	0

Figure 15.9 Incentive structure for A_i.

The matrix in Table 15.2 displays the individual key performance parameters, along with the threshold performance levels, objective performance levels, minimum incentives and maximum incentives. This data will be incorporated into the acquisition solicitation, on which basis the performance-based contract will be negotiated.

TABLE 15.2   Tableau with KPP Incentive Allowances and Performance Levels

KPP	Performance Threshold	Performance Objective	Minimum Incentive (%)	Maximum Incentive (%)
MTBF	226 h	350 h	0	43% (0.43)
MTTR/$\overline{M}_{ct}$	2.83 h	2.25 h	0	36% (0.36)
MTBM$_{pt}$	237 h	400 h	0	8% (0.08)
$\overline{M}_{pt}$	2.09 h	1.5 h	0	12% (0.12)
A_i	0.98763	0.99361	0	1% (0.01)
				100% (1.00)

Allocation of Human Resources for Individual KPPs

It is advisable that the technical team provide estimates for allocation of human resources that indicate levels of effort anticipated by the procuring agency for dedication to work on the individual key performance parameters. This would provide insight to potential contractors on the degree of emphasis on the individual program elements. Figure 15.10 portrays the type of human resource data that could be incorporated into the solicitation, as well as the resultant contract.

Key Performance Parameter	Allocation of Resources
MTTR	23.5 Man-Years
MTBF	20.0 Man-Years
$\overline{M}_{pt}$	13.5 Man-Years
MTBM$_{pt}$	7.0 Man-Years
A$_i$	3.0 Man-Years
Total	67.0 Man-Years

Figure 15.10 Anticipated allocation of human resources.

Order of Importance of the Key Performance Parameters

Figure 15.11 portrays the order of importance of the KPPs, which parallels the essentiality values shown in Figure 15.2. This data is included in the contract solicitation document, but are not required for inclusion in the final contract. As with the resource allocation data set forth in Figure 15.10, this section conveys to the potential contractors relative essentiality of the individual key performance parameters. The Federal Acquisition Regulation does not permit inclusion of the essentiality values (Figure 15.2) in the contract solicitation, so the potential contractors need to infer from the Figure 15.11 data the necessary insight on how to best develop a responsive and competitive proposal for accomplishment of the specified performance requirements.

KPP	Order of Importance
MTBF	1st
MTTR	2nd
MTBM$_{pt}$	3rd
M$_{pt}$	4th
A$_i$	5th

Figure 15.11 Key performance parameters—order of importance.

Highlights of Contract for Performance-Based Systems Engineering

The essential contract program requirements, which normally emanate from negotiations based on the solicitation, include a description of the system; statement of work; special provisions setting forth the cost, base fee, and incentive fee considerations; and a schedule of the performance metrics. The work will be accomplished pursuant to a CPIF contract, which should incorporate the guidelines of FAR 16.405-1.

Description of the System

The system consists of five modules, for which the threshold (acceptable) performance metrics are described by Table 15.3.

Statement of Work

The following summarizes the major features of the technical effort:

1. The Contractor shall conduct a Development and Demonstration Program to enhance the key performance characteristics of System XYZ so as to establish the Systems Engineering baseline preliminary to Production and Deployment phase of the System.

2. The System will be a continuously operating system over a service life of 87,600 operating hours (10 years).

3. The Mean Time Between Failure (MTBF) shall be a minimum of 226 hours with an objective of 350 hours.

4. The Mean Time to Repair (MTTR) shall be a maximum of 2.83 hours with an objective of 2.25 hours.

5. The Mean Time Between Preventive Maintenance ($MTBM_{pt}$) shall be a minimum of 237 hours with an objective of 400 hours.

6. The Mean Preventive Maintenance Time ($\overline{M}_{pt}$) shall be a maximum of 2.09 hours with an objective of 1.5 hours.

7. Inherent Availability (A_i) shall be a minimum of 0.98763 with an objective of 0.99361.

TABLE 15.3 Threshold (Acceptable) Performance Parameters in Solicitation for Contract Proposals Based on Systems Configuration and Predicted Technical Attributes

System XYZ

Module	MTBF	λ	$MTBM_{pt}$	f_{pt}	$\overline{M}_{ct}$	$\overline{M}_{pt}$
1	800	0.00125	1000	0.001	2.0	3.0
2	1000	0.001	1100	0.00091	2.5	2.5
3	1200	0.00083	1200	0.00083	3.0	2.0
4	1400	0.00071	1300	0.00077	3.5	1.5
5	1600	0.00063	1400	0.00071	4.0	1.0
Σ 1–5	226 h System	0.00442 System	237 h System	0.00422 System	2.83 h System	2.09 h System

Note: Threshold A_i is based on threshold MTBF of 226 and M_{ct} (MTTR) of 2.83.

Special Provisions

The following summarizes the financial cost, fee, and incentive considerations:

1. The allowable program costs (excluding fees) under the contract is $10,000,000.

2. The base fixed fee is $500,000. *Note: Negotiated fixed fee is 5% of initially negotiated program costs.*

3. All allowable costs exceeding $10,000,000 shall be subject to a 60/40 sharing ratio. *Note: The sharing ratio reflects "Government share/ Contractor share." The Contractor shall absorb 40% of cost "overrun."*

4. A total of $1,000,000 has been designated for the performance incentive pool, in addition to the consideration established for allowable costs and base fixed fee.

5. The maximum sum for the fixed fee and incentive fees authorized pursuant to this contract shall not exceed $1,500,000.

6. This minimum fixed fee and technical incentive fees allowable under this contract is zero.

Schedule of Performance Metrics

The schedules of incentives applicable to the key performance parameters, which are incorporated into the contract are as follows:

1. Mean Time between Failure (MTBF), as detailed in Figure 15.5.

2. Mean Time to Repair (MTTR), as detailed in Figure 15.6

3. Mean Time between Preventive Maintenance ($MTBM_{pt}$), as detailed in Figure 15.7.

4. Mean Preventive Maintenance Time ($\overline{M}_{pt}$), as detailed in Figure 15.8.

5. Inherent Availability (A_i), as detailed in Figure 15.9

Evaluation of Contractor's Technical Performance

The contractor's achievements with respect to the individual key performance parameters are summarized in Table 15.4.

TABLE 15.4   Systems Engineering Metrics Resulting from Contractor's Performance

Module	MTBF	λ	$MTBM_{pt}$	f_{pt}	$\overline{M}_{ct}$	$\overline{M}_{pt}$
1	1200	0.00083	1300	0.00077	1.8	2.5
2	1400	0.00071	1600	0.00063	2.3	2.2
3	1450	0.00069	1650	0.00061	2.5	1.8
4	1800	0.00056	1950	0.00051	3.0	1.4
5	1900	0.00053	2100	0.00048	3.0	1.0
Σ 1–5 System	301 h System	0.00332 System	333 h System	0.003 System	2.45 h System	1.87 h

Note: Based on system MTBF and $\overline{M}_{ct}$ (MTTR) the achieved A_i is 0.99193.

Evaluation of Incentives Authorized Based on Achieved KPP Levels

The summaries of achievements by the contractor on individual key performance parameters and technical incentives authorized are detailed in the following by Figure 15.12 (MTBF); Figure 15.13 (Mean Corrective Maintenance Time/MTTR); Figure 15.14 ($MTBM_{pt}$); Figure 15.15 (Mean Preventive Maintenance Time); and Figure 15.16 (A_i). The scalar performance values and corresponding incentives are based on the same performance-incentive structures previously described in Figures 15.5, 15.6, 15.7, 15.8, and 15.9.

System Program: *XYZ*
Contract: *XXXXXXXX*

Key Performance Parameter: MTBF
Threshold: 226 h
Objective: 350 h
Maximum Incentive: $430,000 (43%)

Performance Achieved	Authorized Incentive (%)	Authorized Incentive ($)
≥ 350 h	43%	$430,000
≥ 340 h ≤ 349 h	37%	$370,000
≥ 330 h ≤ 339 h	34%	$340,000
≥ 320 h ≤ 329 h	31%	$310,000
≥ 310 h ≤ 319 h	28%	$280,000
→ ≥ 300 h ≤ 309 h	25%	$250,000
≥ 290 h ≤ 299 h	22%	$220,000
≥ 280 h ≤ 289 h	19%	$190,000
≥ 270 h ≤ 279 h	16%	$160,000
≥ 260 h ≤ 269 h	13%	$130,000
≥ 250 h ≤ 259 h	10%	$100,000
≥ 240 h ≤ 249 h	7%	$ 70,000
≥ 230 h ≤ 239 h	4%	$ 40,000
≥ 226 h ≤ 229 h	1%	$ 10,000
≤ 226 h	0	0

MTBF Achieved: 301 h   Authorized Incentive: $250,000 (25%)

Figure 15.12   MTBF performance and incentive achieved by contractor.

System Program: *XYZ*
Contract: *XXXXXXXX*

Key Performance Parameter: MTTR/$\overline{M}_{ct}$
Threshold: 2.83 h
Objective: 2.25 h
Maximum Incentive: $360,000 (36%)

Performance Achieved	Authorized Incentive (%)	Authorized Incentive ($)
≥ 2.83 h	0	0
≥ 2.78 h ≤ 2.82 h	3%	$ 30,000
≥ 2.73 h ≤ 2.77 h	6%	$ 60,000
≥ 2.68 h ≤ 2.72 h	9%	$ 90,000
≥ 2.63 h ≤ 2.67 h	12%	$120,000
≥ 2.58 h ≤ 2.62 h	15%	$150,000
≥ 2.53 h ≤ 2.57 h	18%	$180,000
≥ 2.48 h ≤ 2.52 h	21%	$210,000
→ ≥ 2.43 h ≤ 2.47 h	24%	$240,000
≥ 2.38 h ≤ 2.42 h	27%	$270,000
≥ 2.33 h ≤ 2.37 h	30%	$300,000
≥ 2.28 h ≤ 2.32 h	33%	$330,000
≥ 2.25 h ≤ 2.27 h	35%	$350,000
≤ 2.25 h	36%	$360,000

MTTR/$\overline{M}_{ct}$ Achieved: 2.45 h   Authorized Incentive: $240,000 (24%)

Figure 15.13 MTTR/$\overline{M}_{ct}$ performance and incentive achieved by contractor.

System Program: *XYZ*
Contract: *XXXXXXXX*

Key Performance Parameter: MTBM$_{pt}$
Threshold: 237 h
Objective: 400 h
Maximum Incentive: $80,000 (8%)

Performance Achieved	Authorized Incentive (%)	Authorized Incentive ($)
≥ 400 h	8%	$80,000
≥ 378 h ≤ 399 h	7%	$70,000
≥ 356 h ≤ 377 h	6%	$60,000
≥ 334 h ≤ 355 h	5%	$50,000
→ ≥ 312 h ≤ 333 h	4%	$40,000
≥ 290 h ≤ 311 h	3%	$30,000
≥ 268 h ≤ 289 h	2%	$20,000
≥ 246 h ≤ 267 h	1%	$10,000
≥ 237 h ≤ 245 h	0.5%	$ 5,000
≤ 237 h	0	0

MTBM$_{pt}$ Achieved: 333 h   Authorized Incentive: $40,000 (4%)

Figure 15.14 MTBM$_{pt}$ performance and incentive achieved by contractor.

System Program: *XYZ*   **Key Performance Parameter: $\overline{M}_{pt}$**
Contract: *XXXXXXXX*   **Threshold: 2.09 h**
Objective: 1.5 h
Maximum Incentive: $120,000 (12%)

Performance Achieved	Authorized Incentive (%)	Authorized Incentive ($)
≥ 2.09 h	0	0
≥ 2.00 h ≤ 2.08 h	1%	$ 10,000
≥ 1.95 h ≤ 1.99 h	2%	$ 20,000
≥ 1.90 h ≤ 1.94 h	3%	$ 30,000
→ ≥ 1.85 h ≤ 1.89 h	4%	$ 40,000
≥ 1.80 h ≤ 1.84 h	5%	$ 50,000
≥ 1.75 h ≤ 1.79 h	6%	$ 60,000
≥ 1.70 h ≤ 1.74 h	7%	$ 70,000
≥ 1.65 h ≤ 1.69 h	8%	$ 80,000
≥ 1.60 h ≤ 1.64 h	9%	$ 90,000
≥ 1.55 h ≤ 1.59 h	10%	$100,000
≥ 1.50 h ≤ 1.54 h	11%	$110,000
≤ 1.50 h	12%	$120,000

$\overline{M}_{pt}$ Achieved: 1.87 h   Authorized Incentive: $40,000 (4%)

Figure 15.15 $\overline{M}_{pt}$ performance and incentive achieved by contractor.

System Program: *XYZ*   **Key Performance Parameter: A_i**
Contract: *XXXXXXXX*   **Threshold: 0.98763**
Objective: 0.99361
Maximum Incentive: $10,000 (1%)

Performance Achieved	Authorized Incentive (%)	Authorized Incentive ($)
≥ 0.99361	1.0%	$10,000
≥ 0.99276 ≤ 0.99360	0.8%	$ 8,000
→ ≥ 0.99191 ≤ 0.99275	0.6%	$ 6,000
≥ 0.99106 ≤ 0.99190	0.4%	$ 4,000
≥ 0.98763 ≤ 0.99105	0.2%	$ 2,000
≤ 0.98763	0	0

A_i Achieved: 0.99193   Authorized Incentive: $6,000 (0.6%)

Figure 15.16 A_i performance and incentive achieved by contractor.

Cost Performance Evaluation Determined by the Contracting Officer

The initially negotiated program cost was $10 million, but the total program cost incurred is $10,500,000. The cost "overrun" of $500,000 is subject to the 60/40 sharing ratio incorporated in the special provisions of the contract. Inasmuch as the originally negotiated base fixed fee was $500,000, the contractor's 40 percent share must be deducted from the initially negotiated base fixed fee of $500,000, i.e.,

Final Fee = Initial Base Fee − Cost Overrun)(Contractor Share %)

Final Fee = ($500,000) − (500,000)(0.40)

Final Fee = $300,000

Computation of KPP Incentives Plus Base Fee

The following summarizes the final profit posture of the contractor, which reflects the performance on the individual key performance parameters as well as the cost performance. The maximum technical incentives plus base fixed fee provided for in the contract totaled $1,500,000.

Performance Parameter	Incentive Payment Earned By Contractor
MTBF	$250,000
$MTBM_{pt}$	240,000
$MTTR/\overline{M}_{ct}$	40,000
$\overline{M}_{pt}$	40,000
A_i	6,000
Final Fee (Cost Performance)	300,000
TOTAL	$876,000

In this hypothetical scenario, the contractor earned 58.4 percent of the maximum allowable for technical incentives and base fee.

Benefit of Incentivized Program for Long-Term Logistics System Support Costs

A comparatively modest investment of incentive funding during the pre-production phases of system development could offer significant cost reductions over the projected service life of the system or equipment. The hypothetical system discussed in this chapter presumed continuous operation over a period of ten years (87,600 h) service life. The results reflected in Table 15.5 emphasize significant reduction in both maintenance actions and maintenance man-hours per system unit when comparing threshold (acceptable) performance levels to those achieved by the contractor. The total cost benefits are largely governed by the economy of scale, i.e., number of units to be fielded during the total life cycle of the system. This perspective must be considered when analyzing the effectiveness of the investment in incentive funding for a performance-based logistics engineering program. The analysis provided in Table 15.6,

TABLE 15.5 Summary Analysis of Systems Engineering Results: 87,600 hr (10 yr) Service Life

Contractor Performance Evaluation

Comparing Achieved KPP Metrics to Threshold KPP Metrics

Total Maintenance Actions/System Unit

KPP Metric	Threshold	Achieved	Threshold Nr. Actions	Achieved Nr. Actions	Achieved vs Threshold
MTBF	226 h	301 h	388	291	Reduction of 97 (25%)
MTBM$_{pt}$	237 h	333 h	370	263	Reduction of 107 (29%)

Total Maintenance Hours/System Unit

KPP Metric	Threshold	Achieved	Threshold Nr. Hours*	Achieved Nr. Hours**	Achieved vs Threshold
MTTR	2.83 h	2.45 h	1,098	713	Reduction of 370 (35%)
$\overline{M}_{pt}$	2.09 h	1.87 h	773	492	Reduction of 281 (36%)

* *Multiply*: 2.83 (Threshold MTTR) by 388 Actions; 2.09 (Threshold $\overline{M}_{pt}$) by 370 Actions
** *Multiply*: 2.45 (Achieved MTTR) by 291 Actions; 1.87 (Achieved $\overline{M}_{pt}$) by 263 Actions

TABLE 15.6 Analysis of Economic Benefits Based on Data Provided n Table 15.5

Comparing Achieved KPP Metrics to Threshold KPP Metrics-Economic Analysis

Corrective Maintenance Actions (Based on MTBF)

System Life Cycle Population: 100 Units   Service Life per System (87600 h)   Material Cost per Action: $50

Threshold Values

$$\text{Nr. Units} \times \text{Mct Actions/Sys} \times \text{Matl Cost/Act} = \text{Cost}$$
$$(100)(388)(\$50) = \$1,940,000$$

Achieved Values

$$\text{Nr. Units} \times \text{Mct Actions/Sys} \times \text{Matl Cost} = \text{Cost/Act}$$
$$(100)(291)(\$50) = \$1,455,000$$

Net Saving
$$\$485,000$$

Preventive Maintenance Actions (Based on $MTBM_{pt}$)

System Life Cycle Population: 100 Units   Service Life per System (87600 h)   Material Cost per Action: $50

Threshold Values

$$\text{Nr. Units} \times \text{Mpt Actions/Sys} \times \text{Matl Cost/Act} = \text{Cost}$$
$$(100)(370)(\$50) = \$1,850,000$$

Achieved Values

$$\text{Nr. Units} \times \text{Mpt Actions/Sys} \times \text{Matl Cost} = \text{Cost/Act}$$
$$(100)(263)(\$50) = \$1,315,000$$

Net Saving
$$\$535,000$$

Corrective Maintenance Man-Hours (Based on MTTR)

System Life Cycle Population: 100 Units   Service Life per System (87600 h)   Cost per Man-hour: $100 (1 Technician)

Threshold Values

$$\text{Nr. Units} \times \text{Mct Hours/Sys} \times \text{Labor Cost/Hour} = \text{Cost}$$
$$(100)(1098)(\$100) = \$10,980,000$$

Achieved Values

$$\text{Nr. Units} \times \text{Mct Hours/Sys} \times \text{Labor Cost/Hour}$$
$$(100)(713)(\$100) = \$7,130,000$$

Net Saving
$$\$3,850,000$$

Preventive Maintenance Man-Hours (Based on $\overline{M}_{pt}$)

System Life Cycle Population: 100 Units   Service Life per System (87600 h)   Cost per Man-hour: $75 (1 Technician)

Threshold Values

$$\text{Nr. Units} \times \text{Mpt Hours/Sys} \times \text{Labor Cost/Hour} = \text{Cost}$$
$$(100)(773)(\$75) = \$5,797,500$$

Achieved Values

$$\text{Nr. Units} \times \text{Mpt Hours/Sys} \times \text{Labor Cost/Hour}$$
$$(100)(492)(\$75) = \$3,690,000$$

Net Saving
$$\$1,107,500$$

Total Savings   \$5,977,500

which compares the threshold and achieved performance metrics, underscores the potential long-term logistics cost savings accomplished by the contractual incentives. The $876,000 in technical incentives and base fee earned by the contractor resulted in potential life cycle savings of $5,977,500 for a population of 100 system units. Based on the economies of scale, the financial benefits should be significantly increased for a larger population.

Logistics Systems Management and Operations

16

Logistics Facilities

General

Facility location and specification, especially with respect to those facilities dedicated to production logistics, product distribution, and customer support, are becoming the province of the logistics manager. This phenomenon is attributable to the predominantly logistic nature of the governing economic factors and the related social and political considerations. As this book is aimed at the novice logistician, the purpose of this chapter is to establish a conceptual threshold from which to proceed to the more advanced aspects of facility technology and analysis.

Facility decision considerations

The facility planner is faced with four major areas of consideration which influence facility requirements analyses and decisions:

1. Purpose, or mission, of the facility
2. Principle of locational gravitation
3. Economic factors
4. Sociopolitical criteria

These considerations will be individually addressed in terms of their constituent elements and how these individually interact with other elements to form the basis of decision making.

Purpose of Facility

The types of facilities which fall within the purview of logistics analysis, as categorized by their purpose or mission, are (1) production facilities, (2) distribution centers, (3) service centers, and (4) inventory storage points.

Production facilities

The mission of the production facility is to receive raw materials, working stock, and fabrication components and convert them into finished products for subsequent transfer to the customer. The location of the manufacturing facility relative to the sources of raw material for working stock and source vendors reflects the optimum trade-off among the factors of transportation from the production point to raw materials sources, transportation from the production point to the market zones, required site manufacturing resources, production material inventory requirements, and order processing and communications.

Distribution centers

Distribution centers receive consignments of finished products from the production points and store them in sufficient quantities to replenish the stocks of the retail outlets within their assigned market support areas as needed. Distribution centers may be either wholly owned by the manufacturer or primary corporation or operated by another enterprise under franchise or contract with the producer of the goods. This intermediate-level type of activity is, on a smaller scale, akin to that of a military depot, but with its mission focus being the consumer market zone. Distribution centers provide (1) replenishment support to retail outlet inventories of the product, (2) stockage of spare components and repair parts, (3) wholesale transactions to retailers, and (4) intermediate maintenance capabilities beyond the scope of retail service activities. Distribution centers typically support product-related activities within geographic areas or defined market zones.

Service centers

This type of logistics facility is similar to a distribution center, but generally not as comprehensive in its scope of activities. It is typically wholly owned by the producer to support the needs of customers within its defined market zone. The focus is on the needs of the retail-level customers who use the company's products rather than on providing the depot-type operations of distribution centers.

Inventory storage points

This type of facility serves as a storage warehouse for (1) goods received from the production point for subsequent transhipment to distribution centers and service centers in defined market zones and (2) raw materials and working stock to support production schedules. Inventory storage points may be either producer-owned warehouses or public warehouses. There are considerations that favor company-owned warehouses rather than public warehouses, and considerations which favor public warehouses over company-owned warehouses; these are discussed in depth in Chap. 17.

Principle of Locational Gravitation

In a generic sense, *locational gravitation* embodies those forces and factors which influence whether an industrial activity is located close to its source of input (e.g., source of raw materials for production) or close to the recipients of its output (e.g., market areas in which there are active and potential customers). This principle applies especially to analysis for determining the location of a production facility relative to production materials sources, distribution centers, service centers, inventory storage points, and market centers. From a logistics perspective, the primary analytical factors are defined by the economics of *transfer* (transportation) and *production*. There are cases in which social and political issues arise in the decision process and, therefore, must be incorporated and given weight in the analytical equation.

Transfer-production trade-offs

Locational gravitation results from the pressures of the production and transfer costs associated with the manufacture and distribution of a product. Increases in production costs (e.g., labor, taxes, utilities, facility costs) which might tend to result from closer proximity of the production site to the market centers as a result of the metropolitan area cost levels may be offset by lower costs of transportation of the finished products to the distribution outlets. Production locations in areas remote from population centers may enjoy lower production costs, but with higher costs of transferring finished products to the market centers. Product transfer costs must consider the economic effects of the production location relative both to the market centers and to the sources of raw materials and working stocks. The rules of logistics economy suggest that whether the production site is located close to the market center or close to the source of production materials is a function of the ratio of the weight of the materials used for the production process to the weight of such materials in the finished product—i.e., how much of the weight of input materials is consumed in the production process. The higher the weight ratio of materials input to materials retained in the finished product, the greater the likelihood that the production site will be located in close proximity to the production materials source. On the other hand, the closer the input-output materials ratio is to parity, the greater the likelihood the production location will be in convenient proximity to the market centers. Liquor distilleries and precious metal extraction plants are examples of facilities in which significant percentages of input materials are consumed in processing the finished products; hence, the production sites are typically located close to the materials sources. The other end of the spectrum is typified by a jewelry manufacturer, whose products are processed by assembly of lightweight components, and who would probably locate production facilities in convenient proximity to the market centers.

In essence, the economic principle of locational gravitation is governed by the costs of transporting raw materials to the production facility relative to the costs of transporting finished products to the marketplace.

Effects of Economic Factors on Location

The predominant factors affecting the selection of the location of a production facility are the transfer and production costs. Each candidate production site is subject to analytical scrutiny through application of these factors to it in relation to the production materials sources and market centers.

The goal of the analyst is to determine the optimum location in terms of distance (or transport zones) between the materials sources and market centers, giving consideration to

1. Costs of transfer of raw materials and working stocks from the sources to the production plant
2. Costs of transfer of finished products from the plant to market centers
3. Production cost (labor, taxes, utilities, facility costs, etc.) at the plant site

Figure 16.1 portrays a hypothetical case graphically. It depicts cost profiles determined by the effect the distance between the production materials sources and the market center would have on the aforementioned transfer and production costs. For this example, unit production costs would increase in direct proportion to the decrease in distance between the plant location and market centers. Costs for transfer of raw materials to the production site increase significantly as the plant is moved further from the materials source. The costs for shipment of finished products from the plant to the market increase in proportion to the distance from the market centers. The cost value on the *total cost* curve is equal to the sum of the individual values at corresponding distance

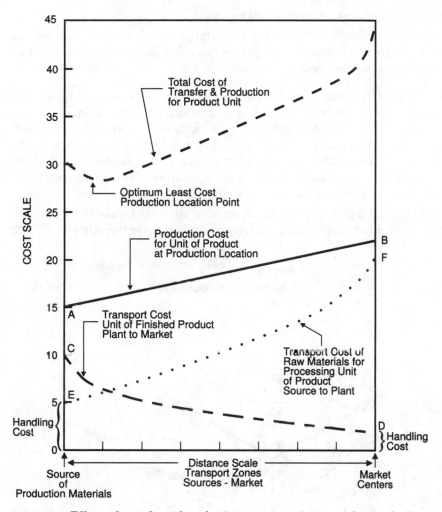

Figure 16.1 Effects of transfer and production costs on optimum production facility location.

scale points on the transfer (plant to market and materials source to plant) and production cost curves. In this case there is an initial downward trend in total costs, but they then start to increase in proportion to the distance away from the production materials source. The optimum location would be the point on the total cost curve where the slope (rate of change of cost relative to distance) is equal to zero, as indicated in Figure 16.1.

Impact of logistics on commercial locations

The logistics analyst evaluating candidate commercial facility locations is aided by two frequently employed quantitative and graphic analysis techniques, *isotims* and *isodapanes*.

Isotims. The elementary isotim method, as illustrated by Fig. 16.2, is based on production costs and transportation costs. An isotim is a cartographic display of the candidate site location as a central point surrounded by concentric dotted circles. The production costs are noted by the center point. The intervals between the dotted circles, called *blockets*, define the transportation costs per prescribed distance interval (usually 100 mi) from the facility location point.[1]

Figure 16.2 portrays isotims describing the situation of three competing firms. Each has a plant located equidistant from the other two. Each

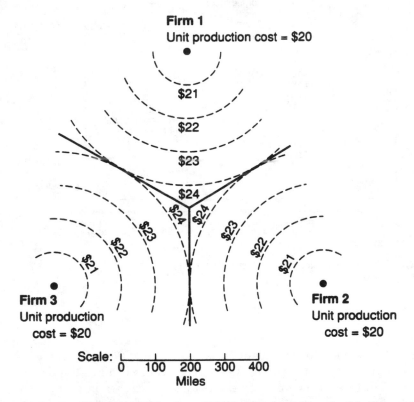

Figure 16.2 Market areas for three firms with equal costs of production and transportation. Total cost = AB + CD + EF (Reference Figure 16.1).

---

[1]100 mi = 160.93 kilometres (kms); this also applies to *blocket* intervals shown in Figure 16.2 and Figure 16.3.

has the same production cost per unit ($20), and transportation costs per unit are constant and equal for all firms ($0.01 per mile per unit).[2] The latter are indicated by the cost belts radiating from each production site. In this case, the boundaries of the markets served by each plant (assuming completely rational economic behavior and a price policy which bases the price charged on production cost plus cost of transportation to any given point) will be defined by points at which the price charged by any two or more of the firms is the same. Theoretically, all of these points are situated at the intersections of comparable cost belts of two or three firms (for the example, the boundary intersection value is $24). By connecting these points, market boundaries are formed.

Transportation cost impact on market boundaries.   Decreased transportation costs, production cost, or both for one firm relative to another will enlarge the market area of the first at the expense of the second. This is

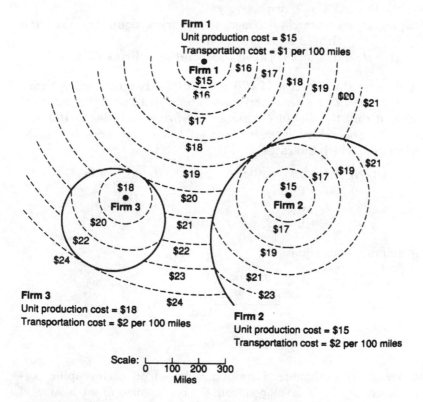

Figure 16.3   Market areas for three firms having unequal costs of production and transportation.

---

[2]$0.01 per mile per unit = $0.0063 per kilometer (km) per unit.

the situation shown for the three firms in Fig. 16.3. Firm 1 benefits at the expense of firm 3, enlarging its market area to completely encircle a small area surrounding firm 3's plant.

If the diagram were extended further, would firm l's market area also eventually encircle firm 2's? The following is a formula for the determination of the point at which a theoretical market boundary between two competing fixed facilities will fall. It requires a knowledge of production costs at the two competing points, the transportation rate per unit of distance paid by each competing firm, and the distance between competing fixed facilities.

$$p_1 + r_1 d_1 = p_2 + r_2 d_2$$

where p_1 and p_2 = production costs of two firms, 1 and 2, respectively, at competing fixed facilities

r_1 and r_2 = transportation rates per unit per mile[3] paid by firms 1 and 2, respectively

d_1 and d_2 = distances between the market boundary line and firms 1 and 2, respectively

$d_1 + d_2$ = distance between the facilities of firms 1 and 2

Assume a production cost of $20 per unit for two competing firms. Assume further that the transportation rate is $0.10 per mile per unit[4] for each and that the distance between the firms' facilities is 100 mi.[5] Where will the boundary between them fall? Common sense is verified by the formula, which indicates

$$20 + 0.10d_1 = 20 + 0.10d_2 \qquad \text{and} \qquad d_1 + d_2 = 100$$
$$0.10d_1 = 0.10d_2$$
$$d_1 = d_2$$
$$d_1 - d_2 = 0$$

Adding simultaneous equations,

$$d_1 - d_2 = 0$$
$$\underline{+d_1 + d_2 = 100}$$
$$2d_1 \quad = 100$$
$$d_1 \quad = 50$$
$$d_2 \quad = 50$$

Isodapanes. The *isodapane* technique is based on cartographic portrayal of boundary lines developed from the confluences of the isotims of a multiplicity of locations. Figure 16.4 illustrates a typical isodapane

---

[3]mile = 1.6093 kms
[4]$0.10 per mile per unit = $0.063 per km per unit
[5]100mi = 160.93 kms

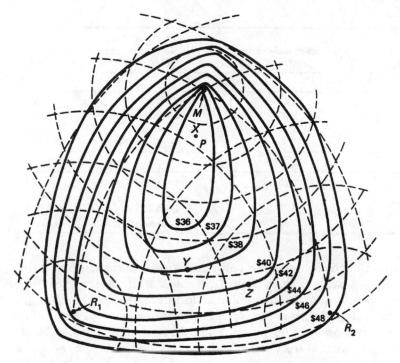

Figure 16.4 A construction of isodapanes showing the interrelated influence of transportation and production costs on the selection of a production site. Each isotim (broken lines) around points R_1, R_2, and M represents $4 of transport cost per blocket. Values for isodapanes (solid lines) represent the total transportation costs from any point on a given isodapane to points R_1, R_2, and M.

construction. The isodapane patterns in the figure show the interrelated influences of transportation costs governing the selection of a key facility location such as a distribution center. Determining the isodapane that describes the optimum location entails consideration of the interaction of the distribution center with the market zone and production plants and is constructed on the basis of the following:

1. *Market zone.* Point M represents the market center.
2. *Production plants.* Points R_1 and R_2 denote the locations of production plants supplying finished marketable goods to the distribution center.
3. *Distribution center.* Point P pinpoints the optimum location for the distribution center, determined on the basis of transportation costs from P to R_1, R_2, and M.

The isodapane lines are lines of indifference defined by a common transportation cost factor. Each isodapane line describes collinear loca-

TABLE 16.1

Isotim leg	Blocket scale value (100 mi)[6]	Transport cost for isotim leg ($4/100 mi)[7]	Leg total
	Point Y		
R_1 to Y	2.5	4 × 2.5	10.00
R_2 to Y	2.25	4 × 2.25	9.00
M to Y	4.75	4 × 4.75	19.00
		Summary Total	$38.00
	Point Z		
R_2 to Z	1.25	4 × 1.25	5.00
R_1 to Z	3.5	4 × 3.5	14.00
M to Z	5.25	4 × 5.25	21.00
		Summary Total	$40.00

tion points that have the same cumulative transportation cost value, incorporating transport costs from R_1, R_2. and M, irrespective of where any of the points may lie on the line of indifference. For this example, the blocket value of each of the R_1, R_2, and M isotims is $4 per 100 mi. The transportation costs for points Y and Z, for example, would be calculated on the basis of isotimetric scale values, determined as shown in Table 16.1

Point Y is located on the $38 isodapane line. Computation of isotimetric blocket costs for any location point on the same line would also result in transportation costs of $38, considering the point relative to R_1, R_2, and M.

Point Z is logically located on the $40 isodapane line. As was the case for point Y, any collinear point would result in the same transportation cost factor.

Optimal location. As depicted in Fig. 16.4, the optimal location point for the distribution center, point P, falls within the area circumscribed by the $36 isodapane line, which was determined by the same computational methodology described above.

Sociopolitical Criteria

There are intangible factors involved in the facility location decision which transcend economic pragmatism and technical requirements. Decision planners are inevitably obligated to address social and political issues which may either reinforce or contradict the conclusions drawn from the quantitative cost-effectiveness analyses. These generally in-

---

[6]100 mi = 160.93 kms
[7]$4 per 100 mi = $2.49 per 100 kms

clude such issues as quality of life, workforce dynamics, community attitudes, political indicators, the environment and ecology, and the absorp-
tive capacity of the local infrastructure. These qualitative criteria could
favor a specific location when the quantitative analyses result in cost-
benefit indicators favoring other candidates.

Quality of life

Opportunities for recreational activities, availability of high-caliber school
systems, access to colleges and institutions offering advanced education,
and favorable climate are typically important considerations for employees and their families. Such inducements aid in recruitment and in
maintaining a stable, productive workforce.

Workforce dynamics

Community stability fosters minimum turnover of employees and retention of skilled personnel, and thereby reduces expenditures for retrain-
ing. Community stability is a prerequisite to workforce stability and
cultivating a base of loyal, productive employees.

Community attitudes

An indifferent or antigrowth sentiment on the part of local government
officials could challenge the analytical conclusions favoring a candidate
site on the basis of economic effectiveness. Lack of popular endorsement
and support for a new corporate resident could degenerate into community hostility. Such a situation could erode employee morale and reduce
the prospects for a stable, skilled workforce.

Political indicators

An assortment of active political constituencies in a community could
portend turbulence in the relationships between the company and the
community. One group holding governmental power might enthusiasti-
cally promote the introduction of a new commercial facility. On the other
hand, a shift in the political tides could bring to power a different political party that was not favorably disposed to the new facility and was
inclined to impose hostile tax structures, zoning regulations, and other
measures inimical to the interests of commercial activities. Notwithstanding such adverse potential, many communities are composed of
diverse political groups that are united in their desire to promote commercial development, a scenario which would ideally serve the interests
of the new facility.

   The political ambience must therefore be thoroughly assessed with
respect to sensitivities that could affect the long-range corporate interests.

Environment and ecology

Among the qualitative issues concerning a candidate host community
are the possible effects of the activity on the social and geographic envi-
rons, e.g., local customs and activities, community values, atmosphere,
land areas, waterways, flora, fauna, the ecological scenario, and the
interaction of the populace with their social surroundings. The decision
planner would be well advised to prepare an environmental impact
study addressing all such pertinent issues before making the location
decision.

Absorptive capacity of local infrastructure

It is essential that there be available or that provision be made with the
host community to accommodate the required physical support facilities and services, such as accessible road networks, traffic control, links
with local transportation, water sources, power utilities, contiguous
land area for expansion, sewers, and disposal services. The decision
planner should prepare a comprehensive checklist of the detailed infrastructure requirements and verify the availability of those local facilities and services necessary to sustain the long-term operations and
mission of the facility.

Packaging, Handling, Storage, and Transportation

General

The packaging, handling, storage, and transportation (PHS&T) functions play critical roles in both the military and commercial market sectors. Military items must be packaged to withstand the most severe climatological conditions as well as the impacts and shock levels characteristic of materials handling and movement in zones of contingency operations. Military packaging and handling specifications must take into account conditions of austere operational facilities, shelf-life constraints, and climatic extremes anywhere in the world. The military logistician is additionally faced with a multiplicity of material staging patterns, pallet configurations, movement modes, and lines of communication in the contingency zone, which must all be anticipated by the specified packaging procedures. Military item packaging is dominated by priorities of the military missions, tempered, as appropriate, by cost-effectiveness considerations. The crucial determinants for commercial packaging are predominantly overall cost minimization and product profitability. There is a large body of literature, in the forms of standards, specifications, agency regulations, etc., governing military PHS&T procedures and practices, which is available to the student of logistics on request, as required. Accordingly, this chapter will emphasize those PHS&T principles related to commercial logistics.

Packaging

The concept of commercial product packaging has assumed multifarious purposes beyond its original, traditional one of conveying products from the production line to the hands of receptive consumers. Packaging designers should be concerned with what the package can do for the product in terms of protection, economy, convenience, and promotion. The basic purpose of product packaging is to provide for product security, transportability, and storability, with the added utility of serving as a medium of communication from the producer to the purchaser. Corporate packaging engineers in the contemporary commercial context are called upon to accommodate a multiplicity of issues and concerns, derived primarily from governmental regulations and market dynamics.

Packaging objectives

The major considerations defining the functions and purposes of product packaging are

1. *Protection.* Design characteristics which shield the package contents against temperature extremes, humidity, aridity, and peculiar climatological elements; shock, compression, puncture, and vibration impacts during movements; and likelihood of pilferage during shipment.

2. *Labeling and identification.* Markings which clearly identify the product, note product safety statements, and assure the purchaser of the standards of quality attributable to the producer and its reputation for achieving levels of expectation.

3. *Sales promotion.* Incorporation of a method of display and distinguishing markings which entice the consumer to purchase the product.

4. *Consumer security.* Tamperproof features which serve to assure the purchaser that, if there is no indication of violation, the product is in the same form as when it was released from the production line.

5. *Storability.* Features which serve to ensure the optimum shelf life or interval between receipt by the distribution or retailer and issue to the user.

6. *Transportability.* Characteristics of design which accommodate the standard pallet and container utilization configurations for shipment from the producer to the distribution centers and redistribution to the retail activities.

7. *Environmental factors.* Characteristics of the product and product packaging which provide for ease of disposal or recycling after use, so as to conform to ecological and environmental regulations.

8. *Communication to the consumer*. Labeling to promote sale of the product, and incorporation of markings and liability disclaimers which portray warranty stipulations and, if appropriate, warnings of potential hazards associated with use of the product.

Packaging design

The design of a package is the result of the amalgamation of inputs from the corporate staff as necessary to accomplish the objectives of product packaging. Figure 17.1 provides a checklist summary of major design considerations. The primary contributors to this process include the following areas of expertise:

1. *Production*. This element must address whether the current production capability is adaptable to the product configuration (i.e., size, shape, and structure), the effects of the configuration on the ability of production to meet customer orders, and whether new production facilities might be necessary.

2. *Purchasing*. The procurement activity evaluates whether the product conforms to existing sizes and types, prospects for standardization, costs for alternative packaging, prospects for vendor conformance, provisions for storage of empty containers, and effects on inbound transportation costs.

- Protection against rough handling and resulting shocks.
- Protection against dirt, dust, moisture, and other contaminants
- Prevention of pilferage and tampering
- Safeguarding of hazardous materials
- Simplification and more efficient piling, warehousing, and inventory control through use of standard shipping units
- Definition of unit of sale
- Provision of product identity and definition of quality, quantity, size, color, etc.—an information function
- Assistance to marketing by means of "miniature billboards," which are provided by attractive shipping containers
- Building of customer goodwill through convenient, reusable containers
- Relationship to environment in terms of disposability and multipurpose uses
- Facilitation of production, marketing, and logistics functions

Figure 17.1 Checklist of major packaging design considerations.

3. *Engineering*. The form of the package must provide the required strength. The package specifications should incorporate the latest material technology and minimize the materials required per product unit. It should also be determined if the optimum package configuration necessitates redesign of the product itself. The safety engineer should specify applicable warnings or safety precautions to be identified through special markings.

4. *Materials handling*. The package should, ideally, permit manual handling without physical harm. Adequate markings for picking should be included in the design. The packaging should accommodate the standard quantities in which items are picked; afford ease of order picking from broken packages; and facilitate packing, repacking, closing, and reclosing. The package design should also be adaptable to standard modular load configurations.

5. *Storage*. The packaging should be constructed so as to discourage inventory pilferage. It should not require special storage equipment and should have optimum stacking characteristics in terms of strength, dimensions, and surface properties in order to minimize cubic storage space requirements.

6. *Marketing*. The marketing staff should specifically address the following considerations:
   a. Need for special use instructions
   b. Purchaser expectations regarding quality and variety
   c. Customers' ability to handle and store the product
   d. Ease of disposal
   e. Environmental compatibility

   Based on an assessment of consumer trends, the marketing staff should determine if the package requires redesign for product promotional campaigns.

7. *Traffic management*. The package design should optimize the density of the product in package form, protect the product from damage during shipment, and minimize vulnerability to in-transit pilferage. The package could invite the most advantageous carrier rates by being designed to conform to the dimensions of standard carrier equipment and providing for easy return of empty, reusable containers.

8. *Legal counsel*. The role of the legal staff is to ensure that descriptions are marked or affixed on the packages and products and are using the appropriate wording to clearly convey warranty terms, safety requirements, warning statements, user precautions, and other product stipulations that may be required by public law and standards of trade.

Packaging design interlace with pallet configurations. Economy and efficiency in storing, staging, and carriage of packaging products dictate that individual product cartons be configured to the capacity of larger bulk carton packages used for wholesale lot shipments and that the bulk package dimensions be configured to allow maximum use of the load capacities of conventional pallet patterns. Figure 17.2 illustrates the basic block, brick, row, and pinwheel pallet configurations used in the transportation industry.

Economic significance of packaging

The cost significance of packaging is reflected by industry estimates that for typical commercial goods, packaging averages 6 percent of the

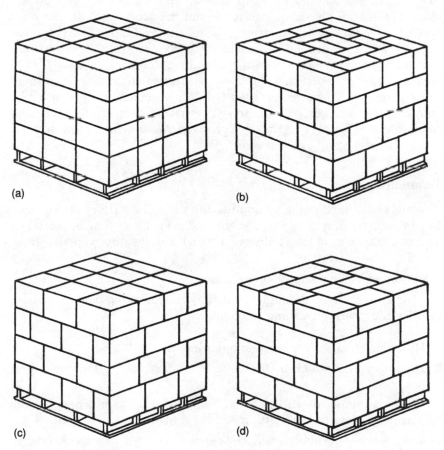

Figure 17.2   Basic pallet configurations. *(a)* Block. *(b)* Brick. *(c)* Row. *(d)* Pinwheel.

product value. Certain factors tend to influence this statistic; e.g., for certain canned food products such as sardines and kipper filets, the containerization technology used in processing the goods at the sources of supply inevitably inflates the cost of packaging relative to that of the contents. Notwithstanding the disproportionate packaging costs, the popularity of these products in worldwide consumer markets has not diminished. Proliferation of brands of like or similar products, as occurs when corporate conglomerates market similar items under a multiplicity of brand labels, also tends to increase unit packaging costs. Variation in packaging design for the same product elevates the unit packaging cost. This situation is typified by many breakfast cereal producers, who frequently change the motif of the package for a particular product for promotional purposes and special events. In such instances, marketing strategy prevails over engineering economics considerations. Variations in configurations and shapes of packages and containers govern the dimensions of the bulk shipment packages, and may not result in the most cost-effective system for palletizing, shipment, and breakdown at destinations. Many producers of products that are well established in consumer markets will opt to maintain consumer identifications and loyalty based on association of their products with unusual package conformations or unique images rather than risk loss of consumer recognition by changing the packaging design. In such cases, the concern for continued consumer loyalty outweighs any perceived cost savings from streamlining the packaging configuration.

Materials Handling

The materials handling subdiscipline enhances the efficacy of the logistics process by providing an efficient and responsive flow of materials within the context of the facilities for which the systems were designed.

Principles of materials handling systems

The logistics planner should be guided by the following basic principles relative to material handling technology:

1. Handling should be reduced to a minimum.
2. Distances over which materials are handled should be as short as possible.
3. Routes of materials should be on the same level as much as layouts permit in order to avoid lifting and lowering.
4. Once started in motion, materials should be kept moving as long as possible.

5. Mechanical and automatic means of materials handling should be used wherever routes of travel and work volume justify the investment.

6. Materials handling equipment should be standardized to the greatest extent possible.

7. Gravity flow (the least expensive form of energy) should be incorporated whenever practical.

8. In mechanized systems, maximum investment should be in movement rather than stationary equipment.

9. In equipment selection, an effort should be made to minimize the ratio of dead weight to payload.

Materials handling methods

The basic methodologies of materials handling systems are manual systems, mechanized systems, and automated systems.

Manual systems.   Manual methods are appropriate in a situation in which there is a large variety of types of items, predominantly in small packages which permit manual handling, and there is no cost-benefit rationale for mechanized or automated handling systems.

Mechanized systems.   Mechanized systems are appropriate for activities involving larger shipments for which equipment such as forklifts and overhead cranes would have a high degree of utility. Mechanized systems offer optimum efficiency through their ability to accommodate palletized and containerized shipments.

Automated systems.   Automated materials handling systems are used mostly in privately owned facilities, where the system is dedicated to specific cargo handling requirements and where the functions of the system can be preprogrammed. They are appropriate for materials handling requirements which are repetitive, are frequently occurring, have a high volume throughput rate, and typically involve complete containers.

Combination systems.   Given the potential for achieving cost-effectiveness, some facilities with diverse types of cargo throughput might find it advantageous to employ a combination of manual, mechanized, and automated systems. The type of system would, in such cases, be determined by the needs of individual departments and the cost benefits afforded by using a specific type of system to meet these departmental needs.

Principles of pallet utility

Pallets are used in the transportation industry to facilitate the economic and efficient storage, staging, and movement of cargo. Pallets are more cost-effective where the cargo density per cubic measurement is maximized through the use of bulk packaging cartons which dimensionally conform to one of the four basic stacking configurations (block, row, brick, or pinwheel) shown in Fig. 17.2. There is a variety of pallet sizes available for consideration by the logistics planner in developing loading schemes. The most common pallet size is 40 in × 48 in; other commonly used pallet sizes are 48 in × 48 in, 40 in × 32 in, and 36 in × 48 in. The appropriate pallet size to use in moving products is governed by the specification of the cargo bed of the carriers' vehicles.

Figure 17.3 depicts a typical pallet which accommodates both forklift vehicles and hand pallet trucks. The forklift trucks can capture the pallet load from any of the four sides of the pallet; the hand trucks have access on the two length sides.

The variety of pallet patterns derived from the four basic configurations can accommodate a multiplicity of bulk carton designs. Figure 17.4 illustrates an array of possible pallet patterns that can be developed from the basic block, row, brick, and pinwheel configurations.

Economic utility of materials handling systems

The economic advantage of a materials handling system is manifested by its contributions to the logistics service system, such as

1. Increased efficiency of materials flow
2. Improved facilities utilization
3. Improved safety and working conditions
4. Facilitation of logistics service functions

A well-planned and well-controlled materials flow pattern ensures that the required items are in the right place at the right time to support operations of the facility.

As with any logistics service function, the level of efficacy of a materials handling system is enhanced by effective integration of the equipment into the design of the total logistics system. Materials should be organized and staged according to building location, bays within the building, and designated stack bins so as to afford ease of retrieval and conveyance to the designated reception points. Computerized techniques, such as use of bar codes for item identification and preprogrammed flow patterns, can help ensure accuracy in material dispatching and internal routing of the items to the appropriate destinations.

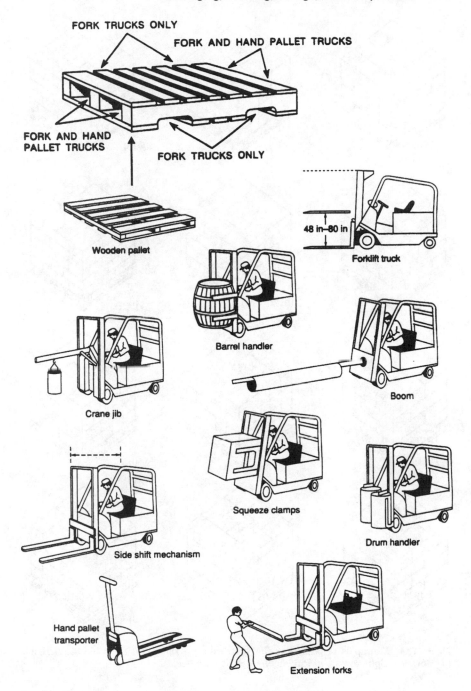

FORK TRUCKS ONLY

FORK AND HAND PALLET TRUCKS

FORK AND HAND PALLET TRUCKS

FORK TRUCKS ONLY

Wooden pallet

48 in–80 in

Forklift truck

Barrel handler

Crane jib

Boom

Side shift mechanism

Squeeze clamps

Drum handler

Hand pallet transporter

Extension forks

Figure 17.3   Typical uses of materials handling equipment.

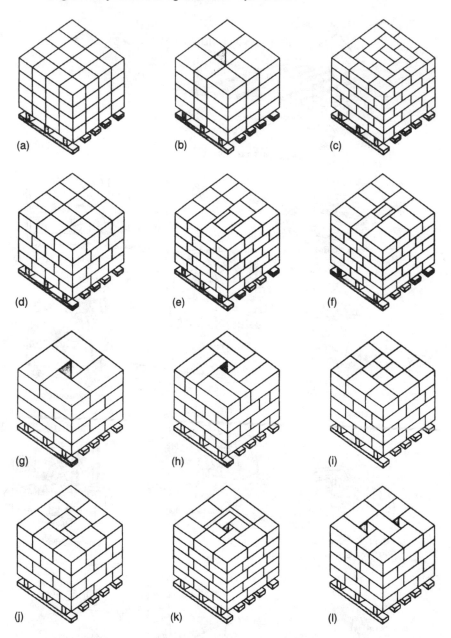

Figure 17.4 Examples of possible pallet patterns. *(a)* Block. *(b)* Split block. *(c)* Brick. *(d)* Row. *(e,f)* Split row. *(g–k)* Pinwheel. *(l)* Split pinwheel.

Warehousing and Storage

Warehouse storage facilities serve as staging areas for working materials to support production lines; storage points for finished products, replacement components, and repair parts to replenish the stocks at product service centers, distribution centers, and retail outlets; and inventory repositories to accept and assume custody of outputs of the production plants. A warehouse and storage activity is described in terms of its legal status, type of ownership, physical functions, types of commodities stored and handled, and functional role within the logistics system.

Legal status of warehouses

Commercial storage facilities are legally defined as either *public warehouses* or *private warehouses*.

Public warehouses. Commercial facilities defined as public warehouses are broadly regulated by the Uniform Commercial Code. Unlike the bodies of regulations and statutes promulgated by Federal agencies and congressional legislation, the Uniform Commercial Code essentially defines the legal relationships of the public warehouses with their customers and provides specifications on contracting procedures. It does not provide for federal economic regulation of warehousing like that for common carriers in the transportation industry. However, while not subject to federal economic regulation, public warehouses are under federal jurisdiction with respect to laws, regulations, and guidelines applicable to commercial organizations on matters concerning the environment, occupational safety, labor laws, etc. Public warehouses are subject to laws and regulations that may be imposed by state and local governments with jurisdiction over the community in which the public facility is located. Like common carriers, they are obliged to provide to the public at large an array of services that complies with their charters granted by the local jurisdictions and in accordance with their publicly stated rate structures and fee schedules. As with all enterprises, pricing schedules may be tempered by the degree of competition within the local market zones.

Private warehouses. This legal designation describes commercial storage facilities that are either owned or leased by the user organizations solely for support of their own logistics requirements. Private warehouses, in keeping with the regulatory obligations of their corporate operator, are subject to the state, local, and municipal laws, ordinances, and regulations imposed by their host communities.

Special legalities regarding consignment of goods. In addition to the warehouse's legal status as either a public or private commercial storage facility, there is another aspect of legality relevant to the use of the facility for consignment of certain goods. This concerns use of storage facilities as *field warehouses* or *bonded warehouses*.

Field warehouses. Field warehouses provide property security and are responsible and liable for the goods consigned to them by the owners of the consigned goods. Custody and possession are generally documented by *warehouse receipts,* which authenticate the description, quantity, and location of the consigned goods. A warehouse receipt provides evidence of ownership and is a negotiable instrument frequently used by brokers and transaction agents to authenticate the purchase, sale, or other transfer of ownership of the receipted goods, frequently with neither the purchasers nor the sellers actually seeing the goods.

Bonded warehouses. Bonded storage implies a covenant of trust with a governmental agency for custody of and accountability for the consigned goods. Bonded storage facilities may be either public warehouses or private warehouses. They are used primarily for interim custody of goods subject to taxes or duties until their release for market distribution.

As well as enjoying unique legal status, field warehouses and bonded warehouses provide capabilities that are essential to the commercial and governmental fiscal infrastructure. Both types of storage areas are frequently located at the same facility.

Warehousing operations

Warehousing operations are characterized by their roles in the logistics system, the commodities handled, and the relative attributes of their capabilities and services provided as public or private storage facilities.

Role in the logistics system. The role of a storage facility is typically defined by one of the following categories of service:

1. Interim custody and protection of goods pending their reconsignment, secondary destination shipment instructions, or release to the owners
2. Materials handling, which includes the functions of (1) assembly of inbound goods to be consolidated into specified cargo modules for reshipment to designated terminal points and (2) mixing of carload (CL) and truckload (TL) shipments into customized cargo lots for redistribution to one or more secondary destinations

3. A combination of the storage and materials handling roles

4. Special storage, such as cold or frozen storage, hermetically sealed storage for such sensitive items as medical and pharmaceutical items, and unique bulk storage facilities, such as tank farms for petroleum products

Commodities handled.   Warehousing may be classified, according to the types of cargo handled, as one or more of the following:

1. *Commodity warehouses.* Only certain commodities are stored and handled. Tobacco, lumber, grain, and cotton products are examples of such commodities.

2. *Bulk storage warehouses.* Services are restricted to storing, handling, mixing, and breaking bulk of bulk products such as liquid chemicals, petroleum, and syrups.

3. *Cold-storage warehouses.* Controlled, low-temperature facilities for the storage and handling of perishable products, chemicals, and drugs are employed for preservation purposes.

4. *Household goods warehouses.* Specializing in the storage and handling of household furnishings and furniture, these warehouses are utilized by the household goods carrier industry for both temporary and longterm storage.

5. *General merchandise warehouses.* These facilities are used for the storage of merchandise which does not require the specialized services of the previous types of warehouses. These warehouses handle all types of mercantile items.

Although the above roles are more typical of public warehouses, private warehouses may provide the same capabilities, as required by the owner or lessor of the facility.

Public vis-à-vis private warehousing.   The logistics planner is frequently obliged to determine which type of ownership would best serve the projected needs of the product storage and distribution system. There are conditions favoring each alternative.

Public warehouses are favored in the following situations:

1. Irregular and seasonal demand patterns prevail.

2. Markets are widely dispersed and frequently changing.

3. Inbound and outbound movements utilize several transport modes.

4. Warehousing functions are performed by specialized professionals.

5. Investment in facilities is negligible.

6. National warehouse coverage is attainable.

The commercial organization would best be served by buying or leasing a dedicated storage facility under the following conditions:

1. Direct controls are needed.

2. Specialized facilities, equipment, and layout are required.

3. There is significant product throughput, accompanied by high occupancy.

4. Demand is relatively stable throughout the year.

5. Markets are highly concentrated.

6. Products require some assembling, mixing, or processing prior to delivery to customers.

It might be determined that a combination of both public and private storage facilities within a logistics network would provide the most cost-effective support capability.

Warehouse design

The parameters and specifications for design, construction, or acquisition of a storage facility are generally defined by the purpose of the facility, the required layout, storage and aisle areas, planned stock locations, stock replenishment patterns, and compatibility with planned material handling systems.

The purpose of the facility is its projected role as a routine storage, special storage, assembly, or distribution point.

Layout is determined by how the inbound and outbound cargo flow will interact with the internal facility processing points, workstations, and staging and storage areas.

Storage and aisle areas are governed by and designed to accommodate pallet placement points and required aisle widths.

Stock location must be compatible with the materials picking and item retrieval schemes.

Stock replenishment is related to and influences stock location and facility layout.

Compatibility with materials handling is achieved by initially establishing the facility layout, storage and aisle areas, stock location points, and stockage replenishment patterns preliminary to installation of materials handling equipment.

Economics of warehousing

The location of a commercial storage facility falls into one of three geographic categories:

1. A production-oriented site is located near the production facility; it is used when additional processing of the material is required.
2. A market-oriented site is located near the market centers; it is used when responsiveness and service to the customers are the predominant priorities.
3. An intermediate location may be chosen when neither production nor market considerations prevail, when there is geographic dispersion of production points, when there is broad diffusion of markets, and when the corporate goal is solely to minimize overall distribution costs.

The site location decision is, therefore, based on the application of functional warehouse cost standards to determine the optimum size or volume of each warehouse and the location of the storage sites relative to the production or processing points and the market centers.

Functional warehouse cost standards.   Typical cost criteria include

- Storage costs per square foot (0.093 sq. metre)
- Storage costs per hundredweight
- Storage costs per case or pallet
- Handling costs per hour
- Handling costs per hundredweight
- Handling costs per case or pallet
- Clerical costs per document
- Clerical costs per line item
- Clerical costs per hour
- Cartage costs per vehicle mile (1.6215 km)
- Cartage costs per hundredweight-mile (1.6215 km)
- Cartage costs per vehicle-hour
- Cartage costs per shipment

These factors are integrated with the pertinent utility factors to provide the governing cost-estimating relationships for analyses of the alternatives.

Warehouse volume.   The volume, or workload capacity, of a warehouse directly affects warehousing costs per unit stored. Unit cost varies with method of warehousing, throughout volume, and inventory turnover rate. These variables are affected by the consistency of the volume of activity, the physical size of the inventory, the number of stock locations, the size of the area being served by the warehouse, and inventory issue policy.

In determining the total impact of warehouse volume on logistics distribution costs, effects on *delivery costs* as well as the specific impact of *warehousing costs* must be considered. An increase in volume implies expansion of the service area, which would produce an increase in delivery or local transportation costs as an offset to the reduction in unit storage costs. The logistics analysis must achieve an optimum trade-off which addresses both delivery and warehousing costs. Figure 17.5 provides a graphic portrayal of this cost tradeoff concept.

The total cost of supplying an area from a given location can be determined using the following equation:

$$C = a + \frac{b}{V} + c\sqrt{A}$$

where C = cost (within the warehouse district) per dollar's worth of goods distributed, the measure of effectiveness

V = volume of goods, in dollars, handled by the warehouse per unit of time

A = area in square miles (sq. kilos) served by the warehouse

a = cost per dollar's worth of goods distributed independent of either the warehouse's volume handled or the area served

b = fixed costs for the warehouse per unit of time, which, when divided by the volume, will yield the appropriate cost per dollar's worth distributed

c = the cost of the distribution, which varies with the square root of the area; that is, costs associated with miles (kilometres) covered within the warehouse district, such as gasoline, truck repairs, driver hours, and others

As shown in Fig. 17.5, the optimum capacity of the warehouse is at the point on the total cost function curve at which the slope, or rate of change, is zero. At this point the slope makes a transition from a negative rate of change (cost rate reduction) to a positive rate (cost rate increase).

Warehouse dispersion.   In addition to these volume cost considerations, the number of warehouses and their geographic dispersion have trade-

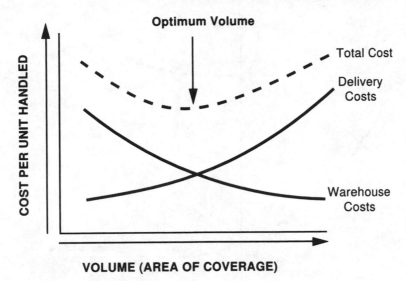

Figure 17.5   Optimum warehouse area coverage.

off effects on the costs and efficiency of the total logistics system. The more prominent effects are evidenced in the relationship of delivery times to the costs for delivery to the customers. Typically, the advantages afforded by an increase in responsiveness to customers are countered by an increase in distribution costs. Figure 17.6 illustrates this relationship.

The major system cost categories are inventory costs, transportation costs, and order processing (customer communication) costs. These costs are marginally sensitive to alternatives ranging from a network of a

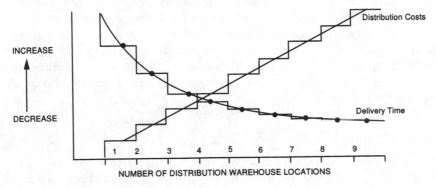

Figure 17.6   Relationship of inventory location and delivery time.

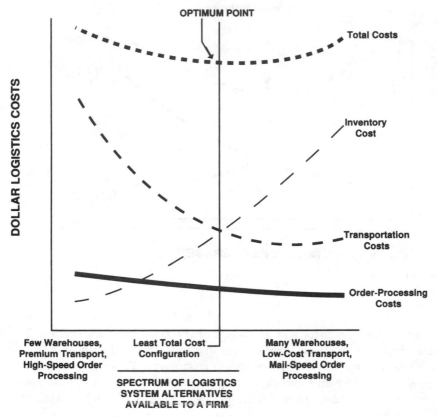

Figure 17.7   Total cost concept trade-offs.

few warehouses to many warehouses. Figure 17.7 portrays this inter-relationship.

For this example, it is apparent that, with an increase in number of warehouses, the cumulative costs of warehouse inventories increase, the costs of transportation to the market centers decrease, because the dispersed locations are in closer proximity to these centers, and the costs of communication with customers decrease minimally. In keeping with economic discipline, the analyst would probably recommend the number of warehouses that produces the optimum (zero slope) point on the total cost function curve.

Transportation

From the perspective of a production facility, transportation is critical to the distribution of material from the sources to the production sites and from these sites to and within the market centers.

To condense the magnitude and complexity of the transportation industry for tutorial purposes, this chapter will provide highlights on its legal forms, operational modes, auxiliary users, transfer facilities, intermodal movement, regulatory and rate structures, and economic considerations.

Legal forms of transportation

The legal forms of carriers in the transportation industry are *common carriers*, *contract carriers*, *exempt carriers*, and *private carriers*. Figure 17.8 portrays the legal structure in terms of public and private carriers and their subsidiary functions. Each legal category has a unique role within the logistics system.

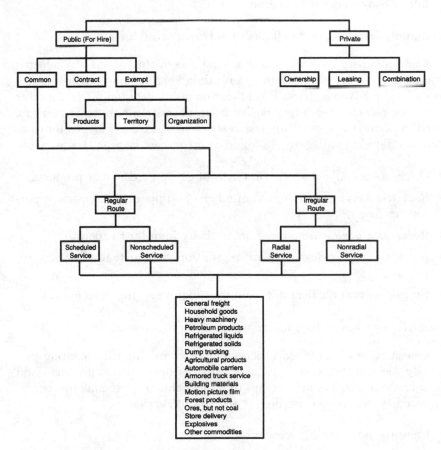

Figure 17.8 Transportation legal and regulatory structure.

Public (for hire) carriers.

Common carriers.   Carriers in this legal classification are available to all users at published rates. All tariffs are approved by the cognizant regulatory agencies. When so legally classified, a common carrier

- Agrees to serve all who apply without undue preference, prejudice, or discrimination.
- Provides adequate facilities and services at reasonable rates.
- Delivers goods entrusted to its care with reasonable dispatch.
- Publishes rates and charges.
- Accepts liability for goods tendered to it.
- Assumes fairly intensive economic and safety regulations.
- Requires a certificate of public convenience and necessity from a regulatory body in order to operate.

Common carriers operate all modes of transportation.

Contract carriers.   Carriers in this legal classification perform selected transportation functions. Rate differentials for the same type of service are allowed; however, these rates must be published and made a matter of public record. Regulatory bodies issue permits for contract carriers, but the permits are generally less restrictive than those issued for common carriers. When legally classified as a contract carrier, the carrier

- Offers a specialized service in terms of equipment and/or products.
- Restricts service to a limited number of shippers (one to seven persons or firms).
- Receives a permit from a regulatory body in order to operate.
- Assumes only modest economic regulation, but fairly intensive safety regulation.
- Enjoys geographic flexibility and latitude in serving customers.

Contract carriers include motor, water, and air carriers.

Exempt carriers.   Transportation companies in this classification primarily move unprocessed products, such as agricultural products and fish. Exempt carriers are exempt from economic restrictions by regulatory bodies. A carrier classified as an exempt carrier

- Requires no economic regulation.
- Can enter or exit the market freely.

- Is so classified because of the products carried, territory served, or organization formed.

- Negotiates its own rates and can discriminate.

- Assumes no obligations to provide regular service or to provide a particular quality of service.

- Assumes fairly intensive safety regulation.

Exempt carriers generally include motor and water carriers.

Private carriers.   Carriers in this classification are operated by the producer or distributor of the cargo. A private carrier is not legally for hire by outside organizations. A carrier classified as a private carrier

- Provides transportation service only for the firm; the carrier and the owner of the goods are identical.

- Requires no economic regulation.

- Assumes fairly intensive safety regulation.

- Transports only goods that are incidental to the primary business of the firm.

- Enjoys geographical flexibility and latitude in serving customers; it can go anywhere.

Private carriage involves ownership, leasing, or a combination thereof, and can include all modes of transportation.

Public carriers vis-à-vis private carriers.   There are differences in the corporate advantages offered by public carriage and by private carriage. These must be considered by the logistics planner on a situational basis.
    Public (for hire) carriers offer advantages such as

- Expert, professional service
- Extensive national and international geographic coverage
- Well-defined duties and responsibilities
- Shipper protection through economic and safety regulations
- Assumption by the carrier of operating and transportation management problems
- The required investment being made by the carriers, not the shippers
- Flexibility to use other modes and other carriers

- Transportation charges which can be competitively economical
- Extended services through intramodal and intermodal coordination
- Dedicated equipment for particular shippers

Private carriage through corporate ownership or leasing can provide user advantages such as

- Improved scheduling and delivery
- Greater control over service and products
- Flexibility in operations and territory
- Reduced loss and damage claims because goods are not mixed with those of other shippers
- Transportation rates which can be more cost-effective in light of the owner's situation
- Specialized equipment and service
- Improved customer relations—greater interest in customers by drivers
- Use of equipment for advertising purposes—rolling billboards
- Coordination of inbound and outbound shipments
- Less expensive packaging and reduced inventories through improved transit times

Operational modes of transportation

The U.S. transportation system includes five major operational modes: rail, highway, water, air, and pipeline. Each of these five modes has varying attributes with respect to governing legal forms, cost structures, types of ownership, and service features. Table 17.1 shows the array of distinguishing capabilities characteristic of each mode. The following highlights the key attributes:

TABLE 17.1 Transport Modes

Operating characteristic	Individual rankings within operating characteristic				
	Transportation mode				
	Rail	Highway	Water	Pipeline	Air
Dependability	3	2	4	1	5
Speed	3	2	4	5	1
Frequency	3	1	4	5	2
Availability	2	1	4	5	3
Capability	2	3	1	5	4
Economic efficiency*	3	4	2	1	5
Energy efficiency†	3	4	2	1	5

TABLE 17.1 *(continued)*

Operating characteristic	Ranking of modes according to operating characteristic				
	Highest 1	2	3	4	Lowest 5
Dependability	Pipeline	Highway	Rail	Water	Air
Speed	Air	Highway	Rail	Water	Pipeline
Frequency	Highway	Air	Rail	Water	Pipeline
Availability	Highway	Rail	Air	Water	Pipeline
Capability	Water	Rail	Highway	Air	Pipeline
Economic efficiency*	Pipeline	Water	Rail	Highway	Air
Energy efficiency†	Pipeline	Water	Rail	Highway	Air

*Based on cost per ton-mile.
†Measured in BTUs per net ton-mile.

Relative Characteristics of Operating Modes

Mode	Legal Forms	Service Features
Rail	Almost exclusively common	Mass movements, long distances, low unit cost High capabilities Extensive coverage and large number of accessorial services
Highway	Common Contract Exempt Private	Accommodates all types of goods Door-to-door service Extensive geographical coverage Speed, dependability, and frequency
Water	Common Contract Exempt Private	Mass movement of bulk commodities at the lowest cost High capabilities Advanced technology for loading and unloading
Pipeline	Almost exclusively common	Mass movement of liquids and gaseous materials at very low cost High capabilities Greatest dependability Pricing flexibility
Air	Common Contract Private	Premium service Reduced packaging and handling requirements Reduced costs for other logistics components

In terms of system utilization, the following approximates the apportionment of the total cargo movement within the U.S. traffic system.

Distribution of Cargo Traffic by Individual Modes
(Estimates based on analysis of currently available data)

Mode	Utilization
Rail	38%
Water	25%
Highway	15%
Pipeline	21%
Air	1%
	100%

It is evident that the bulk cargo carriage capacity of water and rail, which respectively rank first and second in the operating characteristics of capability, together account for almost two-thirds of the cargo moved within the United States.

Inherent advantages of individual modes.   The logistics manager and the traffic manager must assess the relative advantages of the various transportation modes, given the specific transportation needs and the modal operating characteristics. The following sections highlight the inherent advantages of each mode.

Rail.   The primary advantages of rail transportation are the following:

- Mass movement of goods—large capabilities
- Low unit cost of movement
- Dependable form of transport
- Long-haul movements
- Fairly extensive rail system network—coverage to major markets and suppliers
- Numerous ancillary services—switching, in-transit privileges, storage, etc.
- Goods transfer to other carriers
- Specialized equipment

Highway.   The trucking industry is the monarch of the highway transportation mode and is a healthy contributor to the national economy. The highway mode offers cargo customers

- Flexibility—can go anywhere
- Speed—3–5 days delivery to any point in the continental United States
- Frequency—hourly and daily pickup and delivery service

■ Convenience—loading and unloading at the shipper's and receiver's places of business

■ Goods transfer to other carriers

■ Equipment diversity

Water. Water transportation is relatively slow, but large cargoes of materials can be carried at economic rates. Advantages of this mode of movement include

■ Mass movement of bulk commodities—large capabilities

■ Very low unit cost

■ Movement of low-unit-value commodities, such as sand, gravel, or shell, which otherwise would have limited distribution

■ Long-haul movements

Pipeline. This mode is unsurpassed in dependability, inasmuch as it is rarely affected by the climatological phenomena and other schedule hindrances that can affect the other modes. Advantages offered by the pipeline system include

■ Mass movement of liquid or gas products

■ Lowest unit cost of movement

■ Large capacity and volume of throughput

■ Most dependable of all the modes

■ Long-haul movements

Air. Air cargo is a premium transport service used when speed is paramount and cost is a secondary consideration. The advantages of air cargo movement are

■ Highest movement speed

■ Overnight service to any point in the continental United States

■ Frequent service to major cities

■ Increasing capabilities—up to 200,000 lb (90,718 kgs) in a single aircraft

Auxiliary users. This category of transportation organizations exists to provide companies with economic advantages and the most responsive services that meet their specific needs from the various elements and modes of the transportation system.

Major auxiliary users include United Parcel Service, Federal Express, DHL, the U.S. Postal Service (parcel post and Express Mail), freight forwarders, shippers' associations, and freight brokers. Some third-party providers of logistics services may also be regarded as auxiliary users. Auxiliary users do not own or operate the equipment used for the line-haul movement of their loads; instead, they emphasize the collection of freight at its origin and distribution at its destination. While many cargo movements may involve more than one carrier, auxiliary users assume complete responsibility for the movement of a shipper's freight regardless of the number of carriers or modes involved.

Freight forwarders provide a major service to their customers by consolidating small shipments into larger ones for long-distance movement. While providing lower line-haul rates and faster service for shippers, they retain a portion of the differential between vehicle-load and less-than-vehicle-load rates to defray the expenses of their operations and to earn a profit.

Shippers' associations, while similar in function to freight forwarders, are actually voluntary organizations composed of members with mutual commercial interests who use the service to take advantage of the economies of consolidation. These associations' members are usually firms shipping similar items between common origins and destinations.

Third-party providers of logistics services exhibit some of the characteristics of auxiliary users. They provide an increasing variety of services to their client firms (usually manufacturers), including warehousing, order processing, data transmission, and shipping (including the purchase of transportation).

Transfer facilities

This type of activity has been regarded by some as an element of materials handling. For tutorial purposes, transfer facilities will, by virtue of their role in the logistics system, be addressed here as an adjunct to the transportation modes.

Transfer facilities can be best characterized by the nature of the freight or items they are designed to accommodate. They include terminals for assembling, sorting, transferring, or distributing packaged freight; terminals for handling and transferring containers used in coordinated transportation services; specialized bulk commodity terminals dedicated to particular commodities ranging from grain to oil; and carrier yards, terminals, and ports for combining or separating carrier equipment.

Packaged freight terminals. Carriers that transport small shipments, such as those engaged in highway or air operations, may utilize terminals to consolidate, mix, or deconsolidate packaged freight shipments.

Small-shipment or package sorting facilities typically employ mechanical devices and computer-controlled sorting mechanisms. Their automated sorting equipments is normally located at centralized hub points on their route systems so that they can gain maximum productivity from their investments.

Container terminals.   The growth of coordinated transport, or the use of two or more modes in the transportation of a single shipment, has led to the use of large containers that are compatible with two or more modes of carriage. The transfer of a 40 ft × 8 ft × 8.5 ft (26,580 kgs payload potential) container from one mode to another typically requires the use of a crane capable of lifting up to 38,480 kgs. The larger terminals located in the United States and other international ports are capable of transferring large containers from trucks to ships to trucks, etc.

Bulk commodity terminals.   Transportation, storage, and handling of commodities in unpackaged form (e.g., grain, coal, petroleum) has become a sizable economic activity. It requires facilities for assembling, transferring, and breaking bulk commodities in great quantities at high speeds. Such facilities often have the ability to transfer such commodities between vehicles of two cooperative transportation modes. As there are no stipulated restrictions on forms of ownership, bulk commodity terminals may be owned and operated by carriers, shippers, or public organizations, such as port authorities.[1]

Carrier equipment transfer facilities.   Rail, highway, and water carriers operate terminals for the transfer of cars, trailers, and barges, respectively. The development of the Interstate Highway System and the improvement of the tributary roads have enabled motor carriers to haul heavier loads and more trailers per powered vehicle between terminals. It is at the carrier equipment transfer terminals that such trailer combinations are assembled and disassembled for handling over more restricted highways.

Intermodal transportation

Intermodal transportation systems are joint, point-to-point, and through transportation services which involve two or more movement modes.

---

[1]Based on a 1999 LATTS (Latin American Trade & Transportation Study) data involving 42 U.S. coastal and inland ports, estimated U.S. throughout requirements for break-bulk, neo-bulk, dry bulk, and liquid bulk cargo totals 620 million short tons (562 million tonnes) annually {17 million short tons (15.4 million tonnes) per day}. The individual port throughout averages over 400,000 short tons (362,880 tonnes) daily.

The coordination of transport systems so as to optimize available capabilities by integrating modes

- Combines the inherent advantages of several modes.
- Offers cost economies.
- Reduces loss and damage claims through containerization.
- Provides increased flexibility through service extension to more shippers and origin/destination points.
- Affords increased efficiency through reduced handling and storage.

The primary intermodal combinations offering service on a regular basis are truck-rail, truck-water, air-truck, and rail-water. Containerization is an important consideration in these coordinated systems.

Truck-rail.   Truck-rail service may be trailer-on-flatcar (TOFC) or container-on-flatcar (COFC).

Trailer-on-flatcar (TOFC). This combination uses the popular "piggyback" arrangement, in which the conventional highway trailers are placed on rail flatcars.

Container-on-flatcar (COFC). In this method, intermodal containers are detached from the highway trailer chassis and placed on rail flatcars.

Truck-water.   This combination involves intermodal truck and water service, sometimes referred to as "fishyback," which is accomplished by coordination of truck and water transport movements. The roll-on, roll-off (Ro-Ro) method allows standard highway trailers and railcars to be driven directly on and off specially adapted ships via large side or stern doors.

Air-truck.   Inaugurated in 1957 by a single air carrier in cooperation with a number of trucking firms, this service is now widely utilized. It provides feeder and delivery service between major airport hubs and remote communities deprived of adequate air freight service. This service frequently utilizes containers which conform to the upper or lower configuration of the aircraft fuselage and which may be preloaded and unloaded by shippers and consignees at their facilities.

Rail-water.   This type of service has been in operation for many years. The "hydro-train" rail-water service uses specially constructed ships or barges on which strings of freight cars are moved over water to the port nearest their ultimate inland destination. In international rail-water services, the rail portion is often regarded as a "bridge" connected to one or two water movements. The term *land bridge* refers to an overland movement that is preceded and followed by water movement.

Other intermodal characteristics. Ship-barge combinations, known as barge-on-board, are most frequently used in the transhipment of bulk cargoes (e.g., coal, grain, wine, petroleum, and lumber) when the ship cannot be navigated to the origin or destination of the cargo. Barges provide feeder service for container shipments between inland waterway points and major ocean terminals. A variation on this alternative is the lighter-aboard-ship (LASH) system, in which specially constructed lighters (small barges that may be self-propelled or towed) are floated or lifted on and off specially built vessels for movement into ports unable to accommodate the "mother" ship or to a location where such an operation will save the time required to dock and load or unload.

Another intermodal application is described as container-on-airplane, sometimes termed "birdy back." This combination employs aircraft containers, called igloos, which are placed on highway trailers.

Modal combinations involving pipelines typically require bulk storage in transit at a terminal and therefore are not considered intermodal transportation.

Total modal structure. Figure 17.9 portrays the infrastructural relationship of the major modes to their corresponding legal forms, auxiliary user systems, and intermodal coordination.

Transportation regulation

Rather than examining in detail the labyrinth of bureaucratic intricacies affecting transportation regulation, this section will address those aspects that are necessary to provide an initial understanding of how they relate to the logistics system.

Transportation regulatory agencies and their jurisdictions. The major federal state and local regulatory agencies are described in this section.

Federal agencies. Federal agencies are concerned with *interstate commerce*. The major organizations are described in the following.

The Surface Transportation Board

Overview

A singularly significant development during the mid 1990s was the termination of the Interstate Commerce Commission (ICC) and the establishment of the Surface Transportation Board (STB) as the successor to the ICC.

The Surface Transportation Board was established on January 1, 1996, as a decisionally independent, bipartisan, adjudicatory body organiza-

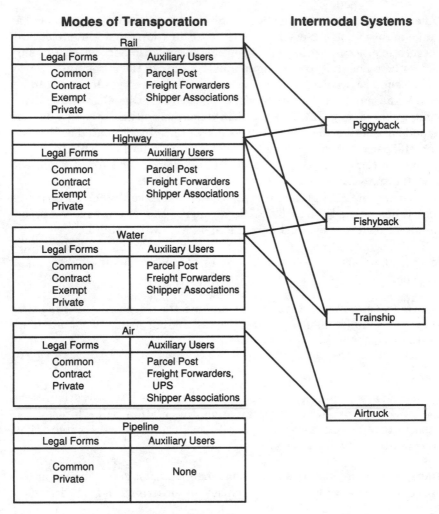

Figure 17.9 Total modal structure: The relationship of modes, legal forms, auxiliary users, and principal intermodal systems of transportation.

tionally housed within the U.S. Department of Transportation (DOT), with jurisdiction over certain surface transportation economic regulatory matters. The new organization was created by a December 29, 1995, Act of Congress (49 U.S.C. 10101, *et seq,* known as the ICC Termination Act of 1995 (ICCTA). The ICCTA terminated the Interstate Commerce Commission effective December 31, 1995; eliminated various functions previously performed by the ICC; and transferred licensing and certain motor carrier functions to the Federal Highway Administration (now handled by the Federal Motor Carrier Safety Administration, FMCSA) or the Bureau of Transportation Statistics (BTS) within the DOT. All other rail and non-rail functions were transferred to the Surface Transportation Board. This legislation represented a further step in the process of streamlining and reforming the Federal economic regulatory oversight of the railroad, trucking, and bus industries that was initiated in the late 1970s and early 1980s.

Table 17.2 lists the former ICC functions retained by the Surface Transportation Board. Table 17.3 lists the ICC functions eliminated by the ICCTA.

The mission of the STB is to adjudicate disputes and regulate interstate surface transportation. In this regard, the Board's general respon-

TABLE 17.2   Functions Retained by the Surface Transportation Board (STB)

<table>
<tr><td align="center">ICC Termination Act of 1995
Retained Functions</td></tr>
</table>

Rail Regulations *(all administered by the Surface Transportation Board)*

    Common Carrier Obligation
    Exemptions
    Rail Mergers
    Line Transfers and Trackage Rights
    Line Sales to Noncarriers
    Labor Protection
    Rate Regulation for Common Carriage, including:
        Public Disclosure of Rates and Service Terms
        Advance Notice of Rate Increases or Changes in Service Terms
        Maximum Rate Reasonableness for Captive Traffic
    Contracts for Transportation of Agricultural Products Requirements, including:
        Filing of Summaries
        Protest and Matching Rights
        Equipment Limitations
    Rail Cost Adjustment Factor (RCAF) Computation
    Reasonableness of Practices
    Rate Discrimination
    Car Supply and Interchange
    Emergency Service Orders
    Competitive Access
    Line Constructions, including Line Crossings
    Line Abandonment, including:
        Financial Assistance
        Rails-to-Trails
        Public Use Provision for Right of Way
    Feeder Line Development Program
    Collective Ratemaking (and Antitrust Immunity)
    Interlocking Officers and Directors
    Recording Liens
    Data Collection and Oversight

(continued)

TABLE 17.2 *(continued)*

Motor Carrier Regulation

Common Carrier Obligation (together with freedom to provide Contract Service)

Exemptions *(by STB and Secretary, DOT, as applicable)*

Registration of Carriers—Trucking and Bus Companies *(Secretary, DOT)*

Insurance Requirements *(Secretary, DOT)*

Rate Reasonableness *(STB)* for:

    Residential Household Goods Moves

    Joint Motor-Water Rates in Noncontiguous Domestic Trade

    Collectively Set Rates

Collective Ratemaking (and Antitrust Immunity) *(STB)*

Pooling *(STB)*

Undercharges *(STB)*

Background Commercial Rules (except informal dispute resolution) *(Secretary DOT)*:

    Owner-Operator Leasing

    Lumping

    Loss and Damage Claims

    Duplicate Payments and Overcharges

Household Goods Operations, including:

    Tariffs (not files) *(STB)*

    Mandatory Arbitration *(Secretary, DOT)*

    Rules for Binding Estimates and Guaranteed Service *(Secretary, DOT)*

    Agent-Van Line Relations *(Secretary, DOT)* with Antitrust Immunity (STB)

Performance Standards *(Secretary, DOT)*

Preemption of Intrastate Regulation

Data Collection and Oversight *(Secretary, DOT)*

Freight Forwarders—Registration and Insurance *(Secretary, DOT)*

Brokers—Registration and Insurance *(Secretary, DOT)*

Bus Company Mergers *(STB)*

Bus Company Through-Route Requirements *(STB)*

Mexican Carriers *(Secretary, DOT)*, including:

    Registration and Insurance

    Enforcement of NAFTA Restrictions

Regulation of Water Carriage

Tariff Filing and Rate Reasonableness Requirements for Noncontiguous Domestic Trade *(STB)*

Regulation of Pipelines *(all administered by STB)*

Common Carrier Obligation

Rate Regulation, including:

    Public Disclosure of Rates and Service Terms

    Advance Notice of Rate Increases or Changes in Service Terms

    Reasonableness

Reasonableness of Practices

Rate Discrimination

Intermodal Regulation *(STB)*

Rail-Water Connections for Noncontiguous Domestic Trade

TABLE 17.3   Eliminated ICC Functions and Provisions

ICC Termination Act of 1995
Eliminated Functions and Provisions

Rail Regulations

Tariffs
Rate Investigation and Suspension Process
ZORF (Zone of Rate Freedom)
Minimum Rate Regulation
Contract Rate Regulation for Nonagricultural Commodities, including
    Filing Requirements
    Percentage Limitations on Equipment Commitments
    Complaint and Matching Rights
Joint Rate Surcharge and Cancellation Supervision
Recyclables Special Rate Standards
Labor Protection for:
    Class III Carriers
    Noncarrier Line Acquisitions
    Feeder Line Cases
Intermodal Ownership Restrictions
Commodities Clause Restrictions
Interlocking Officers and Directors Restrictions for Class III Carriers
State Certification and Regulatory Authority
Securities Issuances
Valuation
Passenger Train Discontinuances

Motor Carrier Regulation

Licensing
    Need-Based Criteria (except for certain subsidized traffic)
    Permanent Licenses (must now be reviewed)
Contract Carriage
    Limitations and Restrictions Removed
Control and Transfer Transactions for Trucking Companies
Collective Activity
    Permanent Approval (must now be renewed)
Tariff Filing (except for noncontiguous domestic trade)
Rate Regulation (except for residential household goods moves, noncontiguous
    domestic trade, and collectively set rates)
Undercharges (September 1990 cut-off for unreasonable practice provision)
Informal Dispute Resolution in the following areas:
    Owner-Operator Leasing
    Lumping
    Loss and Damage Claims
    Duplicate Payments and Overcharges
    Household Goods and Auto-Driveaway Carriers
Intercity Bus Route Discontinuances
Intrastate Preemption (Sunset Rate for Hawaii Intrastate Regulation)

Regulation of Domestic Water Carriage

All regulation, except for noncontiguous domestic trade (Residual jurisdiction
    retained for preemption purposes)

(continued)

TABLE 17.3   *(continued)*

Regulation of Pipelines
    Tariff Filing
    Rate Investigation and Suspension Process

Organizational
    Interstate Commerce Commission (ICC)
    Joint Boards
    Rail Services Planning Office
    Rail Public Counsel

sibilities include the oversight of firms engaged in transportation in interstate and foreign commerce to the extent that it takes place within the United States, or between or among points in the contiguous United states and points in Alaska, Hawaii, or U.S. territories or possessions. Surface transportation matters under STB jurisdiction generally include railroad rate and service issues, rail restructuring (mergers, line sales, line construction, and line abandonment) as well as labor matters related thereto; certain trucking company, moving van, and non-contiguous ocean shipping company rate matters; certain intercity passenger bus company structure, financial and operational matters; and certain pipeline matters not regulated by the Federal Energy Regulatory Commission. The STB is headed by Board Members appointed by the President and confirmed by the Senate. In the performance of its functions, the STB is charged with promoting, where appropriate, substantive and procedural regulatory reform in the economic regulation of surface transportation, and with providing an efficient and effective forum for the resolution of disputes, through the granting of exemptions from regulations where warranted, the streamlining of its decisional process and the regulations applied thereto, and the consistent and fair application of legal and equitable principles, the Board seeks to facilitate commerce by providing an effective forum for efficient dispute resolution and facilitation of appropriate market-based business transactions. The STB strives to develop, through rulemakings and case disposition, new and better ways to analyze unique and complex problems, to reach fully justified decisions more quickly, to reduce the costs associated with regulatory oversight and, where appropriate, to encourage private-sector negotiations and resolutions.

Prominent Aspects of STB Oversight Responsibility

Railroad rates

Railroads have a legal common carrier obligation to provide rail service upon reasonable request. They can provide that service under rates and service terms agreed to in a confidential contract with the shipper or under openly available common carriage rates and service terms. Railroads are required by the Code of Federal Regulations to file with the STB summaries of all contracts for transportation of agricultural products. A railroad's common carriage rates must be disclosed upon request (and published for agricultural products and fertilizers). Advance notice must be provided for increases in carriage rates or changes in service terms. If a railroad does not have a rate in place to move a shipper's traffic, it must promptly establish a rate and service terms upon the shipper's reasonable request. The STB maintains regulations governing the establishment, disclosure, publication, and notification requirements for common carriage rates.

Amtrak

The STB has limited but significant regulatory authority over the National Railroad Passenger Corporation (Amtrak). Specifically, the STB has authority to ensure that Amtrak may operate over the track of the nation's railroads, and to adjudicate disputes between Amtrak and the individual freight railroads concerning shared use of tracks and other facilities and to set the terms and conditions of such use if Amtrak and the freight railroad fail to reach a voluntary agreement. In 2004 Congress amended existing legislation to vest the STB with authority to direct service of commuter rail operations in the event of a cessation of service by Amtrak. The STB has coordinated with the Federal Railroad Administration, Amtrak, and the commuter and freight railroads to explore what would be involved should the need arise to exercise that authority. It is further noted that when a railroad cannot move an Amtrak train over its normal route and Amtrak requests movement over an alternative route of another carrier, the STB may issue an emergency rerouting order to permit uninterrupted passenger service.

Motor carriage

The STB has limited oversight responsibilities over motor carriage. The functions not specifically assigned to the STB are administered by the FMCSA or the BTS. The STB may approve agreements by motor carriers to participate in bureaus that can collectively set through-routes and joint rates, set rates for the transportation of household goods, establish uniform classifications and mileage guides, and engage in certain other collective activities. STB approval effectively confers immunity from the antitrust laws for these collective activities, but results in STB monitoring of the activities conducted under the approved agreements. Pursuant to legislative actions enacted in 1999, the STB must conduct a periodic review to determine whether the approvals for existing motor carrier bureau agreements should continue.

Water carriage

The STB has authority to regulate rates for transportation in the non-contiguous domestic trade, which consists of domestic transportation to or from Alaska, Hawaii, American Samoa, the Northern Mariana Islands, Guam, the Virgin Islands, and Puerto Rico. Carriers engaged in the noncontiguous domestic trade are required to file with the STB tariffs containing their rates and service terms for this category of transportation (except that tariffs are not required for transportation provided pursuant to contracts between carriers and shippers, or for services provided by freight forwarders). Carriers have the option of filing printed tariffs or filing their tariffs with the STB electronically. In either event, these tariffs are available in the STB offices for review by the general public.

Pipeline carriage

The STB regulates interstate transportation of commodities other than water, gas, and oil. Pipeline carriers must promptly disclose (in either written or electronic form) their rates and service terms upon request. Additionally, pipeline carriers must provide at least 20 days' notice before a rate increase or change in service terms may become effective. It is the responsibility of the STB to ensure that pipeline rates and practices are reasonable and nondiscriminatory.

Federal Maritime Commission

The Federal Maritime Commission (FMC) was established as an independent regulatory agency on August 12, 1961. The FMC is charged with administration of the regulatory provisions of the shipping laws. The FMC is also responsible for regulation of ocean-borne transportation in the foreign commerce of the United States. Specifically the FMC executes the following significant roles in its mission:

- Monitoring activities of ocean common carriers, marine terminal operators, conferences, ports, ocean transportation intermediaries (OTIs) who operate in the U.S. foreign commerce to ensure they maintain just and reasonable practices, and providing oversight of the financial responsibility of passenger vessel operators.

- Maintaining a trade monitoring and enforcement program designed to assist regulated entities in achieving compliance and to detect and appropriately remedy malpractices and violations, as stipulated by the Shipping Act.

- Monitoring the laws and practices of foreign governments which could have a discriminatory or otherwise adverse impact on shipping conditions in the United States.

- Enforcing special regulatory requirements applicable to ocean common carriers owned or controlled by foreign governments.

- Processing and reviewing agreements and service contracts.

- Reviewing common carriers' privately published tariff systems for accessibility and accuracy.

- Issuing licenses to qualified OTIs in the United States and ensure all maintain evidence of financial responsibility.

- Ensuring that passenger vessel operators demonstrate adequate financial responsibility for casualty and non-performance.

Federal Transportation Safety Regulators

Collectively, the U.S. Department of Transportation (created by Congress in 1966), the National Transportation Safety Board, the Federal Aviation Administration, the U.S. Coast Guard, and the Federal Railroad Administration provide safety regulation of the air and surface transportation carriers respectively assigned to their jurisdictions.

State agencies.   State agencies are empowered to regulate *intrastate-commerce*. State governments set up such organizations as railroad commissions and public utilities commissions to exercise economic and safety control over air and surface carriers, as well as other related transportation operations. The Department of Transportation or equivalent agency within each state is generally responsible for planning and operation of transport facilities, except where preempted by federal law.

Local agencies.   Local agencies are concerned with regulation of *local commerce*. Authorities such as city councils, county commissioners, city managers, and municipal transportation authorities may exercise economic control over such enterprises as local taxicabs; limousine services; and bus, subway and streetcar operations; they are also responsible for municipal transportation planning.

Transportation regulatory functions.   Transportation regulatory agencies are primarily concerned with oversight and appropriate jurisdictional administration of the economics and safety of their assigned sectors of the transportation industry.

Miscellaneous rate modifications.   Deregulation has brought about increased use of a system of modified rates, especially with respect to shipper-carrier contracts which provide a full range of carrier services for a particular shipment.

1. *Value rates.* The general rule of carrier liability provides that the carrier is liable for the value of goods lost or damaged while in its cus-

tody. Value may be specified in numerous ways. Released value rates are based on carriers assuming a certain fixed liability, usually stated in value per pound, for goods transported by them. This liability ordinarily is substantially less than the actual value of the goods and results in a lower rate.

2. *Guaranteed schedule rates.* Carriers and shippers may agree on a guaranteed schedule for the movement of a shipment of commodities such as livestock and perishables. If the carrier fails to meet the schedule and a loss is incurred, the carrier may be held liable. A carrier is normally responsible only for exercising "reasonable dispatch," which permits greater latitude than a guaranteed delivery schedule.

3. *Space-available rates.* Shippers who are not affected by delays in the delivery of their freight may take advantage of lower space-available rates. These rates are used frequently by air and water carriers because they provide a way to achieve higher capacity utilization after regular cargos have been loaded.

4. *Measurement rates.* Rates that reflect weight-density ratios are used in ocean transportation, in airfreight, and on certain bulk rail shipments. Ocean rates commonly are quoted at a cost per long ton (2240 lb) or per 40 ft^3 (1.1327 cu. metres), whichever produces the higher charge. The loading of ships requires careful attention to weight and balance. If the cargo weight-density ratio is low, maintaining the ship's stability will require adding nonrevenue ballast (usually seawater). Generally, heavy cargo is not a problem if it is properly stowed at the appropriate deck level. For aircraft, there are absolute weight constraints and the primary consideration thus becomes a matter of fore-and-aft balance.

5. *Agreed rates.* Agreed rates are loyalty incentive rates, granted in exchange for the shipment of a specific percentage share of total annual shipments. When offered by common carriers to all shippers without reference to specific volumes, they have been acceptable to the Interstate Commerce Commission. Contract and exempt carrier rates of a similar nature are established through direct shipper-carrier agreements.

6. *Governmental rates.* Federal Statute and regulations provide that the general rules of rate making need not apply to shipments made by government agencies.

7. *Multimodal rates.* The growth of intermodal transportation services has produced a system of multimodal rates designed to make these services more price competitive, as well as to simplify the documentation and billing for such shipments. Most multimodal rates are quoted by the originating carrier on a single, joint, through rate basis for door-to-door service and are generally based on carrier costs.

Accessorial service and terminal charges.   In addition to route movement rates, carriers are entitled to charges for accessorial services and terminal use.

Accessorial services.   Accessorial types of service, often considered as privileges by the providing carriers, result from special needs of the shippers and competition among carriers. The more significant of such services are described below.

1. *Diversion and reconsignment.* Although diversion means a change in the routing of a shipment and reconsignment means a change of consignee for a particular shipment, the terms are used interchangeably. Diversion is used most frequently in connection with the movement of perishables, grain, and lumber. This service permits the delivery of shipments to one of a number of alternative points, depending upon changes in market conditions or terms of sale, after the shipment has left its point of origin, but before it has reached its originally stated destination. Diversion is also a method of using the carrier as a warehouse. Through diversion, a shipper can route a shipment circuitously so that transit time will be longer than usual and the need for warehousing at the destination will be reduced.

2. *Protective service.* Protective services are offered for goods that could be affected by the environment. Such services include ventilation, refrigeration, and heating. In some situations, carriers provide these services without charge as a necessary element in the transportation of a commodity. In other cases, where the shipper specifies the type of service to be given to a particular shipment, charges are assessed for the additional service.

3. *Special equipment.* Extra charges are sometimes assessed for the use of equipment that meets special shipper needs. Enclosed tri-deck railcars used only to transport new automobiles are an example of such special equipment.

4. *Transit Privileges.* Transit privileges facilitate the flow of goods that require some handling or processing at an intermediate point between origin and destination. Some of the types of transit privileges and the commodities for which they were designed are milling (grain), compressing (cotton), planing (lumber), fabrication (steel), blending (wine), and storage (for commodities produced seasonally and consumed nonseasonally). With the advent of rate deregulation in the railroad industry, the use of in-transit rates has declined sharply, and they have largely been replaced with simple point-to-point rates. This evolution has caused problems for shippers that previously used in-transit rates. With railroads tending toward basing their rates on cost of service, many facilities that once benefited from in-transit rates are now at a disadvantage compared with competing locations which are either raw-material-source- or market-oriented.

5. *Stop-off privileges.* This type of transit privilege permits the shipper or receiver to stop a shipment to partially load or unload it at an intermediate point. With such privileges, shippers are normally allowed

one stop to complete loading and three stops for partial unloading. Both rail and highway carriers offer stop-off services.

6. *Split delivery*. Split delivery is an additional service provided by highway carriers. It permits an unlimited number of segments of a larger shipment to be delivered within the limits of a specific split-delivery area. Special split-delivery charges are made for each partial shipment delivered to a different point; these are the same regardless of the total weight of the shipment.

7. *Pickup and delivery*. Certain carriers commonly quote rates on a terminal-to-terminal basis. Additional charges usually apply for pickup and delivery services. These charges must be noted and included when comparing terminal-to-terminal rates with those of portal-to-portal services.

Terminal charges.   Certain services provided by terminal operations could generate added charges. Such services are most frequently reflected in demurrage, detention, and switching charges.

1. *Demurrage and detention*. The words *demurrage* and *detention* both refer to charges paid to carriers by shippers or consignees for the delay of a car, vessel, or vehicle beyond a specified time period allowed for loading and unloading. The term *demurrage* is used in rail, water, and pipeline transportation. Detention is assessed on highway vehicles and on trailers used in TOFC, or piggyback service.

Pipeline demurrage is charged when the receiver is unable (usually because of insufficient storage space) to receive the quantity delivered to it by a pipeline. For water movement, charges for vessel demurrage are based on the size of the vessel and are usually specified in the applicable tariffs or contracts. The free time allowance is based on the volume and characteristics of the cargo to be unloaded. Detention charges vary widely, depending on the legal form of carriage and the type of equipment involved. Many shippers find it economically advantageous to use demurrage as a supplement to their own warehouse capabilities. A producer with fluctuating storage demands might, for example, plan to use rail demurrage in the form of a "rolling warehouse" capability as a cost-effective alternative to acquiring additional storage facilities.

2. *Switching charges*. Line-haul rail rates usually include switching services to the ultimate shipping destination. Under a reciprocal switching agreement (a frequent practice among carriers), the switching carrier will, within defined switching limits, deliver to its destination a car that it did not handle in line-haul movement, and settlement of switching charges is made directly between the line-haul and switching carriers. When the line-haul carrier does not have access to the destination point (such as a railroad siding) and the destination is closed to reciprocal switching, the shipper or consignee must pay for the switching as charged on a per-car basis.

Utilization of Automated Logistics Information Technology

The efficacy of the logistics system has been significantly enhanced by the introduction and utilization of a number of forms of information technology (IT). Among the more exemplary and useful forms of the new technologies is Radio Frequency Identification, which will be addressed in this section.

Radio frequency identification

Radio Frequency Identification (RFID) has had a dramatic impact on the processes involved in packaging, handling, storage, and transportation within the logistics infrastructure. RFID is an automatic identification method, relying heavily on storing and remotely retrieving data using devices called RFID tags or transponders. An RFID tag is a small object that can be attached to or incorporated into an object, person, or any appropriately designated logistics item. There are several forms of identification, but the most common is to store a serial number that identifies the person or object, and perhaps other information, on a microchip that is attached to an antenna (the chip and antenna together are called an RFID transponder or an RFID tag). The antenna enables the chip to transmit the identification data to a reader. The reader converts the radio waves reflected back from the RFID tag into digital information that can then be transmitted to computers that can make use of it.

Overview of the RFID System

An RFID system may consist of several components comprising tags, tag readers, edge servers, middleware, and application software. The purpose of an RFID system is to enable data to be transmitted by a mobile device, called a tag, which is read by an RFID reader and processed according to the requirements of the specific application. The data transmitted by the tag may provide identification or location information, or other specifics about the item tagged, such as price, color, date of purchase, etc. The utilization of RFID in logistics has expanded dramatically because of its ability to track objects moving within the logistics infrastructure. In a typical RFID system, individual items are equipped with a small, inexpensive tag. The tag contains a transponder with a digital memory chip assigned a unique electronic product (or item) code. The interrogator, an antenna packaged with a transceiver and decoder, emits a signal activating the RFID tag so it can read and write data to it. When an RFID tag passes through the electromagnetic zone, it detects the reader's activation signal. The reader decodes the data encoded in the tag's integrated circuit (silicon chip) and the data is passed to the host computer for processing.

Types of RFID tags

RFID tags can be either *active* or *passive*. Passive RFID tags are the least costly, but have no internal power supply. The minute electrical current induced into the antenna by the incoming radio frequency signal provides just enough power for the tag to transmit a response. Due to limited power, the response on a passive RFID tag is brief and typically consists of just an identification number. Active RFID tags are more sophisticated and have an internal power source. The active RFID tags may have longer range and larger memories than passive tags, as well as the ability to store additional information sent by the transceiver. The smallest active tags currently in use are about the size of a coin and many have practical ranges of tens of meters, along with a battery life of up to ten years. Because passive tags are less expensive to manufacture and have no battery, the majority of RFID tags currently in use are of the passive variety. There are four different tags commonly in use. They are categorized by their radio frequency: low-frequency tags which include 125 kHz (the original U.S. standard) or 134.2 kHz (the international standard); high-frequency tags (13.56 MHz); ultra-high-frequency (UHF) tags (868 to 956 MHz); and microwave tags (2.45 GHz).

Applications of RFID tags

The most common logistical applications for RFID are tracking goods in the logistics chain (also know as the *supply chain*); reusable containers; high-value tools and similar assets; parts moving through a manufacturing production line, etc. Low-frequency RFID tags are used for basic functions such as tracking beer kegs. High-frequency RFID tags are used for such functions as airline baggage tracking and pallet tracking. Ultra-high-frequency RFID tags are commonly used commercially in pallet and container tracking, as well as truck and trailer tracking in shipping yards. Microwave RFID tags are typically used in long-range access control for vehicles.

The global perspective for RFID

There is currently no global public body that governs the frequencies for RFID. In principle, every country can set its own rules for this purpose. In the United States the authority governing frequency allocation for RFID is the Federal Communications Commission (FCC). Low-frequency (125–134.2kHz) and high-frequency (13.56MHz) RFID tags can be used globally without a license. UHF (868 to 956 MHz) tags cannot now be used globally, inasmuch as there is currently no single global standard.

Inventory Management

Overview

Inventory exists to provide the most cost-beneficial material support of its dependent activities. Such activities may take the form of production or wholesale-level support requirements, support of intermediate-level distribution points, operational support of organizational or retail-level activities, or any combination of these three user levels. All inventory managers aspire to fulfill all requests for their stock in a timely manner. Notwithstanding this noble goal, each inventory manager must work within capricious parameters determined by users' needs priorities, order and delivery intervals, and financial constraints. The idyllic scenario of 100 percent responsiveness is impossible in practice. The realistic alternative for the inventory manager is to assign priorities, establish support stockage levels based on plausible probabilities of fulfillment, and establish the most cost-effective inventory support system consistent with these considerations.

Reasons for Inventory

The predominant justifications for inventory investment and material management are defined by the needs of the production lines and responsiveness to the customers. Inventory management systems are designed to

1. *Promote production efficiency*. Timely and effective inventory support can help reduce production costs per unit by providing working stock when needed so as to avoid costly nonproductive time.

2. *Provide a buffer against seasonal demands.* By basing production intervals on the opportune availability of production facilities and schedule time, inventories of seasonal market goods (snow shovels, swimwear, holiday greeting cards, etc.) can be gradually augmented to a level sufficient to meet consumer demand surges during the seasonal market periods.

3. *Assure availability of supplies.* There should be a cushion of critical working stock if the sources of such essential materials are vulnerable to supplier disruptions, such as strikes or sudden shortages of imported critical items (e.g., manganese, tungsten, or other precious metals).

4. *Provide a hedge against price increases.* The key to successful execution of this policy is reliable market intelligence, typically provided by the corporate procurement activity. Advance purchase of sufficient materials to support forecast production schedules can mitigate the economic impacts of projected price increases by key suppliers.

5. *Reinforce stock locations near market centers.* Certain producers maintain networks of dispersed stockage points near or collocated with distribution centers or service centers which support their designated market zones. Such stockage networks ensure responsiveness to customers' needs and foster continued and, hopefully, increased consumer patronage.

6. *Accommodate unanticipated surges in sales.* An impromptu decision by the marketing staff to reduce prices so as to stimulate sales, promote consumer interest in the products, or, for some reason, accelerate the pace of sales because of emergent market factors (e.g., pending product obsolescence as a result of advances in the state of the art) might lead to sudden demands on inventory stocks. The measures available to accommodate this type of situation include provisions for safety stocks, just-in-time (JIJ) inventory replenishment procedures, etc.

Characteristics of Inventory

The total assets physically available to the inventory manager comprise (1) normal on-hand stocks to respond to basic user demands, (2) safety, or contingency, stocks, (3) in-transit (often termed *transportation pipeline*) assets, (4) speculative stocks, and (5) dead stocks.

Basic demand or operating-level stocks are those required for filling orders under conditions of certainty where demand and replenishment time can be predicted with reasonable accuracy.

In-transit stocks are those that are en route from one inventory location to another. They may be considered as part of basic stocks even though they are not available for sale until they arrive at a given destination.

Safety stocks are those in excess of the basic inventory that are held because of uncertainties related to demand for the item or replenishment time, or for disaster or security reasons.

Speculative stocks are created to balance cyclical production, such as foods produced during a short harvest season, or to acquire commodities, such as raw materials, which may change significantly in price or availability.

Dead stocks are those for which there has been no demand over a specified period. If no further demand is anticipated, they are normally scrapped for salvage value.

Material items that are in the production process or in the repair cycle for corrective maintenance or restoration are regarded as assets within the total logistics system, but not normally within the purview of the inventory manager. When items of material are transformed into finished products for market distribution or returned from the maintenance process for reutilization, they are again subject to inventory accountability and control. The inventory manager envisions an environment of systematic stability governed by certainty of events, but must acknowledge that the real world is driven by uncertainty.

Inventory management under certainty

The essential elements of an inventory system are determined by the authorized on-hand inventory level, the reorder level, the order quantity, the order and shipping time (also termed as the procurement lead time), the interval between orders, and the issue rate, which necessarily is a function of the consumption rate. Figure 18.1 shows the theoretical inventory cycle. This cycle is based on governing conditions of certainty and, therefore, an ideal inventory management environment. The inventory manager's "dream world" is characterized by the following conditions:

1. Consumer demands and, consequently, issues from stock occur at a constant rate.
2. Order points are at designated stockage levels which occur at predictable intervals, thereby producing a consistent reorder cycle.
3. The procurement lead time, the interval between order and receipt, is consistently the same.

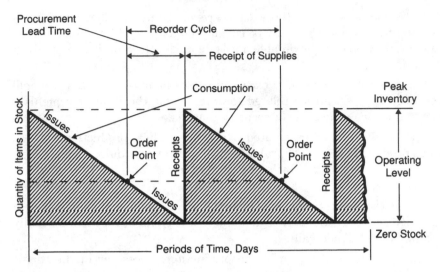

Figure 18.1   Theoretical inventory cycle.

4. Because of the constant demand rate and responsive issue rate, receipt of the items reordered will occur precisely at zero stock level, immediately raising the stockage to the authorized operating level.

5. The peak inventory (authorized stock level), order point, and procurement lead time are such that inventory is always available to fulfill the constant demand pattern, thereby eliminating the need for an additional cushion for contingencies.

Pragmatic logistics managers know that no such dream world exists. The next section describes what inventory managers may expect to encounter.

Inventory management under uncertainty

The basic system elements of consumer demand—issue rates and procurement lead time—are themselves affected by uncertainty, which, in turn, reduces the predictability of the reorder cycle and the ability to sustain peak operating stockage levels. The compounding of these uncertainties portends a stock-out situation, during which condition the inventory manager is unable to fill user needs and must back-order the items. Figure 18.2 portrays a typical scenario under conditions of uncertainty. This case emphasizes the irregularity of the demand pattern and, although replenishment stock was reordered at the designated reorder point, the impact of unforeseen factors that extended the pro-

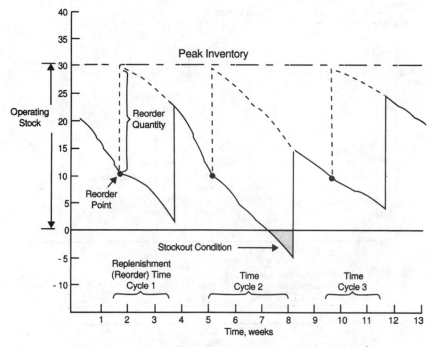

Figure 18.2 Stock-out impact of a longer-than-normal replenishment cycle.

curement cycle. Since the authorized inventory level did not incorporate safety stock, the extended replenishment time coupled with the inconsistent demand pattern produced a stock-out condition.

Figure 18.3 depicts the same scenario, but with safety stock. In this case, when the cumulative demands exceeded the operating stockage prior to replenishment, the safety stockage absorbed the demand impact and assured availability of items for issue.

A major objective of the logistics manager is to seek consistent transit time from carriers. When transit time is regularized, one of the difficult variables of inventory management is eliminated, or at least reduced. Like all inventory resources, safety stock represents a financial investment, with attendant opportunity costs for not having used the same monetary amount to make interest-bearing or profit-assured investments. For this reason, the logistics manager must evaluate the potential consequences of stock-out conditions relative to the financial commitment required to assure a high confidence of the item's being available for issue.

An alternative to maintaining operating-level inventories reinforced by safety levels that has developed in recent years is the practice of

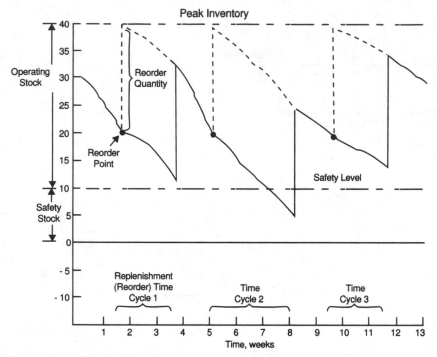

Figure 18.3   Avoidance of stock-out by increasing the reorder point and thereby carrying a large safety stock inventory.

spontaneous replenishment using premium order and transportation methods.

A number of producers have instituted the practice of just-in-time (JIT) inventory replenishment in order to minimize inventories of parts or raw materials. JIT requires that inbound shipments arrive "just in time" to enter the production process and, consequently, places pressure on carriers and traffic departments alike. Although JIT can be an advantageous practice, it has its disadvantages, including the high cost of small lot (inbound) shipments.

Inventory Program Management

The inventory manager is confronted with multifarious factors in determining the optimum inventory management program. This involves evaluation of demand frequencies; the operating-level support posture; make-or-buy considerations; last-in, first-out/first-in, first-out (LIFO/FIFO) policies; stock-keeping unit (SKU) identification and accounting; and the ABC distributive value approach to inventory management.

Establishing inventory levels

Inventory support for new products is fundamentally a matter of responding to marketing strategies, sales forecasts, and customer orders for new units. Inventory management for system components and repair parts poses a different problem. The first phase of inventory management involves the initial provisioning or "lay-in" of components and parts needed for maintenance of the system. This initial provisioning is based on projected estimates for

1. System usage (number of systems in market distribution)
2. Quantity of components and parts per system
3. Usage rates (based on operational failures, induced failures, maldiagnosis, and quality-rejected defectives)

Development of these projections demands close coordination with the product engineers and market strategists. After the initial lay-in, the empirical data that accumulate over a prescribed period (typically, at least one year) should be examined by the inventory manager for indications of demand patterns. On the basis of demand pattern analysis, operating stockage levels and safety levels may be established to accommodate desired fill rates, or levels of confidence based on probabilities of being able to fulfill user requirements.

Demand frequency analysis.   Table 18.1 illustrates atypical summary of demand data based on daily issues at varying demand levels over a typical 250-day operating year (365 days less weekends and holidays).

Table 18.1 indicates that on 30 working days there was zero demand, which denotes a contributory frequency factor of 12 percent for the 250-day period. During each of 62 days, there were two demands, for a con-

TABLE 18.1 Demand Data Summary
(Based on 250 operating days per year)

Demand	Number of days	Total units	Contributory frequency %	Cumulative frequency %
0	30	0	12.0	12.0
1	70	70	28.0	40.0
2	62	124	24.8	64.8
3	40	120	16.0	80.8
4	27	108	10.8	91.6
5	6	30	2.4	94.0
8	11	88	4.4	98.4
16	1	16	0.4	98.8
20	3	60	1.2	100.0
Total	250	610		

tributary frequency factor of 24.8 percent. If the inventory manager desired 90 percent confidence of issue, it would be necessary to maintain at least four items in stock. The cumulative frequency percentage for daily demands of four or less stands at 91.6 percent; for cumulative daily demands of three or less, the cumulative frequency is 80.8 percent. As it is impossible to subdivide a unit of product, the inventory manager must maintain a stockage of four units to assure a 90 percent fill rate. From the standpoint of gross planning estimates, the ratio of total issues to the number of days (610 ÷ 250) reveals an average daily demand pattern of 2.44 units per day.

Operating days' posture. Inventory managers frequently determine cushion stockage levels and reorder points on the basis of projected replenishment intervals. Replenishment intervals comprise days required for order processing, days required for production or source supply of the items, and days required for transportation. In simplistic terms, the reorder point (ROP) would be triggered at a stockage level at which there is sufficient stock on hand to sustain daily demand until replenishment. Such an on-hand quantity would equal the daily usage multiplied by the total number of days in the replenishment cycle. Figure 18.4 illustrates this method.

Figure 18.4 presumes an average daily demand rate of 2.5 units, based on the data in Table 18.1 (the computed daily average of 2.44 rounded to 2.5). Theoretically speaking, if the daily average and the elements of the replenishment cycle are consistent, the net stockage level would be zero at the point of receipt of the reorder quantities. For this type of scenario,

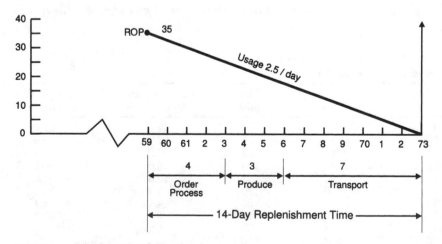

Figure 18.4 Reorder point determination.

the optimum order quantity was calculated on basis of days' operating support posture from the reorder point until receipt of the order.

Make-or-buy considerations

A make-or-buy decision is made by an organization to determine whether materials should be manufactured or purchased. The comparison of the costs involved in making or buying an item is the primary evaluation parameter. The total cost is minimized, based on make-or-buy quantitative criteria.

Although this aspect of material provisioning is traditionally considered part of the procurement process, it is being treated here as an activity that affects inventory management. The criteria applied in the quantitative analysis have much in common with the factors of inventory geometry, and the decisions pursuant to the make-or-buy analysis have a bearing on the issues concerning financial investment in inventory. The issues confronting the logistics manager concern whether it is more cost-beneficial to produce material items in-house than to acquire the items from company vendors and suppliers.

The make-or-buy formula. The same algorithm applies to both the make computation and the buy computation. The following describes a typical analysis.

$$TC_{min} = C_iD + \sqrt{2C_bC_hD\left(1 - \frac{D}{R}\right)}$$

where C_i = item cost
C_b = buying cost or line setup expenses
C_h = holding rate (carrying change for storage)
D = demand during period
R = replenishment rate
TC_{min} = total cost

Consider an item laid out for comparison as follows:

	Make	Buy
Item cost, $	$ 5.00	$ 6.00
Order or setup cost, $	$150.00	$10.00
Replenishment rate/day	25	∞ (unlimited)*
Holding rate per period (% decimal equivalent)	0.005	0.005
Demand per period	12	12

*Supplier's replenishment capacity is considered unlimited, i.e., approaching infinity (∞). Any denominator factor of ∞ makes the ratio equal zero.

$$TC_{\min} = C_i D + \sqrt{2C_b C_h D\left(1 - \frac{D}{R}\right)}$$

$$\text{Make} = (5)(12) + \sqrt{2(150)(0.005)(12)\left(1 - \frac{12}{25}\right)} = \$63.06 \text{ (minimum cost)}$$

$$\text{Buy} = (6)(12) + \sqrt{2(10)(0.005)(12)(1)} = \$73.10$$

For the buy computation,

$$\left(1 - \frac{D}{R}\right) = \left(1 - \frac{12}{\infty}\right) = (1 - 0) = 1$$

The cost considerations are only one of the criteria for the decision. Manufacturing requires investments in machines, people, abilities, and raw materials, plus related overhead and nondirect expenses. It must be acknowledged that a company that makes its own material has better control over schedule, quality, and utilization. On the other hand, buying an item means that the company does not have to invest in those resources and allows the funds to be diverted to more profitable investments.

LIFO/FIFO

FIFO (first-in, first-out) and LIFO (last-in, first-out) are common issue control procedures used in inventory management. FIFO is the issue procedure normally used to assure regular use of items in inventory and avoid obsolescence, excessive shelf life, and deterioration. The use of LIFO and FIFO for financial accounting and valuation of the inventory is largely driven by considerations of tax liability and the financial posture the company prefers to convey to the stockholders and the public at large. For example, in periods of inflation, LIFO would increase material transfer costs, increase the costs of doing business, reduce the profit margin, and minimize tax liability. The monetary LIFO valuation may, therefore, be different from the physical movement (LIFO or FIFO) unless there is identification of specific inventoried items, such as serial-numbered equipment, lots of medicines and pharmaceuticals, etc., in which case issue control and financial accounting must be both LIFO or both FIFO. In certain cases, a pool concept may also be used, in which no control is exerted. Realistically, the inventory manager must be alert to avoid FISH (first in, still here).

Stock-keeping units

A term of reference commonly used to parameterize inventory accountability is the stock-keeping unit (SKU). The SKU relates to a technique many commercial enterprises use to account for, sell, and manage inventories of individual market items. Consider a specific brand of canned beer as an illustrative example: A single can of beer may be an SKU for a small convenience store, a six-pack of beer could be an SKU for a supermarket, a case of beer could be an SKU for a beverage distributor, and a configured pallet load could be an SKU for the brewery. The definition of the SKU for any product is subject to the discretion of the enterprise marketing the items.

An SKU inventory refers to the stock of an individually described SKU, which may contain any number of units. A stock-keeping unit location (SKUL) is the number of units inventoried at one facility. One or more SKUs make up a product line, and the sum of all SKUs at all SKULs is the total inventory for a company.

ABC inventory management approach

The ABC inventory management approach encompasses a variety of practices by which different items (different SKUs) within an inventory are grouped and varying degrees of attention paid to the management of inventories for those various groups of products. Each product group receives a different level of attention on the basis of such factors as (1) rates of sale, (2) value per unit, (3) costs of storage, or (4) the strategic value of the item to customers. Most are concerned with the provision of different levels of coverage, or customer service standards, through the use of safety stocks for various categories of items. The ABC concept parallels the high-value item management concept used in military logistics organizations.

Management of items with high rates of sale.   These items usually offer the greatest opportunity for cost savings and sales benefits from the careful application of inventory control techniques. In contrast, items with slow rates of sale typically are stocked in sufficient quantities to reduce the need for reviews of stock levels to relatively infrequent time intervals.

Management of items with high values per unit.   Where values per unit for individual SKUs vary greatly, the greatest savings in inventory carrying costs may be effected either by the use of a more careful and exacting inventory program for the management of high-value items or by a reduction in inventory coverage levels for them.

Management of items with costly storage requirements.   Products such as frozen foods may be singled out for this type of attention because of the high cost of storing them in expensive freezers. ABC inventory manage-

TABLE 18.2 ABC Concept

	Group A	Group B	Group C
Percent of total inventory investment	65%	25%	10%
Percent of items in total inventory	15%	30%	55%

ment policy may also differentiate SKUs on the basis of cubic footage per unit if storage facilities represent a scarce resource with a high opportunity cost per unit of space.

Management of items with high strategic value to customers. Strategic value to the customer may be exemplified by the question, "When is an automobile not an automobile?" and the answer, "When the automobile is in New York and you need it in Cincinnati." This example illustrates the fact that certain items are highly critical to customers regardless of the monetary values of the items. In the eyes of an airline maintenance manager, a low-cost O-ring used as a seal in the hydraulic system for the landing gear of a passenger aircraft would have a greater strategic value than the more expensive on-board galley microwave cooking unit.

The principle of the critical few. The ABC approach to inventory management is derived from Pareto's principle of the critical few; i.e., a minority will have predominant significance within the defined population. Table 18.2 illustrates the distributive effects when a minority of the SKUs in an inventory constitutes the majority of the investment in inventory resources and the inventory is apportioned according to the relative significance of the constituent groups. The principle of allocation of items in the inventory according to their proportion of the total number of items and their proportionate economic significance, illustrated in Table 18.2, is equally applicable to the use of ABC grids to portray the pertinent attributes of rates of sale, SKU values, and storage costs, as well as strategic values to customers.

Inventory Support of Production Operations

Production material must be at the right place in the right quantity at the right time if a desired level of logistics support is to be achieved. Production inventory scheduling involves the timing, sequencing, and grouping of purchases, production runs, and shipments. The inventory manager must, therefore, be aware of when and in what quantities the goods will be produced, and when the needed raw materials must be on hand.

Logistical prelude to production

An essential prerequisite to production is development of the most cost-effective logistics strategy. Because algorithmic techniques for the determination of economic lot quantities are sophisticated enough to include costs that are not directly related to the production process, such as inventory carrying costs, lost sales impacts, purchase discounts, transportation rate discounts, etc., it is possible to calculate a workable economic order quantity (EOQ). Assuming that such an EOQ can be determined, the next step is to determine *when* lots of particular products should be produced. This can be divided into two parts: (1) the gross production planning of the items and quantities that should be produced during a given time frame, and (2) the detailed scheduling of activities *within* the production process. Because of the logistics implications, the function of gross production scheduling is being assumed by the logistics managers. Inventory managers must coordinate planning on inventory levels and production scheduling with all concerned elements of the corporate staff. This includes the transmission of information on sales projections and decisions regarding customer service standards from the marketing department to the logistics manager. The production department must advise the logistics manager of its production capabilities for various products, the lead times, and the quantity of raw materials required to produce a given quantity of a product. The logistics manager must, on the basis of the stipulated EOQ, inform the production department of which lots of products are to be produced by which dates, and ensure that supplies of working materials are available for production as required.

Production inventory control

Most manufacturers are compelled to plan the production of a number of different products during a given time frame. This requires advance planning to produce relatively constant plant loading, optimum scheduling to meet changing seasonal needs, and allocation of limited production capacity in times of demand surges.

Master schedules.   Advance plans and master schedules are typically developed for production processes on the basis of the following guidelines:

1. Translation of sales forecasts into production schedules for up to six months into the future

2. Development of item-by-item production needs for shorter time periods

3. Reconciliation of production needs with intermediate-term production capacities

4. Development of short-term plant loadings, including the allocation of needed production among plants with overlapping production capabilities

Master scheduling is not normally done for periods of more than one year. In some enterprises, especially those that are not subject to seasonality and that have moderate lead times for materials purchases, master schedules might cover a period of up to six months.

Trade-offs encountered in the master scheduling process are (1) the desire to load to capacity as opposed to retaining some flexibility for last-minute schedule changes, and (2) the desirability, in particular during periods of demand surges in excess of capacity, of scheduling for maximum output as opposed to more frequent production cycles of product groupings which involve retooling of the production lines.

It is nevertheless advisable, even in a short-term production schedule, to be able to accommodate unanticipated manufacturing delays or orders from preferred customers without affecting the master production schedule. The prudent manufacturer never loads its plants to more than 85 to 90 percent of capacity at the time schedules are locked in in order to give the marketing and distribution groups a cushion with which to absorb last-minute orders.

Production sequencing. Items in product groupings that can be produced in the same facility and on the same equipment are normally produced in several cycles during the year. The internal cycle sequence is often designed to minimize the costs of machine setup in switching production-line configurations, which means that once production of an item in a given cycle is completed, further production of that item must await the production of all other items in the cycle.

Production smoothing. If a producer has enough production capacity to meet its annual demand, but not enough to meet peak demands during the year, inventory must be augmented during slack periods in order to meet demand in subsequent peak periods. Inventory levels will thus vary periodically, but the production rate will be stable throughout the year.

Allocation of limited capacity. Plans used by manufacturers for allocating limited capacity include (1) first-come, first-served, (2) items with the fastest production rates first, (3) items found on the smallest orders first, (4) orders from the most profitable customers first, and (5) orders from the oldest, more loyal customers first. The policy most frequently followed is probably the last, particularly when production is scheduled on the basis of individual customer requests rather than forecasts of distribution center replenishment orders.

The experience of a number of enterprises indicates that ability to quote accurate, realistic delivery times is largely a function of

1. An established plan for apportioning demand
2. Conformance to the prearranged plan
3. The degree of allowance for slack in plant loading
4. The efficiency of coordination among the marketing, production, and logistics activities concerning changes in sales forecasts or production capabilities

Production resources planning

There is extensive literature available on a variety of inventory management practices, referred to as MRP I (materials requirements planning), MRP II (manufacturing resources planning), JIT (just-in-time), and the Japanese concept called *kanban*. Although these are related to production scheduling, only two of them (MRP I and *externally oriented* JIT) are true concerns of logistics managers.

Kanban is an internal production control technique for minimizing the capital invested in in-process parts inventories. This prevents accumulation of components that are defective, and reflects the deep concern of Japanese management with quality control and the cost savings possible from minimizing rejects. It should be noted that kanban is a tool of the production manager and not a direct concern of the logistics manager, whose responsibility does not extend to the production line.

Kanban is frequently termed JIT. One definition of JIT is *external* movements of material inbound to a plant.

Materials requirements planning (MRP I). In simplified form, MRP I consists of (1) the preparation of a master production schedule for some period into the future, (2) the preparation of a bill of materials for each item to be produced, (3) the "explosion" of the units on the master production schedule into a component-by-component requirement schedule, in terms of both quantities and dates needed, and (4) the scheduling of component inventory replenishment according to necessary lead times and economic order, buying, or shipping quantities to conform with the requirement schedule rather than average demand over time.

For a multistage manufacturing process, the calculations become overwhelmingly complex. Although MRP I is a relatively simple concept, it requires massive amounts of data that cannot be processed in a timely fashion by manual methods and typically demand the use of a computer. MRP I systems can produce significant inventory savings, but their development requires considerable capital investment and execution by skilled and astute system managers.

Manufacturing resources planning (MRP II). Manufacturing resources planning (MRP II) embodies planning issues related to plant capacity. Although MRP II is not the direct responsibility of the logistics manager, it involves the interests of logistics management when inputs on production materials availability are solicited to enable decisions concerning production planning. Decisions regarding investment in added production facilities or reduction in plant capacity are normally made by corporate management. The resulting plant capacity will govern the decisions of the logistics manager concerning the material provisioning required to support the projected production schedule.

Just-in-time (JIT) scheduling. The Japanese concept of externally oriented JIT is that shipments of materials for the production lines should arrive at the plant "just in time," as needed. The goal of JIT is to minimize inventories of production materials, either purchased or manufactured in other company plants. The JIT concept has recently received a considerable amount of management attention and emphasis in the United States. In the view of the author, this emphasis has resulted in misdirected and costly experimentation in efforts to achieve the logistical nirvana the Japanese seemed to have achieved with JIT.

If a company and its vendors have historically been sloppy with respect to management of inventory levels, placement of orders, and control over order and shipment times, the managers responsible need to be disciplined. The prospect of having a responsive and intelligent inventory management system should not be confused with a perceived need for implementation of JIT. JIT means exercising stringent control over suppliers and carriers to produce a continuous flow of small quantities into the production plant so as to minimize the amount of working inventory on hand. Although it may be possible to control internal company production operations, it is another matter to be able to ensure that suppliers and transportation carriers continuously ship and deliver many orders of small quantities of goods within narrowly defined time frames. While it is recognized that stock-outs of materials required for production are to be avoided at any cost within reason, it must be underscored that JIT requires a very high level of carrier reliability, suppliers preferably located close to the plant, frequent deliveries, smaller shipments at premium transportation cost rates, and an increase in order processing costs, all of which are likely to be incompatible with reasonable EOQ calculations. Any disruption in the system is likely to result in a stock-out situation and close the production line, as no safety stocks are maintained.

Logistics managers considering JIT should be able to answer the following questions in the affirmative.

1. Does the production plant have access to rapid, responsive, and efficient transportation?
2. Are the carriers reliable?
3. Do the carriers schedule frequent deliveries?
4. Is there a high potential for labor stability among the network of suppliers (e.g., a low probability of strikes or work stoppages)?
5. Is high-speed communications technology which provides links with the suppliers available in the plant?

It is incumbent upon the logistics manager to consider these issues carefully and conduct a cost-benefit trade-off analysis of JIT vis-à-vis enhancing the existing controls and stockage levels prior to commitment to the JIT system.

Quantitative Applications to Inventory Management

The elementary quantitative methods used for analytical application to inventory management encompass considerations of cost optimization, essentiality, and confidence based on predictability of demand patterns. The following are quantitative models for periodic total inventory costs, economic order quantities, and inventory levels for consumable and reparable components.

Total inventory cost

Inventory-level decisions would be relatively simple if users organized their requirements so that an equal number of all items were needed each day and if the time required to schedule, produce, and transport replenishment stock were known and constant. Inasmuch as such conditions are virtually nonexistent, the model should be utilized to incorporate real-world variations. The two major factors affecting the order quantity of a product are the cost of placing an order and the cost of carrying inventory. The cost of placing an order or setting up a production run is assumed to be constant regardless of the size of the order; the cost per unit will therefore decrease as order size increases. As the order size increases, it will take longer to deplete the ordered quantity and the average inventory level will be higher, which results in higher carrying charges.

The total periodic inventory cost, considering all of these contingent elements, is

$$TC = \frac{QUR}{2} + \frac{AD}{Q}$$

where TC = total inventory cost for the specified period (in dollars)
  Q = quantity ordered (in units)
  U = average cost per unit (in dollars)
  R = periodic inventory carrying rate as a percentage of product cost
  A = setup or ordering cost (in dollars per order or setup)
  D = period demand (in units per period)

For example, assume the following information, based on a period of one year:

Q = 50 units
U = $50 per unit
R = 0.004 (0.4 percent per year)
A = $125 order placement cost
D = 500 units per year

$$TC = \frac{(50)(50)(0.004)}{2} + \frac{(125)(500)}{50}$$
$$= 5 + 1250$$
$$= \$1255 \text{ per year}$$

The economic order quantity formula

The economic order quantity (EOQ) is derived by calculus, by determining the first derivative of TC with respect to Q. The sequential calculus process is described below:

$$TC = \frac{QUR}{2} + \frac{AD}{Q}$$

$$TC = \frac{QUR}{2} + (Q^{-1})(AD)$$

$$d(TC) = dQ\frac{(UR)}{(2)} + dQ(-1)(Q^{-2})(AD)$$

$$d(TC) = dQ\left(\frac{UR}{2} - \frac{AD}{Q^2}\right)$$

$$\frac{d(TC)}{dQ} = \frac{UR}{2} - \frac{AD}{Q^2}$$

Let $d(TC)/dQ = 0$; i.e., let the rate of change of TC with respect to Q (the slope) equal zero to determine the optimum value of Q.

$$0 = \frac{UR}{2} + \frac{AD}{Q^2}$$

Then

$$\frac{AD}{Q^2} = \frac{UR}{2}$$

and

$$Q^2(UR) = 2(AD)$$

$$Q^2(UR) = \frac{2AD}{UR}$$

$$Q = \sqrt{\frac{2AD}{UR}} = \text{economic order quantity } (EOQ)$$

Therefore

$$EOQ = \sqrt{\frac{2AD}{UR}}$$

Figure 18.5 graphically illustrates the characteristics of the EOQ mathematical relationship.

Example

$A = \$25$, order placement cost
$D = 200$ units, annual demand
$U = \$100$ per unit
$R = 0.005$ (0.5 percent carrying rate)

$$EOQ = \sqrt{\frac{2AD}{UR}}$$

$$= \sqrt{\frac{(2)(25)(200)}{(100)(0.005)}}$$

$$= 141.42, \text{ or } 141 \text{ units}$$

Note: In many EOQ relationships, A is symbolized by C_p (cost to purchase) and UR is represented by C_h (cost to hold based on carrying rate × unit cost). In such formulas,

$$EOQ = \sqrt{\frac{2C_p D}{C_h}}$$

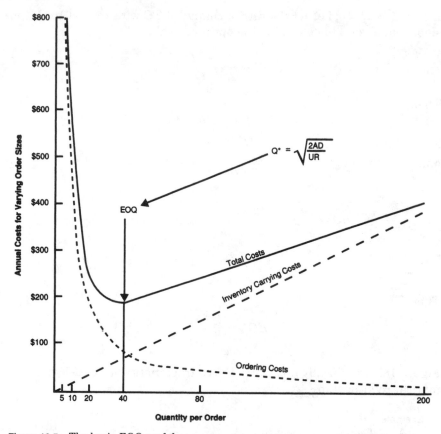

Figure 18.5   The basic *EOQ* model.

The principle of finite replenishment

The *EOQ* concept concerns the customer's perspective on inventory economics. The principle of finite replenishment, also known as the cost-optimum production run, is essentially the producer's *EOQ* formula and involves the same basic constituent factors. The *finite replenishment formula* is described as follows:

$$EOQ = \sqrt{\frac{(2)(C_{setup})(D_{yr})}{(C_h)(UV)\left[1 - \dfrac{D_{yr}}{(OD_{yr})(DR_{day})}\right]}}$$

where C_{setup} = setup costs
D_{yr} = demand per year
C_h = holding rate (% decimal equivalent)
UV = unit value

$$OD_{yr} = \text{operating days per year}$$
$$DR_{day} = \text{delivery rate per day}$$

Example Determine the optimum finite replenishment quantity for a product described by the following attributes:

$$C_{setup} = \text{setup costs} = \$90$$
$$D_{yr} = \text{demand per year} = 300{,}000 \text{ unit yr}$$
$$C_h = \text{holding rate} = 0.22 \ (22 \text{ percent})$$
$$UV = \text{unit value} = \$0.10$$
$$OD_{yr} = \text{operating days per year} = 250 \text{ days}$$
$$DR_{day} = \text{delivery rate per day} = 2500 \text{ units}$$

$$EOQ = \sqrt{\frac{(2)(90)(300{,}000)}{(0.22)(0.10)\left(1 - \dfrac{300{,}000}{250 \times 2500}\right)}}$$

$$= \sqrt{\frac{54{,}000{,}000}{(0.022)(0.52)}}$$

$$= 68{,}704.49 \text{ or } 68{,}704 \text{ units}$$

The sparing model

Determination of spare parts inventory quantities is a function of the probability (confidence factor) of having a spare part available when required, the component failure rate, the composite reliability, and the quantity of the parts used on the system. The formula used for determination of spare parts availability is

$$P = \sum_{n=0}^{n=s} \frac{(R)[-(\ln R)^n]}{n!}$$

where P = probability of having the particular spare part available when required
  S = number of spare parts carried in stock
  K = quantity of the spare parts used
  R = composite component reliability; $R = e^{-k\lambda t}$
  $\ln R$ = natural logarithm of R
  n = designator for quantitative sequence of individual spares, e.g., $n = 0$, $n = 1, n = 2, \ldots, n = s$.

Figure 18.6 portrays a spares nomograph which greatly simplifies determination of quantities of spares and which is a preferable alternative to the formula. The scalar structure of the spares nomograph is

  Scale (1) K = number of parts
  Scale (2) λ = component failure rate (expressed as failures per 1000 operating hours)

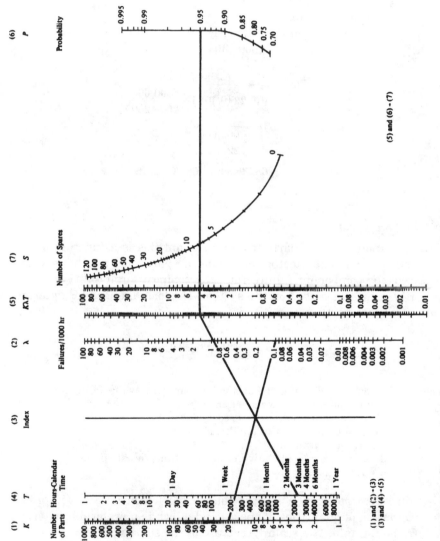

Figure 18.6 Spare-part requirement nomograph.

Scale (3) Index line used for plot purposes

Scale (4) T = reorder cycle (expressed as hours or calendar units)

Scale (5) $K\lambda T$ = composite reliability, a product of scalar elements 1, 2, and 4 (hours)

Scale (6) P = probability of part availability, also expressed as confidence value

Scale (7) S = number of spares required to be on hand to assure availability at specified confidence level P.

Example To use the spares nomograph, assume the following:

$K = 20$

$\lambda = 0.1$ (failures per 1000 hours)

$T = 3$ months, based on 24-hour-per-day system operation (2160 hours)

$P = 0.95$, desired probability (or confidence) of having spares available when needed.

Sequential plot steps:

1. Connect the scale 2 point where $\lambda = 0.1$ to the point at scale 1 where $K = 20$ with a straight line, thereby intersecting the index, scale 3.
2. Plot a straight line from the point on scale 4 where $T = 3$ months or 2160 hours, through the point where the index line intersects the line previously plotted during Step 1, to scale 5. This line will join scale 5 at the point where $K\lambda T$ is 4.32.
3. Plot a straight line from this point, 4.32 on scale 5, to the 0.95 probability point P on scale 6. This line will intersect scale 7 at the point where $S = 8$.
4. The value of S determined by the intersection of the step 3 plot line with scale 7 is the required spares level. For this case, the required level is eight spare parts.

Compared to use of the complex algorithm, the utility to an analyst of the spares nomograph, along with a straightedge and pencil, is obvious.

Economic Implications of Inventory

The economic focus of the inventory manager of a commercial enterprise is the costs of inventory and the profitability of the inventory policy.

Inventory costs

The predominant elements of inventory costs are (1) order processing or production setup costs, (2) investment costs, and (3) warehousing costs, as well as the subtle and elusive cost effects associated with inventory risks and stock-outs.

Order placement or setup costs.   Order placement costs are generated at a distribution center and in the plant purchasing activity. The equivalent type of cost in the production process is the cost of setting up a machine or a production line to produce an item.

Inventory investment.   Investment in an inventory item includes the purchase price, if it is acquired from an outside vendor, or manufacturing and transfer costs, if it is acquired from another corporate activity, plus the cost of inbound transportation. To this must be added any costs of transportation to distribution centers where other inventories may be located for subsequent shipment to customers.

The most significant impact of inventory investment takes the form of *opportunity costs*, i.e., what the equivalent monetary amount would earn if it were invested in interest-bearing financial market accounts or similarly lucrative opportunities.

Warehousing costs.   Warehousing costs, or inventory holding costs, include storage (facility) costs, property taxes, and insurance costs, but not the costs of moving goods into and out of a warehouse.

Unless capacity limits are being reached, the incremental cost of storing more inventory may be very small. Once capacity limits are reached, the incremental cost could markedly increase.

Inventory risk.   Costs of spoilage, damage, obsolescence, and pilferage are difficult to analyze and vary significantly by type of product.

Spoilage or damage can result in the total loss of a product (as with perishable produce), the reworking of defective products, or the reallocation of a product from one distribution point to another. In the case of a total loss, not only the product cost but the associated transportation and storage costs of the product are lost. In the rework process, costs incurred are those of extra transportation (two extra shipments, to and from the plant), potential loss in rework, the cost of rework itself, extra storage, and loss of customer goodwill. Concealed spoilage or damage may, unfortunately, not be recognized until a disgruntled customer complains.

Obsolescence is often difficult to define. A product may be obsolete when it has been superseded by a competitive item or when it has not been sold from a distributor or retailer's inventory for some time, even though it continues to be sold in other market territories. Most firms identify obsolete merchandise by the frequency of orders for such items. An item for which no order has been received at a distribution center for 90 days might be considered dead stock at that center and become a candidate for reallocation to other distribution centers where the product is an active market item.

Inventory losses from pilferage have two adverse effects. First, they represent a financial loss. Second, the pilfered goods may continue to be carried in the inventory as phantom assets and noted as available for sale until the loss is discovered.

Stock-out costs. "Hard" stock-out costs that are comparatively easy to assess include the costs of duplicate order processing, extra communication, and premium transportation costs. It is not abnormal to find such costs amounting to double or triple the costs for processing of a routine stock order.

The "soft" costs of stock-outs, which are seldom analyzed and nearly impossible to measure, are those of lost selling time (a sales representative often must make special efforts to maintain customer goodwill) and lost profit.

Inventory profitability indicators

The ultimate goal of all commercial enterprises is profit. As a member of the company team, the inventory manager has this same corporate objective. The primary indicator of inventory profitability is the stock-turn rate, which reflects the marginal utility of the initial inventory investment. If, for example, the stock-turn rate for a given product is 12, that would indicate that the monetary mount invested in the initial lot was "sold" a dozen times during a 12-month period. A profit margin of only 2 percent per individual lot sold would therefore generate 24 percent gross profit on the original lot investment during the year. After the initial investment of seed money, all subsequent stock replenishments are financed by sales revenues. The longer the active market period for that product, the greater the gross profit ultimately achieved as return on the initial investment.

The stock-turn rate may be expressed in a variety of ways, e.g.,

$$\text{Stock-turn rate} = \frac{\text{total yearly demand}}{\text{order quantity}}$$

or

$$\text{Stock-turn rate} = \frac{\text{total yearly demand}}{2(\text{average inventory} - \text{safety level})}$$

or

$$\text{Stock-turn rate} = \frac{\text{total yearly sales less returns}}{\text{monetary value of order quantity}}$$

Another method frequently used by corporate distribution centers is to express stock turn in number of days, i.e., the number of operating days for complete turnover and replenishment of the inventory. For example, a stock turn of 5 days would, assuming 250 operating days per year, translate into an annual turnover of 50 times. Typical high stock-turn enterprises are supermarkets, which traditionally earn small gross profit margins on sales revenues. On the other hand, the stock turn for an automobile dealer would probably be lower, but, understandably, gross profit margins on product sales would be higher.

19

Personnel Management

Overview

Effective exploitation of logistics technology is contingent upon prudent and efficient utilization of human resources. The ultimate criterion for the corporate organization is the achieved efficacy of the human-machine interface. Human efficiency in an organization is a function of managerial enlightenment and expertise. In this regard, management of logistics personnel is governed by the elements of (1) definition and understanding of the corporation's mission, its products, and its organization as a goal-oriented activity; (2) definition of the logistics mission, functions, and organization within the corporate structure; (3) development of the essential logistics human resources; (4) motivating commitment and productivity from the human resources; and (5) having the ability to redefine and systemically adjust as corporate roles and objectives are redefined.

Corporate Mission

The mission of any commercial enterprise may be broadly stated as to provide a range of goods in the form of systems, products, or services to a defined market and earn a profit for its efforts. The firm endeavors to maintain profitability by providing such goods at competitive prices and by cultivating, sustaining, and continually expanding its customer base within the parameters of its productivity.

Organization of the enterprise

An organization is a logical combination of constituent activities structured so as to achieve stated goals. It consists of a number of persons or groups with specific responsibilities who are united for the same purpose. As a process, organizing the enterprise includes dividing the work into individual jobs as necessary to achieve the company objectives, and instituting means of coordinating the efforts of the individual workers. The result of this process of amalgamation is the *organization*, which defines the participant workers and the network of relationships between them.

The nature of the systems, products, and services marketed by the enterprise can influence the legal form of the organization, as established within the governing jurisdiction, and its structural form.

Legal forms of organization.   In legal terms, any individual or group operating a business for profit is a sole proprietorship, a partnership, or a corporation. The sole proprietor owns all assets, is entitled to all profits, and assumes all losses, risks, and debts of the business. A sole proprietorship is well suited to a small enterprise such as a small "mom and pop" retail activity, but it can also prove successful in much larger operations, as clearly demonstrated by Howard Hughes.

A *partnership* is a voluntary "association of two or more persons to conduct, as co-owners, a business for profit." Many large organizations such as architectural firms, doctors' clinics, and law offices are partnerships, though many are switching to the corporate form now that it is permitted for professionals.

A *corporation* was defined by Chief Justice John Marshall in the Dartmouth College case of 1819 as "an artificial being, invisible, intangible, and existing only in contemplation of the law." The law treats the corporation as a legal person, a business entity that can sue or be sued, hold and sell property, and engage in business operations. A corporation is chartered under the laws of a particular state. The corporate legal form is suited to large, complex organizations with extensive financial investments. A corporation is owned by stockholders and managed by professionals who may or may not also be owners.

Although sole proprietorships make up about 69 percent of all businesses, they account for only about 25 percent of sales. Partnerships involve 5 percent of enterprises and generate 7 percent of sales. Corporations constitute 25 percent of all organizations but account for 67 percent of total sales. Other forms of organization account for about 1 percent in each category.

Logistics and service activities are governed by the legal form of their parent organization.

Structural forms of organization.   The most common form of organization is the *bureaucracy*. According to a translation from Max Weber's *Bureaucracy*, "The fully developed bureaucratic mechanism compares with other organizations exactly as does the machine with nonmechanical modes of production, precision, speed, unambiguity, knowledge of the files, continuity, discretion, unity, strict subordination, reduction of friction and of material, and personal costs—these are raised to the optimum point in the strictly bureaucratic administration."

While the bureaucratic structure is most frequently adopted for commercial activities, it has been found that most organizations have distinct cultural characteristics, which have spawned informal, internal structures of authority. The informal organization, in which the division of duties and the authority structure do not necessarily reflect stated procedures, may evolve along with the formal structure. One of the objectives in developing a formal structure should be to produce a system which will function as planned and to ensure that any informal structures that emerge aid rather than impede achievement of the organizational objectives. In addition to the bureaucracy, other structures which are useful for specific purposes include the committee, overlapping/integrative, matrix/grid, and project/task/goal structures.

Organization charts

An effective means of depicting an organization and communicating its infrastructural functional interfaces is through the use of organization charts. This section will portray each organizational structure from a top management perspective, with which the subsidiary logistics management organization must be integrated and functionally supportive.

Bureaucratic organization.   Figure 19.1 depicts a simplified version of the classic and predominant bureaucratic organizational form. The figure

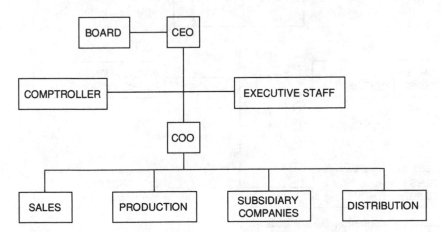

Figure 19.1 Bureaucratic organization.

portrays a typical line-staff organization with an emphasis on functional lines of control.

According to the recognized logistics management expert and author Joseph Patton, "A line organization is made up of those whose work contributes directly to the achievement of a fundamental goal. The staff are those who assist the line in some way, either by providing services or developing plans, giving advice, or auditing performance" (*Logistics Technology and Management,* 1986). The distinction between line and staff in business is similar to that in the military.

Committee organization.   The purpose of a committee is to focus on a specific issue or a group of related issues. Members of a committee, whether they are standing members or ad hoc members, should be knowledgeable about the issues addressed and qualified to represent the interests of their parent organizational elements. A committee aims to reach a consensus on issues rather than to make a decision. Although committees are not normally authorized to make decisions, they are empowered to collectively make recommendations on the issues at hand which, if the individual members have served effectively, will be approved by their respective superiors and result in a multiparty accord on the final decisions. Figure 19.2 portrays a basic committee organization.

Experience has confirmed that a smaller committee is more effective than a larger group; the prospects of agreement on issues tend to diminish as the size of the committee increases.

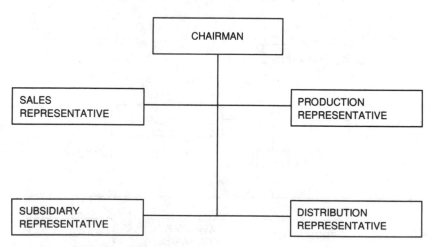

Figure 19.2 Committee organization.

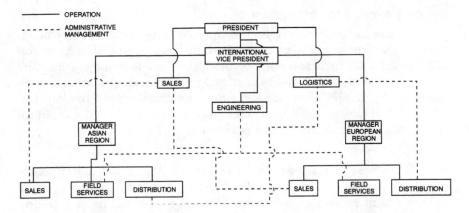

Figure 19.3 Integrative organization.

Integrative organization. The typical commercial version of the overlapping integrative organization resembles a combined or joint military command and control structure (such as that established for the North Atlantic Treaty Organization or pursuant to similar international treaty agreements). Such organizational structures are characteristic of multinational corporations. In such an organization, a single headquarters executive authority controls an organization with a permanent cadre of functional corporate departments and regional branches in globally dispersed areas. Each regional operation is conducted by a combination of indigenous personnel who are technically qualified and familiar with the locale and corporate assignees from home-based organizational elements. In such a scenario, the detailed personnel are "owned" and under the administrative direction of their parent departments, while being under the operational control of the regional manager. Figure 19.3 portrays a simplified version of an integrative organization.

The integrative organization has been described as an example of the functional matrix organization concept being grafted onto a multinational corporate structure. The proponents of such a structure declare that this type of organization, as typically applicable to a multinational corporation, exploits the technical expertise and corporate memory of the assignees coupled with the regional experience and knowledge of the indigenous employees.

Logistics Mission

The corporate logistics mission and organization are inherently derived from the nature of the goods marketed by the enterprise and the form and structure of the parent organization.

Factors governing role of logistics

The primary factors which influence the role of logistics in an organization are the type of business transacted; the importance of logistics relative to the total costs of doing business; the need to manage trade-offs between the important logistics cost categories of transportation, inventory holding, and prospects for lost sales; the complexity of the logistics network; the nature of the overall corporate strategy; and the capabilities of the logistics management personnel.

Type of business transacted.   In organizations that do not manufacture the goods they market to the public, logistics assumes a greater relative importance in corporate operations. Thus, in a wholesaling or retailing organization, it may be called operations and be collocated with accounting and sales.

Importance of logistics costs.   Logistics costs of transportation, inventory holding, and lost sales may be of minimal importance in a firm that manufacturers high-technology items, but they could be as much as 40 percent of the cost of doing business in a firm producing basic raw materials or agricultural products. Studies have established a degree of correlation between the hierarchal level of the most senior logistics executive and the significance of logistics in a firm's total cost posture. A primary contributory factor influencing the importance of the logistics management staff is the importance of the need to manage trade-offs among critical logistics cost categories.

Need for trade-offs among logistics cost categories.   Logistics cost trade-offs contribute to the complexity and sensitivity of the logistics manager's tasks. This situation is typified by considerations such as the following:

1. Should production line resupply be accomplished by JIT procedures, with the attendant low quantity purchase cost penalties, nonroutine order processing, and premium transportation costs, or should stock levels be based on economic order quantities, with routine order processing, large quantity purchase discounts, and routine transportation?

2. Should predicted stockage levels be augmented by safety stockage to assure a high confidence of availability, or should the investment in inventory be reduced despite the associated risk of stock-outs and lost customers?

The more pervasive this dilemma is within the corporate structure, the more significant the logistics cost issues and the greater the importance of the logistics manager.

Complexity of the logistics network. Logistics management entails the coordination of a network of suppliers and market centers, few of whom may be selected by the logistics manager. Service may be provided through a system of collection points, plants, mixing points, and distribution warehouses. Each additional network point increases the network's complexity as well as the possibility of error in analysis or management of the network, which diminishes control over logistics costs. Such loss of control may ultimately lead to management's focusing on the logistics cost situation and paying heed to the logistics functions.

Nature of corporate strategy. Goals for the logistics function in a firm may occasionally shift from cost minimization to an emphasis on customer service, reflecting management's desires to achieve the highest profits in the long run. When such a shift occurs, the importance of maintaining a direct line of communication between those managers setting corporate strategy and those managing logistics operations suggests the desirability of a direct reporting relationship to top management from the logistics manager.

Division structure and reporting relationships reflect corporate strategy and thereby influence the extent to which common logistics facilities, such as private transportation fleets and warehouses, are used jointly by two or more business divisions and how the logistics services are apportioned and paid for.

Capabilities of logistics managers. A lack of capacity or talent in the logistics management of a company may dictate that it adopt a strategy which reduces the complexity of the management task, delegates logistics responsibility to other functional managers, or otherwise minimizes the impact that any single logistics manager or department could have on total corporate performance. Conversely, evidence of strong logistics management expertise may induce top management to develop a strategy based largely on its logistics capability.

Definition of logistics functions

The following generically define the major functional areas of logistics. Some of these tend to be more prominent in military logistics than in commercial logistics; some are accorded greater emphasis in the commercial sector than in military logistics.

- Product maintenance
- Inventory management
- Logistics technical data
- Product packaging
- Support and test equipment
- Materials handling
- Warehousing and storage
- Transportation management
- Facilities planning
- Embedded computer resources
- Purchasing
- Logistics training
- Logistics human resource management
- Logistics management information systems
- Logistics engineering

These generic logistics subdisciplines provide a baseline that can be further refined and defined to conform to the specifications of the corporate logistics mission as well as to identify and develop the required human resource skills and position descriptions.

Organization of logistics

The logistics activities must be organized in such a way as to be functionally coordinated with the other company departments in order to enable effective integration of the total resources to achieve the corporate logistics objectives. The portrait of the organization will inevitably vary according to its mission, whether it be a system research and development program, production program, service center operation, distribution center, retail-level activity, etc. To illustrate this concept, Table 19.1 provides a matrix describing the integration of the logistics activities of a typical production facility with the other functional corporate elements, on which basis a functionally integrative and supportive logistics organization could be developed.

The project organization.   Figure 19.4 portrays a typical project (sometimes termed task or goal) organization. This type of organization is not regarded as a permanent organizational entity. It is assigned an ad hoc mission and, as it is totally oriented toward successful completion of its assigned program, is temporary by nature.

TABLE 19.1 Logistics Interface with Other Corporate Functions

Logistics activities	Marketing	Finance	Production	Traffic management	Purchasing	Human resources	Engineering
Requirements forecasting	X		X				
Order processing	X	X					
Finished product transport, warehouse to customer	X		X	X			
Finished product inventory control	X	X	X				
Distribution center warehousing	X		X	X			
Transportation from plant to distribution center	X		X	X			
Packaging and packaging design	X		X		X		X
Production planning	X		X				X
Material storage			X	X	X		
Production material control			X		X		
Facility location planning	X		X	X		X	
Material transportation		X	X		X		
Material inventory control		X	X		X		
Material provisioning		X	X		X		
Customer support and service	X		X				

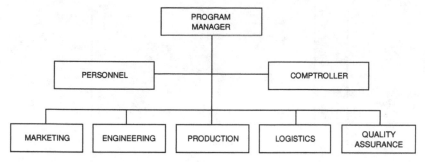

Figure 19.4 Project/task/goal organization.

Members of a project organization are typically specialized professionals with the requisite expertise to focus on the program requirements and achieve results expeditiously.

The matrix organization. The matrix or grid organization (the foundation of the *matrix management* concept) is utilized when a defined project or task requires a specific combination of technical skills or disciplines to support certain work elements of the primary project. The participants may be assigned to the project on either a part-time or a full-time basis. Under a matrix management structure, as illustrated by Table 19.2, the individuals involved are under the functional and administrative management of their parent organizations, but under the operational control of the project managers. The matrix management approach is effective when there are efficient lines of communication and rapport between the project managers and the functional managers of the individuals assigned to the project. Lack of this rapport can have adverse consequences for both the project manager and the assigned individuals.

TABLE 19.2 Grid/Matrix Organization

Functional managers	Program managers			
	Project A	Project B	Project C	Project D
Engineering	3	3	3	3
Logistics	1	2	1	3
Reliability	1	1	1	1
Maintainability	1	3	4	5
Quality	1	2	1	2
Safety	2	2	2	1

The matrix organization, like the project or task organization, is task-oriented. It offers considerable utility in fulfilling the requirements of a specific task when there is limited availability of certain critical skills necessary to accomplish the task objectives.

Logistics under a centralized logistics management structure. In a centralized management organization, top management typically reserves responsibility for the profitability of the constituent activities for itself. In order to provide for profit responsibility, it is often necessary in a centralized organization for all important decisions to be made or coordinated at the top, and so the flow of information must be directed to and from the highest level.

A centralized organization is typically structured along functional lines, with senior executives in charge of finance, marketing, manufacturing, and other major functions reporting directly to a chief executive officer, who is responsible for coordinating major functional decisions.

The basic advantages of centralized management are that it tends to result in more effective use of warehouse and private transport facilities, coordination of the purchase of transportation and other services, and central control over inventories used by two or more business units. An example of a functionally oriented centralized organization is shown in Figure 19.5.

Logistics under a decentralized logistics management structure. In a decentralized management organization, decision-making authority, particularly for profit, is delegated to lower ranks of management within

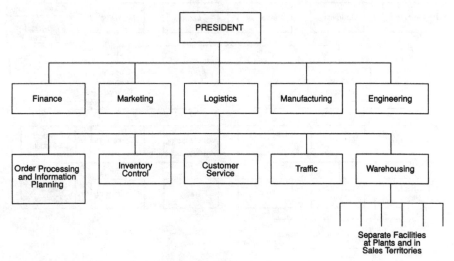

Figure 19.5 Logistics under centralized management.

the organization. Profit responsibility is centered in strategic business units, each with control over its own operating functions, such as marketing, manufacturing, and logistics.

In the decentralized approach to logistics management, two or more divisions or business units might house such groups, as shown in Figure 19.6. Under this approach, decisions with respect to profitability are made at the divisional or business unit level, with corporate officers responsible for advice on such matters as personnel, labor relations, and legal matters, and having line responsibility for the acquisition and allocation of funds to the subordinate elements.

Organization for logistics services. Figure 19.7 provides a functional organizational scheme which describes a typical management services operation for a major hotel chain. The composite shown in Figure 19.7 is based on a study of an international hotel conglomerate conducted in order to identify pertinent logistics elements within the typical management services structure. Such a service organization generally incorporates the functional groups at corporate staff level based on the centralized management concept, whereas line control of the regionalized distribution centers is effected on a decentralized management basis.

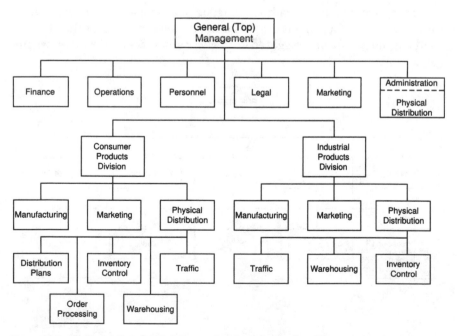

Figure 19.6 Organization for logistics under decentralized management.

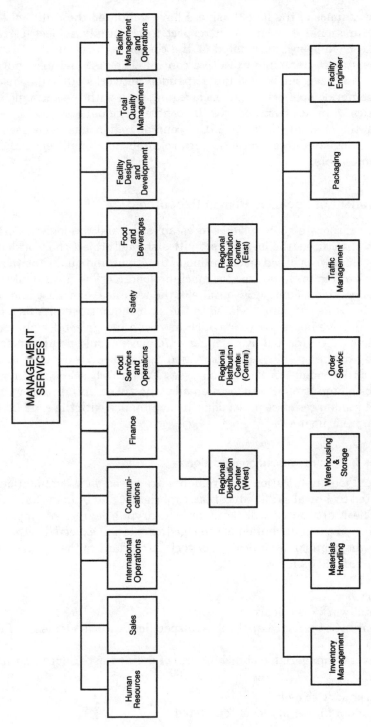

Figure 19.7 Organization of logistics services for a major hotel chain.

The dynamics of the hotel business have validated the utility of this type of functional division. With respect to the study subject, certain functions have been consolidated at the corporate level in order to reduce costs and streamline executive control. The regional distribution centers have been subdivided into expanded networks of similar distribution centers dispersed within each region, each with the substructure of logistics activities portrayed for the central regional distribution center. (For the sake of illustration, the central region has been used to depict the logistics organizational pattern applicable to all regional distribution centers.)

Development of Logistics Human Resources

After the corporate goals have been established and the logistics organization structured, the logistics activities should project the workloads for their employees based on definition of functions and quantitative factors comparable to those used for machine loading. The functional task analyses provide a cumulative profile of the technical expertise and corresponding time estimates related to the skill requirements. A factor of 1800 productive hours per year, which has been recognized for planning purposes as the expected average per employee, should be apportioned according to the specific skills identified and the corresponding levels of effort as determined by the task analyses. From this baseline, the human resources planning effort can focus on allocation of essential skills within the total human resource projections, in compliance with the established budgetary parameters.

Logistics human resources qualifications

Individual job qualification requirements are based on a combination of job-related technical skills, education, experience, and human characteristics. These attributes will be accorded varying degrees of emphasis depending on the contribution to the logistics mission expected of the job. Some of the common qualification criteria attendant to these attributes are summarized as follows:

1. *Experience.*
   a. Total work experience
   b. Years of experience applicable to specific technical elements of the job
   c. Level (managerial, technician, etc.) of defined experience desirable
2. *Education.*
   a. Secondary school
   b. Years of university level completed

c. University degree (B.S., B.A., etc.)
d. Years of advanced degree study completed
e. Advanced degrees achieved (master's, doctorate, etc.)

3. *Professional certification.* Certain engineering and technical professional organizations conduct certification programs to accredit by a process of qualification, examination, and peer review those of their members who have achieved and demonstrated a prescribed level of expertise and experience relative to their professions.

There are certain cases in the development of logistics job requirements in which it is appropriate to stipulate that qualified individuals possess pertinent certification credentials, such as Certified Professional Logistician (CPL), granted by SOLE—the International Society of Logistics (also known as the the Society of Logistics Engineers); Certified Configuration and Data Manager (CDM), granted by the National Defense Industrial Association (formerly American Defense Preparedness Association); or Professional Engineer (PE), granted by an appropriate state jurisdictional board pursuant to the results of a standard national examination. Certain types of governmental projects or other programs subject to certain regulatory criteria might require use of certified professionals.

4. *Technical skills.*
   a. System- or product-related technical skills
   b. Perceptual ability
   c. Analytical skills
   d. Verbal communication skills
   e. Written communication skills

5. *Human characteristics.*
   a. Motivation
   b. Intelligence (I.Q.)
   c. Persistence
   d. Innovativeness
   e. "Team player" qualities
   f. Electromechanically inclined talents
   g. Dependability
   h. Work ethic
   i. Learning proficiency

Every enterprise should maintain an active file detailing the skills of its personnel and other resources it might utilize. The repository could be as elementary as the relevant résumés, or the data could be maintained in a computerized data bank.

Skills information supports a multiplicity of needs. Human resources planning should include identification of possible replacements for every significant position in case of promotion, illness, resignation, or other attrition. If such replacements are not already predesignated, training should be provided to develop them. Government contracts or subcontracts may require security clearances, certificates, or particular qualifications which would require the provision of such data. Finally, the skills data bank enables an organization to evaluate itself in terms of availability of personnel with the requisite skills and provides an inventory of talent for advance planning and development of those with professional potential.

Development of position descriptions

The position description is important to both the employee and the supervisor. This document sets forth what the company expects the employee to contribute to the logistics mission; for management, it is a vehicle through which to assess employee performance and establish appropriate compensation levels for the position. The classification of an individual in the organization can be determined on the basis of

1. The major purpose being served (e.g., firefighter, teacher, president)
2. The process used, such as data processing, engineering, or accounting
3. The types of persons dealt with or served, such as production, field, or customer
4. The place where the work is done, such as the loading dock or branch office

Basic elements of a position description should include the title, the reporting relationship, the scope of responsibility, and measure(s) of performance associated with the job. The scope of responsibility is defined by the size and complexity of the various tasks to be performed. It should be described in basic terms without superfluous verbiage. Each task should be described by an action verb and an object noun; e.g., a traffic manager may be responsible for hiring and firing owner-operators, scheduling loads, allocating equipment, or establishing internal transport cost rates. Measures of performance are an integral part of the position description. The criteria for measuring performance and the specific goals to be achieved should be mutually understood and agreed to by both supervisor and subordinate, which would connote the supervisor's approval and the employee's commitment.

Although the format for position descriptions may be standardized, it should not be assumed that within a specific logistics discipline, all positions are similar and therefore should have a standard job description. It is likely that there will be variations in work requirements among positions within the same discipline. The following is an accepted generic outline for the types of data to be included in position descriptions, which can be tailored to reflect specific organizational requirements.

1. Purpose of the position as related to the logistics mission
2. Principal activities and expected results (what is done, how it is done, and why)
3. Complex problems and unusual aspects that might differentiate this position from others
4. Quantitative supporting data, such as number and skills of personnel reporting through the position, budget size, production rates, machine services, and capital equipment responsibility
5. Special education, experience, certification, or skills required

This outline permits the incorporation of descriptive details within each cited category to accommodate specific job requirements. From the data developed pursuant to the outline, coupled with the aforementioned classification guidelines, a position profile should be established for guidance in determining equitable and competitive compensation schedules.

Recruitment and training

Recruiting the right person for the right job is ultimately the responsibility of the functional manager, although he or she can be ably assisted in this process by the personnel or human resources department. The manager must clearly stipulate the job requirements in the form of a well-defined position description accompanied by a short notice succinctly synthesizing the position description for use in advertisements and solicitation notices.

Sources.   The more frequently utilized sources of qualified candidates are, in order of preference,

1. Personal referral—soliciting a "known quantity."
2. Publishing notices in professional journals and newsletters.
3. Advertisements in public media outlets (newspapers, television or radio spots, etc.).

4. Use of professional services agencies (executive search firms, sometimes called "headhunters" or "body snatchers"). These agencies are likely to require a significant service fee, in typical cases equivalent to at least one month's salary for the position.

The human resources department can relieve the functional manager of many of the administrative trivia and procedural details involved in this process, on the presumption that the manager has provided a clear, unambiguous description of the job requirements, compensation levels, and essential candidate qualifications to govern the recruitment effort.

Coordination of the manager and human resources.   Each candidate attracted to the job opportunity should be screened through a coordinated effort involving both the manager and human resources. This is typically a multistep process.

1. Human resources interviews and screens the candidate on the basis of an introductory checklist providing elementary selection criteria. If a screening test is required for any reason by corporate policy, this is the appropriate point to administer the test.
2. If the candidate passes the initial screening criteria, he or she is further evaluated by a knowledgeable assistant or deputy to the functional manager on the basis of a more comprehensive checklist focusing on pertinent technical details relative to the position.
3. If the candidate passes the functional step 2 screening, the manager conducts the final interview.
4. For candidates that survive the manager's screening criteria, human resources validates the application data and confirms the candidate's references, educational credentials, professional certifications, etc.

The complexity of these four basic steps and the time commitment involved will be governed by the technical nature and criticality of the position. Once a selection is made, it is advisable to tender a job offer promptly (a verbal offer should be immediately confirmed in writing), with a condition that the candidate respond by a specified deadline. The shorter the deadline, the less likely the candidate would be to auction himself or herself to other prospective employers.

Training

The training process is a critical adjunct to effective utilization of a product or a process involved in producing goods for the market.

Training within commercial logistics activities. Once selected, the new employee must be developed through a process of counseling and training. It is often advisable to place the new employee under the mentorship of an older employee. The supervisor must also take the time necessary to explain to the new employee what is expected and what support is available. It is not advisable to leave new employees to fend for themselves. A small commitment of time on the manager's part will be repaid many times over by the employee's improved performance. Some commercial organizations conduct formal training for new employees shortly after they join the organization. This training may be accomplished in a classroom environment through programmed instruction, video techniques, computer-assisted instruction, correspondence courses, or a combination of those methods. Informal instruction is normally conducted as on-the-job training (OJTI) rather than formal schooling.

As addressed in the discussion of human factors engineering in Chapter 6, humans learn best when all senses are involved—sight, sound, feeling, taste, and smell. Depending on the importance of the information to the receiver, retention may be negligible or up to 90 percent. Retention is rarely at the level of 100 percent. The second time the information is presented, particularly if it is presented in a different format, a higher percentage will be retained through reinforcement. A third exposure may provide further reinforcement, but beyond that point, retention diminishes. People learn best by hands-on practice. Motor skills such as turning screws or measuring amperage cannot be acquired from books. They should be demonstrated and subsequently performed by the student. One senior worker should teach others, as this will inspire the mentor to refine his or her own skills in order to teach properly.

Two-way interactions are desirable so that questions may be asked and addressed as they arise. Information should be presented and then reinforced. The material taught should be based on what management requires and has decided upon with respect to

1. Difficulty
2. Frequency of occurrence
3. Time involved
4. Cost impact

This premise is especially applicable to maintenance tasks which, by nature, are labor-intensive and expensive in terms of financial resources. If a task is difficult and can be characterized by hands-on interactions, these should be taught—assuming that they will be encountered often enough for the learning to be retained. The rate of development can be accurately established and is very useful in determining workload and predicting system maintainability.

Certain types of training can be acquired only under realistic conditions. Installation of equipment, for example, is best learned by installing products at a customer's location. Preventive maintenance can be best learned on a machine in a customer's operating environment. To learn well under realistic conditions, a newly trained employee should be placed under the supervision of an experienced team leader or field instructor to ensure that procedures will be practiced in the factory or field according to management precepts.

Human relations training can be conducted in more formal settings through the use of role playing. Audiovisual materials such as videotape are particularly valuable, since they allow people to see how they react to situations, as well as giving them the opportunity to see themselves as others see them. The penalties for failure are far less serious in a classroom than in a customer's location. Management development can also involve exposure, training, and reinforcement by experience. Seminars, workshops, and similar programs are especially effective in this area, since they reinforce the learning experience of the work environment.

Management of Human Resources

The primary functions of management are *planning, organizing,* and *controlling.* Whereas the working-level employee accomplishes assigned functional tasks through a process generally defined by the human-machine interface, the manager achieves assigned management tasks through people, a resource that is infinitely more complex, more inscrutable, and much less predictable than machines and technology. The successful manager adroitly exploits the inherent strengths and talents of his or her subordinates, while minimizing the vulnerability to individual weaknesses, so as to maximize group productivity.

Employee motivation

It is axiomatic that workers have individual needs and require motivation to produce up to their individual capacities. A. H. Maslow aptly captured the spectrum of basic human priorities through his description of the hierarchy of human needs. The Maslow hierarchy is depicted in Figure 19.8.

Maslow's articulation of basic human needs in order of importance is applicable to most populations. Maslow theorized that humans normally strive to satisfy the most basic needs first, then, after fulfilling these,

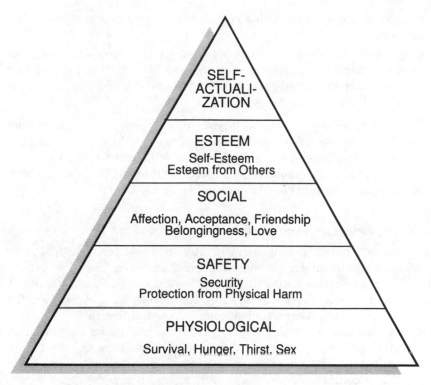

Figure 19.8 Maslow's hierarchy of human needs.

proceed to the second level; the sequence will be continued until achievement of the ultimate level, self-actualization.

The manager's role in employee motivation. The logistics manager has the exclusive authority, responsibility, and opportunity to establish an environment conducive to employee motivation. This embodies certain basic tenets of motivational management.

1. *Employee understanding of job requirements.* This is accomplished through development and employee understanding of a definitive position description.
2. *Valid assessment of employee abilities.* This is a function of logic and accuracy in the screening and evaluation of individual employee qualifications and of supplemental training provided by the organization. The employee must be mentally and physically able to perform the job.
3. *Effective integration of individual ability with job requirements.* If the manager has properly defined the position and accurately evaluated

individual qualifications, this process is a simple fusion of those two elements.

4. *Clear understanding of work objectives.* It is incumbent upon the manager to clearly convey to the employee what will be achieved by accomplishing the stated goals, as well as the consequences of not accomplishing them.

5. *Link employee aspirations to organizational goals.* A manager has greater prospects of success if it is conveyed to the employee that achievement of organizational objectives will foster achievement of personal goals. The employee must perceive that the purpose of a task is compatible with his or her personal interest.

Optimizing human resource productivity

As previously noted, management of human resources, as opposed to machinery and technology, is more artistic than procedural. Two attributes that characterize successful managers are the qualities of being a mentor and an effective communicator. Employees with aspirations for self improvement and career advancement respond favorably to a manager who imparts his or her own expertise to ambitious subordinates (i.e., a good manager is also a good teacher). Effective communication is a critical catalyst to successful enterprise. Hence, the following basic principles of communication are equally as valid in commercial enterprises as in the traditional military environments.

1. A communication from a manager must be legal and within the purview of stated organizational policies and purposes.

2. A communication must be clearly conveyed, as intended by the manager. (What is spoken must be what is meant to be spoken.)

3. A communication from a manager must be clearly understood by and within the comprehension of the employee.

4. The manager's communication must be acknowledged by the employee.

Having established the approach to achieving organizational objectives, the manager should formulate the means to measure and review human resource productivity. This process involves indicators of employee productivity, employee performance appraisal, and a schedule for rewards and recognition for exceptional achievement.

Productivity indicators.   For product processing, the measures of employee output can be parameters such as task elements accomplished or product units processed. It is the manager's responsibility to develop specific

TABLE 19.3 Interrelationship of Logistics Disciplines and Generic Job Categories

Logistics Discipline	Logistics engineering	Technical data specialist	Technical writer	Customer service representative	Cost analyst	Materials handling systems designer and technician	Facility design and planning	Training specialist	Packaging designer and technician	Calibration technician	Logistics data specialist	Logistics MIS data manager	Warehouse technician	Warehouse management and planning	Traffic management and planning	Material control	Production material planning	Inventory management and planning	Material provisioning specialist	Procurement specialist	Contract administration	Maintenance technician	Maintenance management and planning
Maintenance	X	X	X		X			X		X	X	X										X	X
Inventory management	X	X	X	X	X			X						X		X	X	X	X	X	X		X
Packaging	X	X	X		X	X	X	X	X					X		X	X	X	X	X	X		X
Materials handling	X	X	X		X	X	X	X	X				X	X	X	X	X	X	X	X			X
Warehousing and storage	X	X	X		X		X	X					X	X			X	X					X
Transportation	X	X	X	X	X		X	X	X	X					X	X	X	X		X			X
Facility planning	X	X	X		X	X	X	X					X		X		X				X	X	X
Logistics technical data	X	X	X	X	X	X	X	X	X		X	X		X	X		X	X	X		X		X
Logistics human resources	X	X	X	X	X	X	X	X	X	X		X		X	X			X	X		X		X
Logistics training	X	X	X	X	X	X		X	X	X	X	X		X	X			X	X		X		X
Embedded computer resources	X	X	X		X	X	X	X				X		X								X	X
Support and test equipment	X	X	X		X			X														X	X
Provisioning	X	X	X	X	X	X		X	X							X	X						X
Logistics communication	X	X	X	X	X			X				X		X	X	X	X			X	X		X
Configuration management	X	X	X		X		X	X		X		X						X			X		X
Reliability	X	X	X		X	X		X	X			X											X
Maintainability	X	X	X		X	X		X															X
Human factors engineering	X	X	X		X		X	X	X														X
Quality assurance	X	X	X	X	X	X	X	X	X	X	X	X	X	X		X	X	X	X	X	X	X	X
Safety	X	X	X		X	X	X	X	X				X	X				X				X	X

429

definitions of the discriminatory criteria that will enable judgmental evaluation.

Performance appraisal.   Guidelines for supervisory appraisal of employee performance should be categorized according to functional activity and level of assignment within that activity. Employees grouped at each level and within each functional activity should be evaluated on the basis of a stipulated norm. Under this concept, the manager establishes performance criteria for the average employee working within each defined population group. New employees should be evaluated at frequent intervals during the first year (e.g., three, six, and nine months), and subsequently the pattern should be an annual appraisal.

Rewards and recognition.   Employees who achieve normal, expected levels of productivity should be accorded normal increments in compensation. Those who perform in a manner superior to others in their peer group should be so recognized in their appraisal reports and considered for bonuses and award certifications. For those who failed to reflect expected improvements in performance, it is advisable to conduct a special counseling program. Although no financial reward would be in order for subpar performance, the employee should be afforded the opportunity to regain momentum and receive compensation when he or she achieves this.

Personnel performance profiles

This type of data is normally maintained by the human resources department. Personnel performance profiles provide a repository of individual employee skills and qualifications according to functional discipline. They also serve as a human resources data bank for future company programs and to identify candidates for critical vacant positions. Typical data profiles include individual résumés, records of education, certifications, appraisal reports, notes of special bonuses, corporate recognition received, etc. Personnel performance profiles portray the composite expertise of the organization and provide a useful pool of information for management planning purposes.

Interrelationship of logistics disciplines and job categories

Within the purview of logistics, there are a multiplicity of generic job titles which can be correlated with the traditional array of technical disciplines. One of the purposes of this book is to set forth a structure of such interrelationships. Certain job categories are generalist by virtue of their multidisciplinary utility; others reflect a higher focus of specialization. Table 19.3 provides a summary of the job category–discipline interface.

20

Logistics Technical Documentation

Overview

This chapter focuses on the family of system-related documents necessary for logistics support of the system. The primary documentation used by the logistician encompasses engineering drawings, installation drawings, technical manuals, and the spectrum of specifications used in development, production, and quality control of the system. Because of the contributory role of logistics throughout the total cycle of research, engineering development, production, and operation of the system, the pertinent technical documentation governing the evolution of the system during its life cycle has significant implications for the supportability of the system. The logistics engineers participate in defining those design characteristics which are translated into engineering documentation. The operational logistics managers, planners, and technical specialists are bound to use the final product documentation for functional purposes such as maintenance, inventory management, and provisioning, as well as the process of adapting the logistical support system to subsequent design modifications which alter the system design and configuration baseline documentation.

Specifications

In the engineering and logistics context, a specification is a document describing the processes and materials necessary for development, testing, procurement, production, assembly, installation, and logistics support of a system or product. The Department of Defense (DoD) maintains a comprehensive system of specifications and standards (including industry standards), which are identified in the *Department of Defense Index of Specifications and Standards (DODISS)*. Someone desiring further insight into and detailed information relative to the military system is encouraged to consult the specific categories and listings in the DODISS and the procedural details set forth in MIL-STD-961, Defense and Program-Unique Specifications Format and Content.

There are, additionally, such issuances as the federal standards, American National Standards Institute (ANSI) standards, Aerospace Industries Association (AIA) standards, American Society of Mechanical Engineers (ASME) standards, American Society for Testing of Materials (ASTM) standards, American Welding Society (AWS) standards, Institute of Electrical and Electronic Engineers (IEEE) standards, and standards promulgated by the International Standards Organizational (ISO). This section discusses specifications from the perspective of commercial product applications, incorporating, where appropriate and germane, principles of military specifications to strengthen the utility and effectiveness of the commercial applications. In keeping with the systems and life cycle requirements and disciplines the recognized family of specifications that primarily affect logistics applications includes the groups cited below.

1. *Detail specification.* This type of specification sets forth design requirements, such as materials to be used, how a requirement is to be achieved, or how an item is to be fabricated or constructed. A specification that contains both performance requirements and detail requirements is still considered a detail specification.

2. *Item specification.* This is a program-unique specification that describes the form, fit, and function as well as method for acceptance of parts, components, and other items considered elements of the system or equipment.

3. *Material specification.* A material specification describes such raw or processed materials as metals, plastics, chemicals, synthetics, fabrics, and any other material that has not been fabricated into a finished part or item.

4. *Performance specification.* This type of specification states requirements in terms of the required results with criteria for verifying compliance, but without stipulating the method for achieving the required results. A performance specification defines the functional requirements of the system or equipment, its operational environment, as well as interface and interchangeability characteristics. The requirements of a performance specification are frequently incorporated into a detail specification.

5. *Process specification.* This is a type of program-unique specification that provides procedures for fabricating or treating materials and items.

6. *Program-unique specification.* This type of specification describes a system or equipment, item, software program, process, or material developed and produced for use within a specific program, or as part of a single system and for which there is considered to be little potential for use by other systems or equipments.

7. *Software specification.* This is a type of program-unique specification that prescribes the requirements and verification of such requirements for a combination of elements that must function in unison to produce the capabilities needed to fulfill the performance requirements, including hardware, equipment, software, or any combination thereof.

Dual dimensions are a critical considerations in the current context of global technology. When specified acceptable requirements are stated in U.S. measurement units and the global perspective so dictates, the equivalent metric measurement units may be shown in parentheses in order to correlate the U.S. measurement units with their corresponding metric units.

Requirements often found in specifications

The following is a description of requirements that may be part of a specification. These requirements reflect the total scope of specified attributes, which are to be used as applicable. It is, therefore, unusual that a specification would incorporate all of the cited requirements.

1. *Reliability.* This section should state the reliability requirements numerically (with confidence levels, if appropriate). As a minimum, the reliability requirement should constitute a specified reliability value, time associated with the stated reliability, and a desired confidence level. For example, it could be specified that a system or equipment have 90 percent reliability at 2,000 hours of operation with a 95 percent confidence level. This means that the user would want to be 95 percent confident that 90 percent of the system population would survive at least 2,000 hours of operation.

2. *Maintainability.* Maintainability is a significant design parameter which should state quantitative maintainability requirements in such terms as Mean-Time-to-Repair (analogous to Mean-Corrective-Maintenance-Time), Mean-Preventive Maintenance-Time, maintenance man-hours per operational hour, and so on. Maintainability is a measure of the quickness and ease with which a failed system or equipment can be restored to required operational capability. Main-

tainability design parameters center around making a system repairable or restorable as easily, quickly and inexpensively as practicable. A commonly specified logistics-oriented requirement is Maintenance Ratio (MR), which is a measure of intensity of human resources needed to support a system or equipment. MR describes the ratio of maintenance labor hours to usage of the system or equipment.

3. *Availability*. This parameter quantitatively expresses the extent to which the system or equipment shall be in an operable status prior to engagement for the required performance functions. Availability is determined by the criteria of the algorithm stipulated by the users. If quantitative requirements for both reliability and maintainability are specified, this requirement is not applicable.

4. *Environmental conditions*. This paragraph should specify the environments that the system or equipment is expected to experience in shipment, storage, maintenance, operational service as well as other modes which portend environmental impacts. For entities that include software, these requirements should define the environment in which the computer software configuration item (CSCI) would operate, such as the computer hardware or the operating system on which the CSCI must function. This section should include as necessary environmental conditions such as climate, shock, vibration, noise, noxious gases, chemical agents, biological agents, electromagnetic radiation, and so on.

5. *Transportability*. This section should identify requirements for transportability and mobility that are common to all components of the system or equipment to permit operation and logistics support. For example, it could specify that the system or equipment be designed so that, with its packing for transport, each package would be no greater than ___ (volume units) and no more than ___ (length units) high, ___ (length units) wide, and ___ (length units) deep. It should identify all major functional units of the item that, due to operational characteristics, would be unsuitable for normal transportation methods (e.g., oversize, hazardous, or sensitive items).

6. *Materials and processes*. This paragraph should specify both general and specific requirements for materials and processes to be used in the item covered by the specification.

7. *Nameplates or product markings*. This section specifies all applicable requirements pertaining to nameplates or markings, with reference to the governing specifications, drawings, or schedules.

8. *Producibility*. Producibility entails the selection of manufacturing techniques, design parameters, and tolerances that enable the product to be fabricated, assembled, inspected, and tested economically with

assurances of repeatable quality. Product and process characteristics directly relating to safety, performance, durability, or supportability should be correlated and matched to corresponding manufacturing capabilities.

9. *Interchangeability*. This section should specify requirements for the level of assembly at which components should be interchangeable or replaceable.

10. *Safety*. This section should specify requirements so as to preclude or limit hazards to the physical environment as well as personnel and equipment. It also should cite established and recognized standards. The governing criteria should include health and safety considerations, including physical, mechanical, biological, and explosive effects. It is also important to identify those safety characteristics that are unique to the item that could influence the design due to hazards in assembly, disassembly, test, transport, storage, operation, maintenance, or disposal when they are not addressed by standard industrial practices. The applicable "fail-safe" and emergency operating restrictions are critical elements and should be addressed in adequate detail.

11. *Human factors engineering*. This section should specify the pertinent human factors engineering (HFE) requirements, including those of a special or unique nature (e.g., constraints on allocation of functions to personnel, interactions of communications and of personnel with equipment). Special attention should be directed to those designated areas, work stations, or equipment that require concentrated human engineering attention due to the sensitivity of the operation or criticality of the task, particularly where the results of human error would have critical impact. Special considerations should be given to:

    a. Human information processing capabilities.

    b. Foreseeable human errors under both normal and extreme conditions.

    c. Implications for the total system or equipment environment, including training, logistical support, and operational environment.

    It should be noted that the goals of human factors and safety engineering are critically interrelated and demand close coordination between the two disciplines.

12. *Security and privacy*. This section should specify those security and privacy requirements that are essential to the design with respect to the operational environment of the item. This typically concerns security requirements necessary to prevent access to the internal operating elements of the host system or equipment and compro-

mise of sensitive information or materials. Special consideration should be given to:

a. Security and privacy environment in which the item will operate.

b. Type and degree of security or privacy to be provided.

c. Security and privacy risks the item should withstand.

d. The governing policies on security and privacy.

e. Security and privacy accountability that should be provided by the system or equipment.

f. Required criteria for security and privacy certification and accreditation.

13. *Logistics.* This section should include supportability considerations and conditions that will apply to the system or equipment. There should be clear definition of logistics conditions and considerations such as maintenance, computer resources, modes of transportation, supply system requirements, and impact on existing logistics facilities. The following activities, as a minimum, should be addressed for applicability:

a. *Maintenance.* This paragraph should specify requirements relating to

  (1) Utilization of multipurpose support and test equipment.

  (2) Repair versus replacement criteria for reparable items.

  (3) Levels of maintenance.

  (4) Maintenance and repair cycles.

  (5) Accessibility to reparable and serviceable components.

b. *Supply.* This paragraph should specify the limitations on the supply system as a basis for the subassembly and piece part breakout of the system or equipment. It should describe supply elements such as centralized inventory systems use for certain groups of components, inventory stock locations, and types of items stored at those locations.

c. *Facilities and facility equipment.* This section should specify the constraints imposed on the system or equipment by the existing facility resources.

d. *Personnel.* This paragraph should specify logistics human resource requirements, to include:

  (1) Skills and numbers of personnel that should be allocated to the operation, maintenance, and control of the system or equipment.

  (2) Numbers and skills of support personnel needed for each operational mode, both routine and emergency.

e. *Training.* Consideration should be given to the following issues.

(1) Types of training to be used for the system or equipment (e.g., formal technical training program or on-the-job training).

(2) Required capabilities of training devices to be employed, characteristics of such devices, and skills to be developed through the use of the training devices.

(3) Constraints on the length of training time and on training locations.

f. *Interface requirements.* This section should document the external interfaces of the system or equipment hardware and software elements. It should describe each external interface by nomenclature; should designate the interfacing entities (e.g., systems, equipments, configuration items, components, software units) by name, number, version, along with documentary references; and should provide a brief description of each interfacing entity. Where practicable, identifying documentation, such as an interface specification, standard, or drawing should be referenced for each interface.

g. *Computer resources.* Computer resources requirements embody both computer hardware and computer software.

(1) *Computer hardware.* This section should detail computer hardware requirements that must be used by, or incorporated into, the system or equipment. The specifications should include the number of each type of equipment, size, capacity, and other required characteristics of processors, memory, input/output devices, auxiliary storage, communications/network equipment, and other related requirements.

(2) *Computer software.* This paragraph should specify the requirements regarding software that must be used by, or incorporated into, the computer software configuration item (CSCI). Examples include operating systems, database management systems, communication and network software, utility software, input and equipment simulators, test software, and manufacturing software. Item descriptions should provide the correct nomenclature, version, and documentation references of each software item.

14. *Workmanship.* As the predominant basis of the prescribed quality control procedures, this section should detail the workmanship requirements and include the criteria relative to the standards of workmanship desired, freedom from defects, and general appearance of the finished product. Workmanship requirements should be worded so as to provide logical guidelines and rationale for rejection of items found unsuitable for their intended uses.

15. *Product characteristics.* This section should identify specific conditions and properties such as color, protective coating, waviness, surface finish, dimensions, weight, as well as similar attributes that arc necessary for the material to enable the item to perform adequately.

16. *Chemical, electrical and mechanical properties.* This paragraph should define the requirements for composition, concentration, hardness, tensile strength, elongation, thermal expansion, electrical resistivity and other related properties that are necessary for accomplishment of the specified performance of the system or equipment.

17. *Stability.* Stability concerns specified requirements for shelf life and aging that are necessary for the material to perform adequately in its intended use and over its intended life.

Engineering Drawings

Engineering drawings portray the physical design of a system through its natural progression from point of inception to production. As an essential element of the total data package, engineering drawings reinforce the descriptive text in the system specifications and serve as a basis of reference for incorporation into technical manuals which set forth operating and maintenance instructions for the system. Engineering drawings are broadly categorized according to their level of detail, which reflects the phase of evolution of the system. The levels of engineering drawings and their associated data lists may be organized so as to define a conceptual or developmental design, a production prototype, or a limited production design on the highest level of detail included in engineering drawings required for quantity production of the items. These levels are generally categorized as:

> Conceptual design drawings (First Level)
> Developmental design drawings (Second Level)
> Product drawings (Third Level)

There are engineering scenarios for which a combination of levels may be applicable for development or production of the system.

Conceptual design drawings

This class of drawings is normally applicable during the conceptual development and demonstration and validation phases of the system. Engineering drawings and associated lists prepared to this level of detail are used to verify the preliminary design and engineering and confirm that the technology is feasible. These drawings should provide sufficient design information to evaluate the engineering concept and, in some cases, to fabricate developmental models for use in testing of the concepts. Conceptual design drawings should define the design concept in graphic form and provide the appropriate textual data required for evaluation and analysis.

Developmental design drawings

Developmental design drawings are typically applicable to the full-scale engineering development phase and, in certain cases, the production phase of the system life cycle. This type of drawings and associated lists provide sufficient information to permit analysis of a specific design approach and the fabrication of prototype or limited preproduction models. These drawings should include technical data to the level of definitive detail which would permit analytical evaluation of the inherent ability of the design approach to meet the specified system requirements. These selected engineering drawings should include, as applicable, parts lists, detail and assembly drawings, interface control data, logic diagrams, schematics, performance characteristics, critical manufacturing limits, and details of new materials and processes. Special inspection and test requirements necessary to determine compliance with requirements for the item should be defined on the engineering drawings directly or by reference.

Product drawings

Product drawings should reflect the highest level of design maturity to which the system has progressed. This class of drawings is normally effective during the production and follow-on operational phases of the system life cycle. These types of engineering drawings should include details of unique processes, when essential to design and manufacture; detailed performance ratings; dimensional and tolerance data; critical manufacturing assembly sequences; toleranced input and output parameters; schematics; mechanical and electrical connections; physical characteristics, including form and finishes; details of material identification; inspection, test, and evaluation criteria; necessary calibration information; and quality control data. Engineering drawings and lists prepared to this level of detail should, therefore, incorporate design definition that is sufficiently complete to provide the essential design, engineering, manufacturing, and quality assurance requirements necessary to enable the fabrication or procurement of the item from a qualified manufacturing vendor.

These drawings should accommodate the future fabrication of an interchangeable item which would duplicate the physical and functional characteristics of the originally produced item without resorting to additional design engineering effort. Product drawings represent the physical configuration baseline of the product, which should not be modified without a formal engineering change process. When formal changes to the product are approved for implementation, the product drawings should be modified accordingly, with the formal revisions reflected in the document numbers used to identify and manage the drawings.

Installation Drawings

An installation drawing is a special type of drawing which shows the configuration envelope and complete information necessary to install an item relative to its supporting structure or to associated items. An installation drawing may show a specific completed installation. Installation drawings for one-of-a-kind installations may be revised to record the as-installed or as-built condition.

An installation drawing shall include the following, as applicable:

1. Installed item(s) shown in solid lines; other items (e.g., walls and structures) shown in phantom.
2. Interface mounting and mating information, such as the dimensions of locations for attaching hardware.
3. Interface (e.g., cable) attachments required for the installation of the item and its cofunctioning with related items.
4. Information necessary for preparation of foundation plans, including mounting place details, drilling plans, and shock mounting and buffer details.
5. Location, size, and arrangement of ducts.
6. Weight of the unit.
7. Location, type, and dimensions of cable entrances, terminal tubes, and electrical connectors.
8. Interconnecting and cabling data.
9. References to applicable lists and assembly drawings.
10. When not disclosed on other referenced documents, overall and principal dimensions in sufficient detail to establish the limits of space in all directions required for installation, operation, and servicing; the amount of clearance required to permit the opening of doors or the removal of plug-in units; clearances for travel or rotation of any moving parts, including the centers of rotation, angles of elevation, and depression.
11. An installation drawing may include a parts list to establish the requirements for the installation hardware and, if desired, the items being installed.

An installation drawing essentially details the physical juncture of the item with its host system or facility.

Computer Aided Design

Computer aided design (CAD) represents a technological evolution in the development of engineering drawings and is used along with conventional design drawing methods. CAD embodies the use of a wide range of computer tools that assist engineers and other design professionals in their design development activities. It is the primary geometry within the product life cycle process which can involve both software and special-purpose hardware. Current CAD packages range from two-dimensional (2D) vector-based drafting systems to three-dimensional (3D) parametric surface and solid design models. CAD is frequently referred to in other terms, such as CADD (computer-aided design and drafting). All methodologies analogous to CAD are essentially synonymous, although there are some subtle differences in meaning. CAD is a tool to design and develop products and is utilized throughout the engineering process from conceptual design through engineering and analysis of components to definition of manufacturing processes.

Areas of applications of CAD with logistics implications

Computer-aided drafting application.   This generally refers to the actual technical component of a system or equipment project, using a computer rather than a traditional drawing board. The input into this computerized design process may come from specialized calculation packages; preexisting component drawings; graphical images such as maps, photographs, and similar media, hand-drawn sketches, and so on. The operator's task is to use CAD software to blend all the relevant components so as to produce drawings and specifications which can be used to estimate material quantities and project costs and ultimately provide the detailed drawings necessary for fabrication or construction.

Engineering application.   CAD is used in a variety of ways within engineering activities. At its basic level, it is a 2D Wireframe package that is used to create engineering drawings. Due to the technological evolution over recent years, the 3D methodology has been overtaken by 3D parametric feature-based modeling. Component forms are created by either using freeform surface or solid modeling or a hybrid of the 2D and 3D techniques. The individual components are then assembled into a 3D representation of the final product through a process described as the "bottom-up" design approach. These assembly models can be used to perform analyses to determine if the components can be assembled and fit together as well for simulating the dynamic characteristics of the final product. CAD methods and technology have also been developed to accomplish "top-down" design. This approach involves starting with a layout diagram of the product which is broken down into subsystems with increasing detail until the level of single components is reached, with the geometry at each level being associative with the higher level. Detailed design of each of the subsidiary components is then completed before building up to the final product assembly.

Technical Manuals

Technical manuals fall into the category of logistics technical data that provide pertinent technical information pertaining to assembly, installation, operation, routine servicing, preventive maintenance, corrective maintenance, overhaul, and identification of parts and components of the system. Technical manuals are an outgrowth of the full-scale engineering development phase. This element of the total data package synthesizes the final results of the developmental and product engineering efforts in the form of a compendium of instructions governing the operations and maintenance of the system. Technical manuals are generally grouped according to (1) system or equipment manuals, which provide details with instructions used by logistics managers and technical specialists, and (2) consumer booklets, which are oriented toward the typical purchaser and which set forth instructions for procedures and checkpoints that can be utilized by the average consumer without the elaborate technical facilities, tools, and skills required for major logistical tasks.

System or equipment manuals

As part of the increasing emphasis on acquisition streamlining, the Department of Defense has adopted the practices of the commercial sector for development of technical manuals for commercial-off-the-shelf (COTS) equipments acquired for military use. There is a stipulation that, although commercial manuals are acceptable, there is a defined structure of contents and chapter organization that applies to COTS item manuals utilized in the military context. The following guidelines, derived from MIL-DTL-24784, are therefore applicable to manuals prepared for systems and equipments that may have both commercial and military applications.

Structure of technical manuals

Technical manuals for systems and equipment should consider, but not be necessarily limited to the following:

1. Front matter
2. Introduction
3. Preparation for use
4. Installation instructions
5. Principles of operation

6. Operating instructions
7. Maintenance and servicing instructions
8. Reprogramming
9. Preparation for shipment
10. Storage
11. Parts list
12. Illustrations and diagrams
13. Overhaul instructions

Front matter.   This includes a title page, table of contents, and notice of safety precautions where hazards may be present during installation, operation, or maintenance of the system or equipment. The safety precaution notice must be reinforced by danger, caution, or warning statements inserted and prominently visible within the text of those chapters which provide instructions relating to the functions of installation, operation, or maintenance.

Chapter 1, introduction.   Introductory material should include succinct and concise brief statements on the following:

1. Purpose and function
2. Capabilities
3. Performance characteristics
4. Description (e.g., equipment model, dimensions, weight, volume and center of gravity)
5. Power and utility data
6. Environmental information
7. List of items furnished
8. List of additional items required for operation and maintenance, but not supplied with the equipment
9. Tools and test equipment
10. Warranty information
11. Shipping and handling precautions
12. Storage data

Chapter 2, preparation for use.   The chapter shall contain instructions on unpacking and assembly. Inspection for in-shipment damage and instructions on how to handle damaged equipment should also be included.

Chapter 3, installation instructions.   Special installation instructions should include procedures for requirements such as for foundations, ventilation, clearances, plumbing and electrical connections, mountings, wiring runs, initial lubrication, and alignment. This chapter should incorporate reference to the governing installation control drawings, if applicable, which have been previously addressed herein.

Chapter 4, principles of operation.   This chapter shall provide principles of operation data to the technical level commensurate with the intended use of the system or equipment.

Chapter 5, operating instructions.   Operating instructions should include the following, as applicable:

1. Illustrations and explanations regarding functions and uses of all controls and indicators
2. Initial adjustments and control settings
3. Start-up procedures
4. Normal operation
5. Operation under emergency, adverse, or abnormal conditions
6. Shut-down procedures
7. Emergency shut-downs
8. Interface instructions

Chapter 6, maintenance and servicing instructions (preventive and corrective). This chapter shall provide a list of test equipment, special tools and materials needed for maintenance and service. This list should include data on item nomenclature, part and model number, and application. Instructions or illustrations should explain or depict how to make test connections. Actions and normal indications should be defined for each test.

1. *Cleaning and lubrication.* Instructions for periodic cleaning and lubrication defined by calendrical intervals (e.g., monthly, quarterly, semiannually) and operational intervals (e.g., operational hours, mileage). Lubrication and service points should be clearly described or illustrated. The instructions should include procedures applicable during repair, replacement, and reassembly.
2. *Performance verification.* This involves instructions for frequency and step-by-step check procedures required for calibration of test, measurement, and diagnostic equipment needed to restore the system or equipment to serviceable accuracy.
3. *Inspection.* This paragraph provides instructions and schedules for inspection of equipment for damage and wear with emphasis on allowable service limits such as wear, backlash, end play, balance, and length

and depth of scoring. Acceptable service limits are acceptable wear tolerances which will not impair performance. They are not to be confused with manufacturing tolerances.

4. *Troubleshooting.* This concerns identification of malfunctions that might occur during system or equipment operation. Troubleshooting data and fault isolation techniques should include:

   a. Indications of symptom of trouble

   b. Instructions necessary (including test hookups) to determine the cause

   c. Action required or reference to action required to restore the system or equipment to operating condition

5. *Disassembly, repair, replacement, and reassembly.* This paragraph provides sequential procedures for disassembly, repair, parts replacement, and reassembly of the components of the system or equipment. Test adjustment and check-out data after reassembly should be provided. Illustrations, including exploded views, should be used to support each of these functions. Figure 20.1 portrays a typical illustrated parts breakout (IPB).

Chapter 7, reprogramming. A description of reprogrammable memory, reprogrammable theory, program setup, program confidence check-out, program loading, and programming procedures should be provided for systems or equipment having a reprogrammable memory.

Chapter 8, preparation for shipment. Technical manuals should provide instructions for the following, as applicable:

1. Disassembly, removal, and separate packaging of electrostatic discharge devices or fragile components

2. Use of reusable shipping cases or containers

3. Special cradles

4. Mounting

5. Securing

6. Covering and preservation

7. Precautions for shipment

8. Shipment and unloading

Chapter 9, storage. Manuals should provide special instructions, as follows:

1. Indoor and outdoor storage

2. Temperature and other environmental instructions

3. Storage facilities

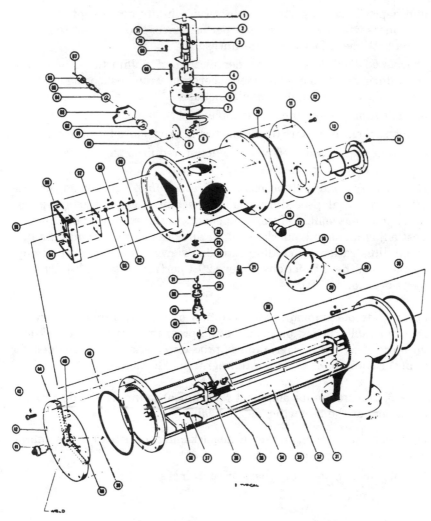

Figure 20.1   Typical illustrated parts breakout.

4. Dunnage
5. Ventilation
6. Revetting
7. Drainage
8. Staking
9. Grounding
10. Covering
11. Preservation

Chapter 10, parts lists. The manual should include a parts list providing positive identification of parts for support of the system or equipment and which shall include actual manufacturer or vendor part numbers or generic descriptions necessary to enable acquisition of replacement parts. Parts list illustrations may be used to clearly identify component parts and parts' interrelationships.

Chapter 11, illustrations and diagrams. Illustrations and diagrams are used to assist in location of all components significant to operation and maintenance of the system or equipment. Symbols used in the documents should be standard or common to the trade or commodity. Where nonstandard symbols are utilized, clear and understandable explanations must be provided.

1. *Illustrations.* The manual should contain illustrations which clearly portray those components identified as having operational and maintenance significance. The illustrations are also used to show configuration of the system or equipment as well as the parts relationship for removal and disassembly procedures.

2. *Diagrams.* As applicable, the following typify the diagrams which should be included in the manual:

   a. Simplified functional block

   b. Locator

   c. Piping

   d. Plumbing

   e. Hydraulic

   f. Schematic

   g. Electrical

   h. Digital

   Although not specifically cited above, other diagrams considered essential to the purpose of the instructions may be incorporated into the manual.

Chapter 12, overhaul instructions. When applicable, the manual shall provide overhaul instructions to return the system or equipment to its originally specified performance capability. As a minimum, the instructions should include the following:

1. List of required support equipment, special tools, and facilities

2. Listing of mandatory parts

3. Preshop analysis, as applicable

4. Step-by-step procedures for performing all functions including disassembly, removal, diagnostics, installation, repair, assembly, in-process testing, adjustment, and inspection

5. Final tests to ensure satisfactory performance of the system or equipment overhauled

Consumer booklets

Consumer booklets, frequently termed "owner's manuals" in the retail trade, are those publications included as part of a purchase of a product, which is typically an electrical, electronic, or mechanical appliance rather than a consumable item. Such manuals are designed to be used by the average consumer and are more likely to be media for customer relations between the producer and customer than documents providing comprehensive operational and maintenance procedures. The typical consumer's booklet serves the purpose of promoting the purchased item and affiliated products offered by the manufacturer, along with providing instructions for routine care and minimal maintenance, oriented toward the perceived limited expertise of the average purchaser. These publications are not in the same category as formal technical manuals used by professional operators and maintainers concerned with the more complex aspects of the system. They are, however, an essential part of the product package which links the product manufacturer or supplier to the customer. The general organization of the typical consumer's booklet, or owner's manual, approximates the following topical outline:

1. *Description of the product*, including an overview of the organization of the booklet.

2. *Unique features of the product,* especially emphasizing implicitly favorable comparisons to its competitors.

3. *Introduction of the user to the product.* This section should include required special warning and precaution notices, preoperational checklists, and data which generally familiarize the user with the product.

4. *Assembly instructions,* if applicable. If this section is used, a clearly defined illustrated parts breakout (IPB) is advisable.

5. *Operating instructions.* This section should provide a clear set of sequential operating procedures, emphasizing the location and purposes of controls and indicators. Warning and caution statements should be prominently stated, where appropriate.

6. *Emergency procedures.*

7. *Routine service instructions,* e.g., periodic lubrication and cleaning.

8. *Owner's assistance available,* typically a local service activity or toll-free network telephone number.

9. *Product specification.* This typically includes a prescribed list of preferred lubricants and cleaning compounds, as well as a list of parts at the "remove and replace" level of competence of the typical customer. The parts list should be supported by a numerically annotated illustrated parts breakout which depicts and clearly identifies the parts on the list.

10. *Listing of service centers.*

11. *Warranty data.*

12. *Index of contents.*

Although the consumer's booklet necessarily excludes logistics-related details, a well-designed owner's manual can be an effective guide which enables the consumer to optimize his or her capability and expertise, while encouraging consultation with the producer's representative when necessary. These factors can contribute to the long service life of the product and enhance the prospects of customer loyalty, satisfaction, and recurring purchases.

Support and Test Equipment

Overview

The generic concept of *support and test equipment* (S&TE) covers a sweeping perspective of all apparatus utilized to sustain and enhance the ability of a system, product, or facility to accomplish its stated function. From the standpoint of logistics, this chapter defines the family of support and test equipment as encompassing all equipment (manual and power-operated) and assistance devices used to support the operation and maintenance of the system or product. This includes the tools, condition-monitoring equipment, diagnostic and checkout equipment, calibration and metrology equipment, maintenance assistance modules, maintenance stands, and handling equipment to facilitate preventive maintenance and corrective maintenance of the prime system. In contemplating the use of S&TE, the feasibility of its use at each level of logistics support within the product distribution system—i.e., the organizational (retail) level, the intermediate (service center, distribution center) level, and the depot (plant or manufacturing center) level—must be addressed. The rationale for incorporating support and test equipment as part of the logistics support subsystem to the prime system must address the stated purpose, economic effectiveness, and quantitative analysis of the utility of the prospective S&TE capability.

Purpose of Support and Test Equipment

The purpose of logistics S&TE is to sustain in a cost-effective manner the designed maintainability parameters of the system, typically specified as mean preventive maintenance time $(\overline{M}_{pt})$ and mean corrective maintenance time $(\overline{M}_{ct})$. When addressing system maintainability in connection with trade-off analysis within the specified reliability parameters, the evaluation must relate the utility of using maintenance personnel hours without S&TE to that of using maintenance technicians assisted by support and test equipment. If the application of S&TE permits a reduction in maintenance labor with a corresponding reduction in life-cycle costs, the logistics manager would be advised to use S&TE; where there is a projected lack of economic benefit associated with support and test equipment, it would be prudent to rely totally on human resources rather than invest in S&TE. With regard to these governing considerations, the following features and stipulations associated with support and test equipment should be addressed by the logistics manager:

1. S&TE equipment must be evaluated for each of the three levels of maintenance (organization, intermediate, and depot).
2. Newly designed equipment should not be considered unless standard equipment is unavailable. It is preferable to use readily available and market-tested standard equipment.
3. The selected support and test equipment must be compatible with the prime, host system.
4. The reliability and maintainability characteristics of the S&TE must be equivalent to those of the prime equipment.
5. Logistics support requirements for the S&TE must be defined, with due consideration of associated maintenance tasks, spare parts, personnel, training, technical data, and facilities.
6. Test and maintenance software requirements must be clearly defined.
7. In principle, S&TE allocation should be based on the minimum, essential capability necessary to support the prime system.
8. Automatic self-test provisions should be incorporated in the prime equipment where cost-effective and technologically feasible.
9. Direct fault signals (visible or audible display devices) should be incorporated in the system to provide indications of malfunctions.
10. Continuous performance-monitoring features should be incorporated in the prime system, where appropriate.
11. System test points should be incorporated in the system design to enable fault isolation beyond the limitations of the self-test features.
12. Test points in the system must be accessible and must accommodate the level of maintenance being accomplished.

13. System test points should be functionally grouped to facilitate sequential testing steps, signal flow, testing of similar functions, or frequency of use, in the case of limited access.

14. Test points should be provided for direct test of reparable items in the system.

15. Test points should be adequately identified, labeled, and illuminated (where possible).

16. At the system level, every malfunction should be detectable by means of a simple "no-go" indication. This feature validates the thoroughness of testing the system.

17. The maintenance software must be designed so as to provide accurate diagnostic data.

Notwithstanding the apparent value of auxiliary support and test equipment for prime system effectiveness, it is reiterated that such investment should be analytically scrutinized in depth with respect to the outlay of resources relative to the expected utility value of the proposed support subsystem.

The Ambiguity Group Concept

Diagnostics is the determinant of the value of the system test capability. The most common basis of reference for determining system diagnostics capability is an appropriately defined ambiguity group. The term *ambiguity group* refers to the level of definition to which a testing unit can isolate a group of components or submodules which includes the cause of the system failure. For example, a testing device which is verified for an ambiguity group of 12 components or submodules is designed to be capable of localizing the source of the malfunction to a specific modular grouping comprising 12 components, one or more of which could be the failed component. Within that grouping, the design engineer has the option of using individual direct test devices to isolate the specific malfunctioning unit or prescribing a set of *maintenance assistance modules* (MAMs) for that purpose.

Maintenance assistance modules

Maintenance assistance modules represent a maintenance practice in which a set of specially marked modules or components identical to those making up a corresponding ambiguity group is used to isolate the faulty module. Each MAM is substituted for the corresponding module in the ambiguity group until the faulty module within the group is identified. The concept of maintenance assistance modules is illustrated by a string of lights connected in series circuit, where if one light in the series fails, the entire string of lights fails to function. By physically interchanging each light bulb in the series with a bulb that is known to be functional until the entire series reilluminates, it is established that a specific bulb, the one whose replacement restored the circuit, is the failed component. There are cases in which MAMs are the simpler and more feasible method of fault isolation within an ambiguity group. As with all design concepts, analysis of the trade-offs between costs and utility should be accomplished before establishing the system testability characteristics. For those systems for which stand-alone self-sufficiency is essential, built-in test elements are mandatory design features, irrespective of the consequent investment in internal testing capability and the increase in system cost.

Diagnostics effectiveness measures

The effectiveness of diagnostic equipment can be measured for utility and reliability. The following are typical elementary methods of assessment:

1. The ratio of types of failures the diagnostic device is able to detect to expected total types of malfunctions:

$$\text{Diagnostic utility} = \frac{\text{number of malfunctions detectable}}{\text{number of potential failures}}$$

2. Reliability of the testing device in detecting malfunctions (as designed) within the ambiguity group:

$$\text{Probability of success of detection} = \frac{\text{number of failures detected}}{\text{number of failures experienced}}$$

   This S&TE reliability value can be expressed as a percentage or decimal equivalent.

3. System-level reliability, which is the probability of success in detecting a malfunction at system level based on system-level fault indicators. This can be computed by determining the product of the reliabilities of the individual testing devices used for measuring ambiguity groups as defined within the system.

Economic Considerations Related to S&TE

Support and test equipment is governed by either the dictated system self-sufficiency or economies of scale and the trade-off of alternatives.

System self-sufficiency demands that built-in test (BIT) capability be incorporated into the design in order to provide the most efficient maintainable design, in terms of accessibility, parts standardization considerations, etc. When S&TE is contemplated as a possible alternative subsystem to the prime system, there are trade-offs to be evaluated in terms of total system reliability (mean time between failures) and maintainability (mean preventive maintenance time and mean corrective maintenance time) relative to the prescribed system operational availability parameters. The greater the incorporation of S&TE into the system design, the less the need for maintenance personnel, but at the expense of increased investment in BIT or auxiliary test devices. The alternative is minimum investment in S&TE, but with increased costs for maintenance labor resources. The following case example illustrates the effects of comparing such alternatives.

Example of S&TE benefits analysis

Statement of scenario. S&TE is under consideration for four collocated primary systems, designed for continuous operation (8760 hours per year). The scheduled maintenance interval $MTBM_{sch}$ is 2190 operating hours. The predicted mean time between failures (MTBF) or projected repair interval is 1000 operating hours. The maintenance man-hour (MMH) cost (basic labor plus facility overhead) is $60 per hour. The total investment cost for acquisition of a complement of S&TE to cover all collocated systems is estimated to be $20,000, with a projected service life of 10 years. The S&TE salvage value is zero.

1. The predicted system maintainability characteristics based on utilization of S&TE are

Mean scheduled maintenance time $\overline{M}_{sch} = 1.0$ h
Mean corrective maintenance time $\overline{M}_{ct} = 1.5$ h

There are three ambiguity groups, with the following fault detection reliability values:

Ambiguity group	Reliability
A	$0.95 = R_A$
B	$0.90 = R_B$
C	$0.85 = R_C$

The system-level fault detection reliability is therefore

$$(R_A)(R_B)(R_C) = (0.95)(0.90)(0.85)$$
$$= 0.72675, \text{ or } 0.727$$

Task descriptions based on use of S&TE require one maintenance technician.

2. The predicted system maintainability characteristics excluding use of S&TE are:

$$\overline{M}_{sch} = 2.0 \text{ h}$$
$$\overline{M}_{ct} = 3.0 \text{ h}$$

Task descriptions without use of S&TE call for two maintenance technicians.

The purpose of the analysis is to determine the more cost-beneficial alternative, use of S&TE or relying totally on maintenance technicians in lieu of investment in S&TE.

Analytical approach.

1. Determine governing system parameters.

Operating hours per year = (number of systems)(annual operating hours per system)
= (4)(8760)
= 35,040 h/yr

System maintenance actions per year:

$$\text{Scheduled maintenance } (M_{sch}) = \frac{\text{total operating hours}}{\text{MTBM}_{sch}}$$

$$(M_{sch}) = \frac{35,040}{2190}$$

$$= 16 \text{ actions/yr}$$

$$\text{Corrective maintenance } (M_{ct}) = \frac{\text{Total operating hours}}{\text{MTBF}}$$

$$= \frac{35,040}{1000}$$

$$= 35.04, \text{ or } 35 \text{ actions/yr}$$

2. Compute annual costs *without* S&TE.

$$\$M_{sch} = (M_{sch} \text{ actions})(\overline{M}_{sch})\left(\frac{\$\$}{\text{MMH}}\right) \text{ (number of maintenance)(technicians)}$$
$$= (16)(2)(60)(2)$$
$$= \$3840/\text{yr}$$

$$\$M_{ct} = (M_{ct} \text{ actions})(\overline{M}_{ct})\left(\frac{\$\$}{\text{MMH}}\right) \text{ (number of maintenance)(technicians)}$$
$$= (35)(3)(60)(2)$$
$$= \$12,600/\text{yr}$$

$$\text{Total annual costs} = \$M_{sch} + \$M_{ct}$$
$$= 3840 + 12,600$$
$$= \$16,440/\text{yr}$$

3. Compute annual costs *with* S&TE.

$$\$M_{sch} = (M_{sch} \text{ actions})(\overline{M}_{sch})\left(\frac{\$\$}{\text{MMH}}\right) \text{ (number of maintenance)(technicians)}$$
$$= (16)(1)(60)(1)$$
$$= \$960/\text{yr}$$

$$\$M_{ct} = (M_{ct} \text{ actions})(\overline{M}_{ct})\left(\frac{\$\$}{\text{MMH}}\right) \text{ (number of maintenance)(technicians)}$$
$$= (35)(1.5)(60)(1)$$
$$= \$3150$$

$$\text{Annual equipment costs} = \frac{\text{acquisition cost}}{\text{service life}}$$
$$= \frac{\$20,000}{10}$$
$$= \$2000/\text{yr}$$

$$\text{Total annual costs} = \$M_{sch} + \$ M_{ct} + \text{annual equipment cost}$$
$$= 960 + 3150 + 2000$$
$$= \$6110/\text{yr}$$

4. Evaluate annual cost differentials.

	A without S&TE		B with S&TE	Net differential A – B
Labor	$16,440		$4110	*$8964
S&TE	0		$2000	–$2000

*The net labor differential is determined by applying the system-level S&TE fault detection reliability, computed as 0.727, to the $12,330 gross difference between the two alternatives (A labor of $16,440 minus B labor of $4110) to calculate the expected value of the potential savings due to S&TE; i.e., the expected value of an objective is always determined by multiplying the probability of achieving the objective by the value associated with the objective. Hence, (0.727)(12,330) = 8963.91, or $8964.

5. *Conclusion:* The expected net savings from using S&TE rather than system diagnostic testing without S&TE is $8964 + (–$2000), or $6964.

Analysis of S&TE Utility

The reader is referred to the supplemental *Instructor's Guide* for an expanded description of the definition of isolation levels of ambiguity; development of an S&TE Testability Profile, and an algorithmic model for evaluation of S&TE testability.

Embedded Computer Resources

Overview

In logistics engineering, computer resources are defined as all computer facilities, equipment, software, firmware, associated documentation, skilled personnel, and supplies needed to operate an embedded computerized subsystem. This technical context includes automated test equipment (ATE) as well as host systems which incorporate embedded software designs. Logistics managers and logistics support technicians are primarily concerned with (1) the fault detection and fault isolation capabilities of the embedded computer elements resident in the prime system, (2) the ability of support technicians to differentiate between hardware and software malfunctions, and (3) the configuration management needed to control software modification during the operational, consumer-use phases of the system.

Embedded computer resources

Those computer resources embedded within a host system fall into the category of firmware. The term *firmware* describes the combination of a hardware device and computer instructions or computer data that reside as "read only" software on the hardware device. Such software cannot be readily modified under program control. This term also applies to read only digital data that may be used by support and test equipment and similar electronic devices other than digital computers.

Influences of Computer Resources on Logistics Engineering

The mnemonics (memory structure) of the computer programs, which have a critical bearing on the instructional routines incorporated in the system and which also have a significant impact on diagnostic procedures, are formulated during the inception of program development. It is incumbent upon the logistics manager during this phase to address the following major considerations relative to computer resources design and development:

1. All computer programs and software required for the prime system should be identified. This includes condition-monitoring programs, diagnostic routines, logistics data processing programs, etc.

2. Computer language requirements, specifications, and requirements for compatibility with other programs should be identified.

3. Software configuration management procedures and quality control provisions should be confirmed.

4. System software requirements for operating and maintenance functions should be identified and developed through system-level functional analysis so as to afford traceability.

5. The software must be complete in terms of scope and depth of coverage.

6. The supporting software must be compatible with the equipment interfacing with the host system.

7. The operating software must be compatible with the maintenance software, as well as with other elements of the system.

8. The computer language requirements for the operating software and maintenance software must be compatible.

9. All software must be clearly described, with supporting documentation including logic function flow, coded programs, etc.

10. The software must be tested, validated, and verified with respect to reliability, maintainability, and performance.

It is underscored that software which is reliable from the outset will be less expensive and quicker to develop, test, and validate. There must be continual emphasis on minimizing the prospects of early errors and on eliminating errors before proceeding to the subsequent phases of development. This consideration reinforces the importance of including logisticians on the computer resources design team on a cradle-to-grave basis.

Organization of Computer Software

Computer software should be organized during the design evolution of the program into one or more computer software configurations (CSCIs), hardware configuration items (HWCIs), or other types of software. Each CSCI is part of a system, segment, or designated prime item and typically consists of one or more top-level computer software components (TLCSCs). Each TLCSC consists of lower-level computer software components (LLCSCs) or units. LLCSCs may consist of other LLCSCs or units. TLCSCs and LLCSCs are logical groupings. Units are the smallest logical entities; they are the actual physical entities implemented in computer code. The static structure of CSCIs, TLCSCs, LLCSCs, and units should form a hierarchal arrangement as illustrated in Fig. 22.1. This hierarchal structure identifies all CSCIs, TLCSCs, LLCSCs, and units.

Software Development Cycle

The total system life cycle encompasses five major phases: concept exploration, demonstration and validation, full-scale engineering development, production and distribution, and operation and customer support.

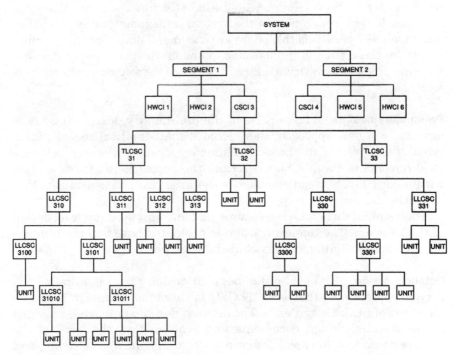

Figure 22.1 Typical CSCI static structure.

The software development cycle consists of six phases: software requirements analysis, preliminary design, detailed design, coding and unit testing, computer software component (CSC) integration and testing, and computer software configuration item (CSCI) testing. *The total software development cycle or a subset thereof may be performed within each of the system life-cycle phases.* Successive iterations of software development tend to build upon the products of previous iterations. It is, therefore, emphasized that the software cycle is effectively a microcosm, a "mini-life cycle" within the system life-cycle process. Figure 22.2 portrays the sequence and flow of the major events and technical review points in the software development cycle, which may occur during any of the five major system life-cycle phases.

The software life-cycle phases

The major phases of the system life cycle are discussed in depth in Chaps. 14 and 15. The following sections describe the major events in the process of software development.

Software requirements analysis.   The objective of the software requirements analysis phase is to completely define and analyze the requirements for the software. Such requirements include the functions the software is expected to perform as part of the system or segment, or as the prime item itself. The results of this phase are documented as approved requirements for the software. If applicable, plans for development of the software are prepared or reviewed at the initiation of the software requirements analysis.

Preliminary design.   The purpose of the preliminary design phase is to develop a design approach; this includes mathematical models, functional flows, and data flows. Preliminary design allocates software requirements to the TLCSCs, describes the processing that takes place within each TLCSC, and establishes the interface relationships within TLCSCs.

The result of this phase is a documented and approved top-level design of the software. The top-level design is evaluated based on the requirements before initiation of the detailed design phase.

Detailed design.   The goal of the detailed design phase is to refine the design approach so that each TLCSC is decomposed into a complete structure of LLCSCs and units. The detailed design approach is provided in the detailed design documents and evaluated on the basis of the requirements and top-level design prior to commencing the coding and unit testing phase.

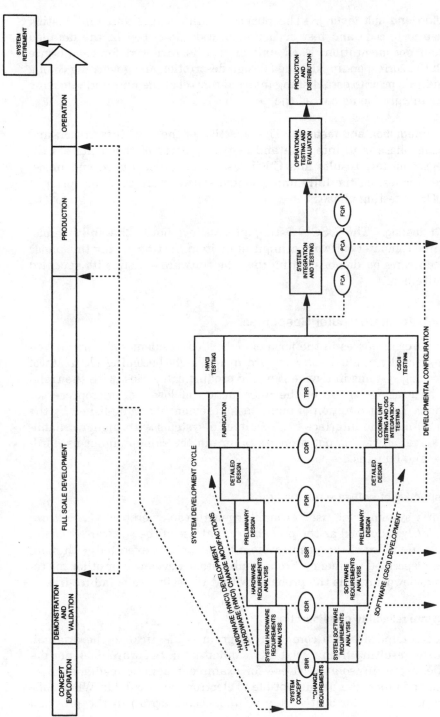

Figure 22.2 Computer program development cycle. *Before production; possible change requirement. **After production. SRR—System requirements review. SDR—System design review. SSR—System specification review. PDR—Preliminary design review. CDR—Critical design review. TRR—Test readiness review. FCA—Functional configuration audit. PCA—Physical configuration audit. FQR—Formal qualification review.

Coding and unit testing. The purpose of the coding and unit testing phase is to code and test each unit of code described in the detailed design documentation. Each unit of code is reviewed for compliance with the corresponding detailed design description and governing coding standards prior to establishing internal control of the unit and releasing it for integration as part of the next phase.

CSC integration and testing. The objective of the CSC integration and testing phase is to integrate and test aggregates of coded units. CSC integration test results and CSCI test plans, descriptions, and procedures for testing the fully implemented software are reviewed prior to the CSCI testing phase.

CSCI testing. The CSCI testing process evaluates the fully implemented computer software configuration item. Testing during this phase concentrates on demonstrating that the software satisfies its specified requirement.

Reliability of Computer Resources

Computer resources in the form of embedded, resident software as well as externally applied software are increasingly becoming elements of operating systems in a wide range of product categories, with the trend accelerating by virtue of the introduction of low-cost microprocessor devices. The assumption of many tasks previously accomplished by the human-machine interface has resulted in systems with higher reliability, since software is not prone to failure in the same manner as hardware and humans.

Perspective of software reliability

From a system-level, user, or similar macroscopic perspective, *software reliability* is defined as the probability that use of the software does not result in failure of the prime system to perform as expected with more than a specified frequency. From a subsystem, developer, or other microscopic viewpoint, it is the probability that the software is failure-free.

Software reliability considerations

As each duplicate of a computer program is identical to the original, failures resulting from variability cannot occur. Software does not degrade, except in isolated cases—for example, magnetic media, such as disks and tapes, can be susceptible to electromagnetic fields. When software fails to perform, it is due to unexpected errors in the program which, consequently, are duplicated in all copies of the program; thus, the copies will also fail under the same circumstances and operating scenario as the original program. This characteristic shows the inherently serious implications of software errors. The software reliability effort is therefore concerned with minimizing the existence of errors by imposing programming disciplines, checking, and testing. The term *software reliability* is not a universally recognized parameter in the same sense as those applicable to hardware reliability, e.g., mean time between failures (MTBF) and failure rate (λ). Whereas reliability with respect to hardware implies a varying quantitative probability value, a computer program either contains one or more errors, in which case the probability of failure in certain circumstances is 100 percent, or it contains no errors, in which case the probability of failure is zero. Because of the lack of conformity of measurement between hardware and software, the commonly used term *rate of failure* has no meaning from the perspective of software reliability. An errorless program will perform indefinitely without failure, as will all duplicates of the program. A program with errors will always fail when executed under the error conditions, as will all copies of that program.

Software failure sources

Software errors or faults can be attributable to the specification, the design, or the coding process.

Specification faults.   Most program errors that arise during the development process are due to faulty specifications. As software is not a physical manifestation, it is difficult to perceive and interpret ambiguities, inconsistencies, or incomplete statements. There are no margins of dimensional variation in software; consequently, the software specification must be logically complete and consistent, and must not include requirements that are not testable.

Design faults.   The software system design must emanate from the specification. The system design is typically a flowchart which defines the program structure, test points, constraining parameters, etc. Errors occur as a result of erroneous interpretation of the specification or incomplete and inaccurate logic. An important feature of software design is *robustness*, the capability of a program to withstand error conditions without adverse effects, such as crashing or looping.

Code generation.   Code generation is a common source of computer program errors, since a normal program involves a large amount of code. Typical coding errors include typographical errors, incorrect numerical values (e.g., erroneous decimal factors, such as 0.1 instead of 1.0), omission of symbols, and inclusion of indeterminate algorithmic citations (e.g., division by ∞, which can force a resultant value of zero).

Categories of computer program errors.   The professionally acknowledged categories of software faults, which are useful for reliability analysis of software programs, include the following:

1. Computational errors
2. Logic errors
3 Data input errors
4. Data handling errors
5. Data output errors
6. Interface errors
7. Data definition errors
8. Database errors
9. Operational errors
10. Documentation errors
11. Trouble report rejections

In addition to these categories, there may be other identifiable causes of errors, such as time limit exceeded, core storage limit exceeded, compilation error, software not compatible with project standards, and code or design insufficiency.

Hardware and Software Interaction

From the perspective of a system user, a system failure has adverse consequences, irrespective of whether the malfunction is a hardware failure or a software failure. In many cases it is not possible to distinguish between hardware and software failures. In software-controlled systems, failures can occur which are difficult to diagnose as attributable to hardware or software because of the precision of the interactions between hardware and software. The software design can minimize such possibilities by incorporating automatic diagnosis and fault indication features. There are other types of failures that are more difficult, especially where the hardware and software interface is less clearly defined. The best resolution of this dilemma is strict enforcement of design discipline and precise formulation of specifications during development of the software.

Comparison of hardware and software reliability characteristics

There are many critical attributes of reliability, relative to which hardware and software are inherently different. The more prominent differences are summarized in Table 22.1.

Quantitative Approach to Evaluating Computer Resource

Patrick D. T. O'Connor, a recognized British reliability engineering expert, in his book *Practical Reliability Engineering*, published a series of software reliability algorithms, including the Poisson model, Musa model, Jelinski-Moranda and Schick-Wolverton models, and Littlewood model. This discussion is an adaptation of the Musa model as described by O'Connor.

The Musa model approach to computer program reliability

The Musa model is useful for the logistics engineer concerned with evaluating computer resources with respect to (1) reduction in the number of inherent faults in a computer program, (2) estimation of testing time necessary to locate and correct system errors in order to achieve an acceptable level of errors in the system, and (3) determination of software system reliability based on the specified program operating cycle and inherent mean time to fault (MTTF).

Reduction in number of system errors before testing to acceptable level of errors pursuant to testing

$$\Delta n = N_0 T_0 \left(\frac{1}{T_1} - \frac{1}{T_2} \right)$$

also expressed as

$$\Delta n = N_0 - N_1$$

where N_0 = number of errors in the system before testing

N_1 = acceptable level of system errors, designated as the specification target parameter

T_0 = inherent MTTF before testing

T_1 = pretest MTTF subject to improvement in order to achieve N_1

T_2 = projected MTTF needed to be achieved in order to reflect reduction in system errors to a level of acceptability, as indicated by N_1

TABLE 22.1   Comparison of Hardware and Software Reliability Attributes

Hardware	Software
Failure can be caused by deficiencies in design, production, use, and maintenance.	Failures are primarily due to design errors, with production (duplicating), use, and program maintenance having a negligible effect.
Failures can be due to wear or other energy-related phenomena. A warning signal may be activated before failure occurs.	There is no wear-out. Software failures occur without warning.
Repairs can be accomplished which can enhance the equipment's reliability.	The only repair to software is by redesign or reprogramming. If this removes the errors and does not introduce new errors, it will result in higher reliability.
Reliability can depend on burn- or wear-out trends; i.e., failure rates can be decreasing, constant, or increasing relative to operational time.	Reliability is not as sensitive as in the case of hardware. Reliability improvement over time may be evident, but not on the basis of an operating time relationship. Such improvement is a reflection of the effort applied to detection and correction of errors.
Failure can be related to the passage of operating or storage time.	Reliability is not time-related as with hardware. Failures occur when a program step or path which is in error is executed.
Reliability is related to environmental factors.	The external environment does not affect software reliability, except insofar as it could affect program inputs.
Reliability can be predicted based on theory from knowledge of design and usage factors.	Software reliability cannot normally be predicted from physical databases, since software reliability depends totally upon human factors during the design process.
Reliability can frequently be enhanced through redundancy.	Software reliability cannot be improved by redundancy if the parallel program paths are identical; if one path fails, the parallel path will have the same error and will likewise fail. It is possible to incorporate redundancy by having parallel paths with different programs written and validated by different teams.
Failures can occur for components of a system in a manner which is largely predictable from the stresses on the components and similar factors. Reliability criticality lists such as those provided by MIL-HDBK-217 are useful for such purposes.	Software failures are not usually predictable from analysis of separate sets of data. Errors are likely to exist randomly throughout the program, and any statement could be the error. Hardware reliability criticality lists are neither useful nor appropriate.

Testing time required to isolate and correct errors so as to improve pretest MTTF to acceptable MTTF level

$$\Delta t = \left(\frac{N_0 T_0}{C}\right) \ln\left(\frac{T_2}{T_1}\right)$$

where Δt = testing time necessary to improve T_1 to T_2
C = test compression factor, equal to the ratio of equivalent operating time to testing time

Factors N_0, T_0, T_1, and T_2 are as described above.

Computer program reliability

$$R_t = e^{-t/T}$$

where R_t = probability of the computer program's completing operation cycle t without failure.
t = specified program run time
T = inherent MTTF of system

Examples of quantitative approach application

Problem scenario.   A computer program is believed to contain 300 errors, and the recorded MTTF prior to testing is 1.5 h. The testing compression factor is established to be 4. The questions to be addressed by the logistics engineer are:

1. How much testing time is required to isolate and repair existing program faults so as to reduce the number of program errors to 10?

2. After effecting the necessary program corrections, what would be the system reliability over a 50-h operating cycle?

Solution

1. Determine T_2, based on Δn algorithmic description.

$$\Delta n = N_0 T_0 \left(\frac{1}{T_1} - \frac{1}{T_2}\right)$$
$$N_0 = 300$$
$$N_1 = 10$$

Therefore

$$\Delta n = 300 - 10 = 290$$
$$T_0 = 1.5 \text{ h}$$
$$T_1 = 1.5 \text{ h}$$
$$T_2 = \text{to be determined}$$

$$300 - 10 = (300)(1.5)\left(\frac{1}{1.5} - \frac{1}{T_2}\right)$$

$$290 = \frac{450}{1.5} - \frac{450}{T_2}$$

$$290 = 300 - \frac{450}{T_2}$$

$$-10 = \frac{-450}{T_2}$$

$$T_2 = \frac{-450}{-10}$$

$$T_2 = 45 \text{ h}$$

2. Having determined T_2, calculate the required testing time Δt.

$$\Delta t = \left(\frac{N_0 T_0}{C}\right) \ln\left(\frac{T_2}{T_1}\right)$$

where $N_0 = 300$
   $T_0 = 1.5$ h
   $T_1 = 1.5$ h
   $T_2 = 45$ h
   $C = 4$

$$\Delta t = \left[\frac{(300)(1.5)}{4}\right]\left[\ln\left(\frac{45}{1.5}\right)\right]$$

$$= (112.5)(\ln 30)$$
$$= (112.5)(3.4012)$$
$$= 382.64 \text{ h}$$

3. Based on the computed value of the upgraded MTTF T_2, calculate the improved system reliability for a 50-h program run time.

$$R_t = e^{-t/T}$$

where $t = 50$ h
   $T = 45$ h (based on step 1, T_2 computation)

$$R_{50} = e^{-50/45}$$
$$= e^{-1.1111}$$
$$= 0.32923, \text{ or } 0.33$$

Conclusions:

1. The required testing time, based on steps 1 and 2, is 382.64 h.
2. The projected computer program reliability, based on step 3, is 0.33.

Equipment Maintenance

Overview

Maintenance is the logistics process that includes all actions necessary to sustain a system at a specified level of performance or to restore it to that level. It is the operational discipline which functionally reflects the parameters of maintainability, a systems engineering design discipline. The maintenance process encompasses the categories of *corrective maintenance* and *preventive maintenance*.

Corrective maintenance consists of those unscheduled actions that are required in order to effect the repairs necessary to restore a failed system to a specified operational capability. Corrective maintenance actions respond to random, unanticipated system malfunctions.

Preventive maintenance consists of those planned maintenance actions accomplished on a scheduled basis through the process of regular inspections, servicing, lubrication, cleaning, on-condition monitoring, periodic replacement of critical components, etc., to avoid certain types of failures. The preventive maintenance process is planned and controlled by the system logistics manager.

The maintenance hierarchy is defined by the three levels of *organizational* (e.g., retail) *maintenance*, *intermediate* (e.g., service center) *maintenance*, and *depot* (e.g., producer's facility) *maintenance*. On each level, maintenance functions are assigned or authorized according to facility capabilities or governing economic effectiveness criteria. All system maintenance strategy and planning emanate from the maintenance concept established at the time of system design and development.

Maintenance Concept

The maintenance concept is typically developed during the early phases of system evolution in order to more effectively integrate logistics with the other system design disciplines. It designates which levels (organizational, intermediate, depot) of maintenance are contemplated for activity in support of the product; the governing philosophy (repair or "remove-and-replace"); the responsibilities projected for each maintenance level; the effectiveness factors, including the predicted reliability and maintainability parameters, which might affect the maintenance process; and the maintenance environment as described by the nature of the market centers, distribution of stockage and service points, and available transportation modes. The maintenance concept is necessarily sensitive to the maintainability design characteristics (discussed in Chap. 3 of this book), especially with respect to forecasting the personnel, skills, support and test equipment, facilities, etc., that may be required to sustain the specified maintainability parameters.

The following considerations pertain to development of the maintenance posture of the proposed system and should be addressed by the logistics manager during formulation of the maintenance concept.

1. The levels in the maintenance hierarchy must be designated and defined.

2. The functions to be accomplished at each maintenance level must be specified and allocated.

3. Repair policies and decision criteria pertaining to the decision to repair, remove and replace, or discard the irreparable modules in the system should be defined for each designated maintenance level.

4. Parameters for turnaround time should be specified for each designated maintenance level.

5. Parameters for pipeline time and in-transit pipeline assets based on realistic order times and shipping times between stockage points and among the locations at each maintenance level should be stipulated.

6. Support and test equipment requirements for each maintenance level must be evaluated and defined.

The analytical methodology and quantitative factors relating to maintenance-level decisions will be governed by the design attributes of the proposed system.

The Maintenance Hierarchy

The maintenance process utilizes a tri-level, hierarchal infrastructure defined on the basis of geographical location, types of facilities, functions performed, and, in general, the economic effectiveness of what an individual facility contributes to the overall logistics posture supporting the product. The three levels of the maintenance process are referred to as the organizational level, intermediate level, and depot level.

Organizational-level maintenance

Organizational maintenance involves retail-level personnel involved on a daily basis with the customer market and operators of the system. At this level, there is limited capability to repair components; consequently, most failed parts are forwarded to the next higher maintenance level for test and repair or discard, as appropriate. It is important that the design of the system reflect modular interfaces and accessibility which take into consideration the maintenance functions designated for and the limitations of the organizational-level activity.

Intermediate-level maintenance

This level of system maintenance is normally performed at product service centers, at distribution centers, or, in certain cases, by mobile maintenance teams with portable maintenance units which can be geographically deployed to augment organizational-level maintenance capabilities. At this level, corrective maintenance usually involves removal and replacement of major modules, assemblies, or repair parts. Scheduled maintenance which requires dismantling and reassembly of equipment may be accomplished at this level.

Depot-level maintenance

Depot-level maintenance is the highest level of maintenance. It is normally located at a production point, original equipment manufacturer's facility, or specialized repair facility (under contract to the prime producer of the item). The skills, support and test equipment, specialized facilities, abundance and array of spare parts, etc., at the depot level exceed those available at the organizational and intermediate maintenance levels. The depot level of maintenance has capabilities for system overhaul, system rebuilding, and calibration of S&TE used at all levels, as well as performance of highly complex maintenance tasks. Depot-level facilities can be designated and located based on specified product lines, regionally defined proximity to service centers or distribution centers, or zonally defined areas focusing on major market centers.

Distinguishing Attributes of Maintainability

The constituent subprocesses of maintenance have inherent unique characteristics which affect their contributions to the total functional process.

Corrective maintenance characteristics

Corrective maintenance is repair. This element of the maintenance process includes all unscheduled maintenance actions which need to be accomplished as the result of randomly occurring system malfunctions. The sequence of actions in the corrective maintenance cycle, as portrayed in Fig. 3.1, includes fault detection and confirmation, fault localization and isolation, disassembly, removal of the failed component, reassembly, and verification and checkout. The repair cycle also contemplates the possibility of "false failures," whereby test and evaluation of a suspected component might indicate that failure never actually occurred; the failure was erroneously signaled as a result of conditions such as faulty diagnostics, erroneous operator interpretation of indicators or signals, etc.

Preventive maintenance characteristics

Preventive maintenance includes all scheduled maintenance actions accomplished to assure that a system remains in condition for a specified level of performance. Preventive, or scheduled, maintenance includes periodic inspections, system or condition monitoring, periodic removal and replacement of critical components, calibration procedures, routine external servicing, lubrication, cleaning, etc.

Determination of reparability of components

The reparability of the modular assemblies, subassemblies, components, and repair parts of a system can be determined on the basis of design characteristics and on the basis of the economics of repair.

Design influences.   Design attributes of a system dictate whether at any level of maintenance an item is categorically nonreparable, partially reparable, or fully reparable.

   Nonreparable item.   A nonreparable item, generally an item constructed by modular fusion or one with a relatively low replacement cost, is one that is discarded when a failure occurs. No repair is accomplished, and the item is replaced by a spare. The failed item is then disposed of as a "throwaway" or reclaimed for other uses. No intermediate-level or depot-level maintenance support is required except for resupply of replacement items. For this designation, maintainability design criteria should provide for high-reliability built-in unit self-test capability to ascertain that a failure has actually occurred before discarding the item.

Partially reparable item.   When a modular assembly is designated as a partially reparable item, certain components of the assembly are identified for removal, replacement, and discard, while other components are considered economically reparable at one or more of the hierarchal maintenance levels and are therefore processed under prevailing reparable item management procedures.

Fully reparable item.   An item within this maintenance category is considered totally reparable at one or more designated maintenance levels and thereby warrants investment in specialized facilities, support and test equipment, specialized skills, specialized technical data, and supporting spare and repair parts inventory.

Economic influences on maintenance-level designation.   Item maintenance responsibilities assigned to the organizational, intermediate, and depot level are determined primarily on the basis of economic effectiveness considerations. Some assignments are based on noneconomic criteria, such as the need for such facilities to be in proximity to market centers to assure prompt response to customer needs. Notwithstanding such possible exceptions, this chapter focuses on the principles of level-of-repair economic effectiveness which govern most locational decisions. The definition of system indenture detail assigned to activities at each level in the maintenance hierarchy reflects the increasing scope and complexity from the lower to higher levels, as shown in Table 23.1.

For those items designated as reparable, either as fully reparable units or as reparable modules within partially reparable units, there are three fundamental maintenance options for the disposition of failed items: (1) repair, (2) referral to the next higher maintenance level, or (3) condemnation and discard in accordance with the governing item condemnation criteria. The following principles pertain to this process:

1. Designation of a level of reparability means that the designated level is the lowest level within the maintenance hierarchy that has the authority and capability to repair the item.

TABLE 23.1

Maintenance level	Maintenance responsibilities
Operational level	Scope of maintenance assignment includes removal, replacement, and repair (as authorized) related to the system and certain subsystems or assemblies. This generally extends to work breakdown structure (WBS) Level 1 or 2.
Intermediate level	All functions assigned to the organizational level plus subassemblies. System indenture complexity usually extends to WBS Level 2 or 3.
Depot level	All functions assigned to the organizational and intermediate levels plus components and spare and repair parts. Indenture detail includes the total system infrastructure extending to WBS Level 3 or 4.

2. Failed items that are designated as reparable at the organizational level are also reparable at the intermediate and depot maintenance levels.

3. Failed items that are referred by the organizational level to the intermediate level and those that are designated for repair at the intermediate level are likewise reparable at the depot level.

4. The depot-level purview encompasses those failed items referred to the depot level plus those items designated as reparable only at depot level.

5. Each maintenance level is provided with item condemnation criteria to be applied in accordance with stipulated factors, conditions, and circumstances prevalent at that level. A condemned item may be discarded for scrap or salvage, or some other appropriate disposition for return of residual salvage value may be made.

These principles are applied in defining the scope of maintenance activity at the designated maintenance levels.

Activity at the Hierarchal Maintenance Levels

In the ideal logistics scenario, all system maintenance actions are responsively accomplished at the organizational level. This necessarily presumes rapid turnaround times and abundance of spare and repair parts stockage, reinforced by an extensive inventory of replacement prime items to compensate for nonavailability during maintenance turnaround and condemnations. The pragmatic system manager acknowledges that such a logistical nirvana is economically prohibitive because of the normal multiplicity and dispersion of retail locations, and therefore is nonexistent. The manager must establish a balance between idealism and pragmatism by trading off costs and maximum support considerations and designing a network of maintenance levels, within which each level of activity contributes to the support structure in the most cost-effective manner achievable.

Organizational-level activity

A maintenance-significant item may be designated as reparable at this level; if it is not possible to repair the failed item in accordance with economic screening factors and the item conforms to the established condemnation criteria, it is discarded. If the failed item is not designated as reparable and is not condemned at the organizational level, it is referred to the next higher maintenance level for evaluation and further disposition. The maintenance policy for the item may stipulate that the organizational level must submit such reparable items to either an intermediate-level or depot-level facility. For items designated as reparable at this level, it is important that the system incorporate highly reliable built-in test (BIT) diagnostic capabilities to preclude the possibility of false failure indications. To the extent that items are designated as reparable at the organizational level, there should be a corresponding stockage of spares and repair parts to support the maintenance policy.

Minimum stockage requirements at organizational level. The minimum-support posture, excluding provision for safety and contingency stocks, for maintenance at the organizational level includes

- Spare and repair parts required to support restoration of the assigned reparable items.
- Stockage of replacements for assigned reparable items based on the item reparable generation rate and local maintenance turnaround time.
- Stockage of replacements for nonassigned reparables which must be referred to higher levels based on the nonassigned reparable generation rate and resupply pipeline time.
- Stockage of replacements for assigned reparable items based on the condemnation rate and resupply pipeline time.

It is underscored that provisions for safety stocks or contingency requirements are in addition to the minimum levels described above.

Intermediate-level activity

Workload at the intermediate maintenance level consists of reparable referrals from the network of organizational maintenance activities plus those items predesignated for repair at the intermediate level by the governing item maintenance policy. Those generated reparables which fail the economic screening and conform to the condemnation criteria are discarded. In keeping with maintenance system discipline, those failed items that are not designated for repair at this level are referred to the depot level.

Minimum stockage at intermediate level.   Minimum support posture at the intermediate maintenance level includes

- Spares and repair parts to support corrective maintenance of assigned intermediate-level reparables
- Stockage of replacements for assigned reparable items based on the rate of reparable generation from organizational-level referrals, the rate of reparable generation within the intermediate-level activity, and the intermediate-level maintenance turnaround time
- Stockage for replacement of nonassigned reparables referred to the depot based on their reparable generation rates and the resupply pipeline time
- Stockage for replacement of assigned reparables based on the condemnation rate and resupply pipeline time

Under certain circumstances, the intermediate-level activity may maintain a specified number of days' stockage of critical items as a cushion to absorb abnormal surges in workload demands.

Depot-level activity

Workload at the depot level is generated by the organizational- and intermediate-level reparable referrals and those uniquely designated depot-level reparables. Reparable items failing the depot economic screening criteria are condemned and discarded for scrap or otherwise disposed of for salvage value. The depot is assigned the responsibilities for system overhaul and rebuilding. Another significant depot function is maintaining sufficient inventory of replacement items and spare and repair parts to support demands from the subsidiary structure of organizational- and intermediate-level activities. The turnaround time patterns of depot-level maintenance notably differ from those of the organizational and intermediate maintenance levels. Depot workload managers typically accumulate reparable carcasses into economical lot sizes before committing to a shop schedule for repairing the specific items. The consequent drawdown of depot assets is offset by an inventory level calculated to support the demand for these assets during the shop schedule intervals.

Unlike the posture at organizational and intermediate levels, the pipeline into the depot is determined by the procurement and administrative lead time required to replenish the total of the system condemnation rates plus the adjustments in logistics system requirements.

Minimum depot-level stockage. Minimum support requirements at the depot level include

■ Spares and repair parts for support of assigned depot-level reparables

■ Stockage replacements for items uniquely reparable at the depot level based on the rates of reparable generation from organizational and intermediate-level referrals, internal depot "job-routed" reparable generations, and the depot turnaround time

■ Stockage of assets required for system overhaul and rebuilding

■ Stocks of reparables based on the aggregate of the organizational-, intermediate-, and depot-level condemnation rates and the procurement and administrative lead time for acquisition of replenishment items from the item vendors and manufacturers

Depot-level maintenance management frequently calls for augmented inventory levels of critical and highly essential replenishment items to provide for demand surges within the logistical support network.

Maintenance Activity Indicators

Performance indicators used to measure the maintenance process fall into the categories of *system supportability indicators* and *facility effectiveness indicators*. System supportability indicators include those descriptive factors which portray the effects and consequences of system operation and downtime with respect to the response capability of the maintenance facility. These indicators are applicable at the organizational maintenance level, inasmuch as this level serves as the critical interface with the systems and is directly responsible for sustaining the functional capability of the systems and responding promptly to the needs of the users. Most system supportability indicators are parametrically defined in terms of system operating time units, although these are frequently converted to calendar time units to be consistent with logistical delay intervals and to accommodate maintenance workload planning, which is normally based on calendar scheduling. Facility effectiveness indicators reflect the implications for the maintenance activity of the circumstances portrayed by system supportability indicators.

These are cited in terms used for calculating potential workload requirements and measuring the productivity of the maintenance activity. Facility effectiveness indicators are typically keyed to calendar time units so that they are adaptable to normal maintenance planning and scheduling time parameters.

System supportability indicators

System supportability indicators essentially characterize demands on the support activity generated by the system. As such, they define and measure those attributes relative to effectiveness in support of the system. The following are system-oriented effectiveness factors (some of which were introduced with maintainability in Chap. 3) derived from system operational and supportability characteristics.

1. *Mean time between maintenance (MTBM).* MTBM is the mean or average time between all maintenance actions (corrective and preventive) and can be calculated as

$$\text{MTBM} = \frac{1}{1/\text{MTBM}_{ct} + 1/\text{MTBM}_{pt}}$$

   where MTBM_{ct} is the mean interval for corrective maintenance and MTBM_{pt} is the mean interval for preventive maintenance. The reciprocals of MTBM_{ct} and MTBM_{pt} are the maintenance rates in terms of maintenance actions per hour of system operation. MTBM_{ct} is equivalent to MTBF, and assumes that a combined failure rate is used which includes primary inherent failures, dependent failures, manufacturing defects, operator- and maintenance-induced failures, etc. The summary maintenance frequency factor, MTBM, is a major parameter in determining achieved availability A_a and operational availability A_o of a system.

2. *Mean time between replacement (MTBR).* This factor refers to the mean time between item replacement and is the governing parameter in determining spare parts requirements. In many cases, corrective and preventive maintenance actions are accomplished without requiring the replacement of a component part. More frequently, however, corrective maintenance involves replacement parts, and this, in turn, necessitates having spare parts available in the supporting inventory. MTBR is a significant factor, applicable in both corrective and preventive maintenance activities involving item replacement, and is a key parameter in determining inventory requirements.

3. *Mean active maintenance time ($\overline{M}$).* This is the mean time for accomplishing all maintenance tasks, both corrective and preventive (a function of $\overline{M}_{ct}$ and $\overline{M}_{pt}$)

4. *Mean corrective maintenance time ($\overline{M}_{ct}$).* This is the average time required to accomplish an action to repair a failed system.

5. *Mean preventive maintenance time ($\overline{M}_{pt}$).* This is the average time required to accomplish a planned, scheduled maintenance action.

6. *Median active corrective maintenance time ($\widetilde{M}_{ct}$).* This factor denotes the value which divides all of the downtime values so that 50 percent are equal to or less than the median and 50 percent are equal to or greater than the median.

7. *Maximum active corrective maintenance time (M_{max}).* This is the value of maintenance downtime below which a specified percentage of all maintenance actions can be expected to be completed (usually specified at 90 percent and 95 percent confidence levels).

8. *Maintenance downtime (MDT).* Maintenance downtime constitutes the total elapsed time required (during which the system is not operational) to repair and restore a system to full operating status, and/or to retain a system in that condition. MDT is based on mean active maintenance time $\overline{M}$ and logistics delay time LDT. The mean or average value of MDT is calculated from the elapsed times for each functional factor and the associated frequencies (similar to the approach used in determining $\overline{M}$).

9. *Logistics delay time (LDT).* Logistics delay time refers to that maintenance downtime that is expended as a result of administrative delays, waiting for a spare part to become available, waiting for the availability of an item of test equipment in order to perform maintenance, etc. LDT does not include active maintenance time, but is nevertheless a major element of total MDT. This factor is mathematically cited as MLDT (mean logistics delay time).

Facility effectiveness indicators

The facility-related factors which are reflective of and sensitive to the effects of system performance and supportability are described below. Facility effectiveness indicators characterize the ability and capacity of the maintenance activity, as reinforced by the total logistics system, to respond to demands generated by users of the system.

1. *Turnaround time (TAT).* This element constitutes the time that it takes for an item to go through the complete cycle from point of receipt

at the maintenance activity, through the restoration process, and into the item inventory ready for reissue.

2. *Reparable generation rate.* This is the rate at which reparables of a particular item are accepted for processing by a maintenance activity, as expressed by the ratio of the number of units accepted to a specified time parameter (e.g., number of reparables per month).

3. *Reparable referral rate.* This is the rate at which reparables of a particular item accepted by an activity, but not authorized for repair at that activity, are referred to higher-level maintenance activities. This rate is expressed in terms of units per specified time parameter.

4. *Reparable restoration rate.* This is the rate at which reparable units of a particular item, designated for support by a maintenance activity, are repaired and returned to inventory for reissue. This factor is expressed in terms of processed units per specified time parameter.

5. *Condemnation rate.* This is the rate at which reparable items are condemned and discarded for scrap or disposed of for salvage by a maintenance activity pursuant to stipulated economic screening and condemnation criteria. This rate is expressed in terms of units per time parameter.

6. *Logistics pipeline time.* Often referred to as pipeline time, this denotes the time, expressed in temporal units (usually days), required for order, shipment, and receipt by the requesting activity of a replacement item. This term is primarily applicable to organizational- and intermediate-level activities.

7. *Procurement and administrative lead time (PALT).* This is the time, expressed in defined temporal units (usually days), for requisition, contractual execution, manufacture, shipment, and acceptance by the requiring activity of items required for replenishment of inventory levels or other needs within the logistics system. This term is applicable to depot- or wholesale-support-level activities.

All the facility effectiveness indicators described are constituent quantitative elements used in evaluation and analysis of the interactions between the organizational, intermediate, and depot levels of the maintenance hierarchy.

Screening for Repair or Discard

The decision to repair or discard a modular component of a system is governed by both economic and noneconomic considerations. Frequently, the final decision is tempered by trade-off analysis of both types of criteria.

Cost differential economic approach

The typical economic decision entails evaluation of the cost of repairing a failed item relative to the cost of discard and replacement. The basic repair vs. discard cost equation is simply stated:

$$\Delta C = C_R - C_D$$

where ΔC = net cost differential
C_R = cost of repair
C_D = cost of discard

If ΔC is positive, the item would normally be discarded based on economic considerations. If ΔC is negative, the item would be repaired.

Example   To determine the appropriate decision concerning whether to repair or discard an item that costs $100 to repair, but is available on the market for $95,

$$C_R = \$100 \quad C_D = \$95 \text{ (replacement unit)}$$
$$\Delta C = C_R - C_D$$
$$-\,100 - 95$$
$$= \$5$$

Recommendation   Discard, as ΔC is positive. If the cost factors were reversed (i.e., $C_R = \$95$ and $C_D = \$100$), ΔC would be negative, thereby suggesting repair as the appropriate policy.

Repair/discard screening methodology for reparable items

The decision for determining reparability of an item is predicated on the steps and methodology described below.

Determination if the item is nonreparable or consumable.   Items can be classified as consumable or nonreparable based on their physical make-up. Examples of these include gaskets, bolts, filters, and other objects whose physical composition does not dictate repair. The logistics analyst would perform the following actions based on the answer to the question: *Is the item nonreparable or consumable?*

If *yes*, then provide repair-level recommendation of discard at the appropriate remove/replace level (Organizational, Intermediate, or Depot).

If *no*, then apply the screening process.

Those items which are disposable by virtue of their consumable characteristics are not in the same category as items which may be considered for discard at a certain level of maintenance due to condemnation and economic screening criteria.

Determinination if there are any preempting factors. The analyst must ascertain if there are preempting factors affecting reparability identified pursuant to a level of repair analysis which would preclude the development of screening criteria. Such preemptive factors include the following considerations:

1. *Safety*. Repair authorization at certain levels would introduce personnel safety hazards.

2. *Security*. Security at certain locations would introduce restraints affecting national security of restricted technology. Security could be compromised due to access by personnel without proper access authority. The system or its components cannot be stored or transported while awaiting maintenance if it retains proprietary data elements or restricted technology.

3. *Warranties*. Product warranties could affect repair-level alternatives.

4. *Calibration requirements*. System requirements may prescribe calibration needs which cannot be delegated to or are beyond the capability of certain repair facilities.

5. *Special skills*. Skill levels required for maintenance cannot be supported for certain repair alternatives.

6. *Specialized training*. Special training requirements cannot be supported for certain repair alternatives.

7. *Product criticality*. A critical component may require highly specialized facilities to meet functional requirements as stipulated in the product specifications.

8. *Special repair facilities*. Repair facilities cannot be established to support specific maintenance alternatives.

9. *Transportation (weight, volume, susceptibility to damage)*. Weight, volume, or handling characteristics preclude establishment of certain maintenance level alternatives. Susceptibility to damage, which cannot be suitably controlled by packaging requirements, discourages transportation and certain maintenance-level alternatives.

10. *Material handling*. Special material-handling requirements prevent consideration of certain maintenance-level alternatives.

11. *Remove/replace decisions*. Accessibility or maintainability constraints indicated by modular remove/replace design features preclude certain maintenance-level alternatives.

12. *Reparability.* Repair of the component is not possible (destructibility) or would degrade the reliability of the component to a non-acceptable level.

If a preempting factor applies, the analyst should exclude from consideration any maintenance level that is affected. If no preempting factors apply, it is essential to proceed with development of item-screening criteria.

Development of item repair/discard screen.   This process is based on the cost threshold concept and considers all components and assemblies which are capable of being repaired. (Nonreparable items such as gaskets, bolts, etc., should have already been screened out and designated as consumables since they cannot be repaired). The cost threshold screen is an analytical technique which uses Mean Time to Repair (MTTR), repair material cost, and estimated labor rates to determine a break-even cost where repair of a specific item is more economical than discard. By using a designated labor rate, the analyst must gather only the normally available information for item cost, MTTR, and average repair material cost for each candidate item to perform the cost threshold screen. The following describe the procedural steps involved in development of the cost threshold screen.

   Step 1: Establish the current procurement cost of the candidate item.

   Step 2: Estimate the MTTR (hours) for the item. The MTTR will consist of all actions beginning after fault isolation and removal and will include verification and screening actions, repair actions, and any calibration actions.

   Step 3: Estimate the cost ($) for repair material for each candidate item.

   Step 4: Determine the cost of the repair materials as a percentage of the item procurement cost. For example, if the repair materials cost $10.00 and the procurement cost of the item is $200.00, the repair material percent is 5 percent (expressed by its decimal equivalent of 0.05).

   Step 5: Use the labor rate ($) established for the maintenance activity under consideration.

   Step 6: For each item, solve the following equation:

$$\frac{(\text{MTTR})\,(\text{Labor Rate})}{\{1 - \text{repair material percent } (>0)\}}$$

If the actual procurement cost of the item is less than the amount calculated based on the above equation, the item should be recommended for discard. It would not be cost-effective to repair the item. The above calculation describes a linear (straight line) relationship and is easily portrayed in a graphic plot by the analyst.

Example   Plots Showing MTTR, Labor Rate, and Repair Materials/Item Cost Ratio

First scenario.   The typical summary in Table 23.2 and the example in Figure 23.1 illustrate a graph for the labor rate of $21.00 per hour. The defined MTTR plot parameters are from 0.5 (the minimum) to 4.5 (the maximum). The lines on the graph are plotted for four material percentage variables (5 percent, 20 percent, 35, and 50 percent). For each specific case, the analyst would actually construct the appropriate labor rate graph based on the governing rate at the maintenance level under consideration and the appropriate material percent line for the item being analyzed. The analyst would then plot the point on the graph for actual item cost and estimated MTTR. If the point falls below the material percent line the item should be coded for discard. If the plot point is above the material percent line it should be recommended for repair.

Labor Rate: $21.00
MTTR Plots: 0.5 and 4.5

Plot scalar points for Repair Materials Cost/Unit Cost Ratio of 5, 20, 35, 50 percent.

$$\frac{(MTTR)\,(Labor\ Rate)}{\{1 - repair\ material\ percent\ (>0)\}}$$

It is noted that the determinant factors in the formulation of the Repair/Discard criteria to construct the graph portrayed in Figure 23.1 are the *MTTR* values, *Labor Rate,* and the *ratio of Repair Materials Cost to Unit Procurement Cost.* The Repair Materials Cost/Unit Cost Ratios of 5, 20, 35, 50 percent are judgmentally selected as these are the more common plot criteria utilized. If these factors are constant, the new procurement cost of any item plotted on this graph helps make the repair-versus-discard decision.

Table 23.2

Repair Materials Cost/Unit Cost Ratio	Mean Time to Repair MTTR (hours)	Calculated Value ($) to be Plotted on Graph
0.05	0.5	11.05 (@ 11)
0.05	4.5	99.45 (@ 99)
0.20	0.5	13.13 (@ 13)
0.20	4.5	118.25 (@ 118)
0.35	0.5	16.15 (@ 16)
0.35	4.5	145.38 (@ 145)
0.50	0.5	21
0.50	4.5	189

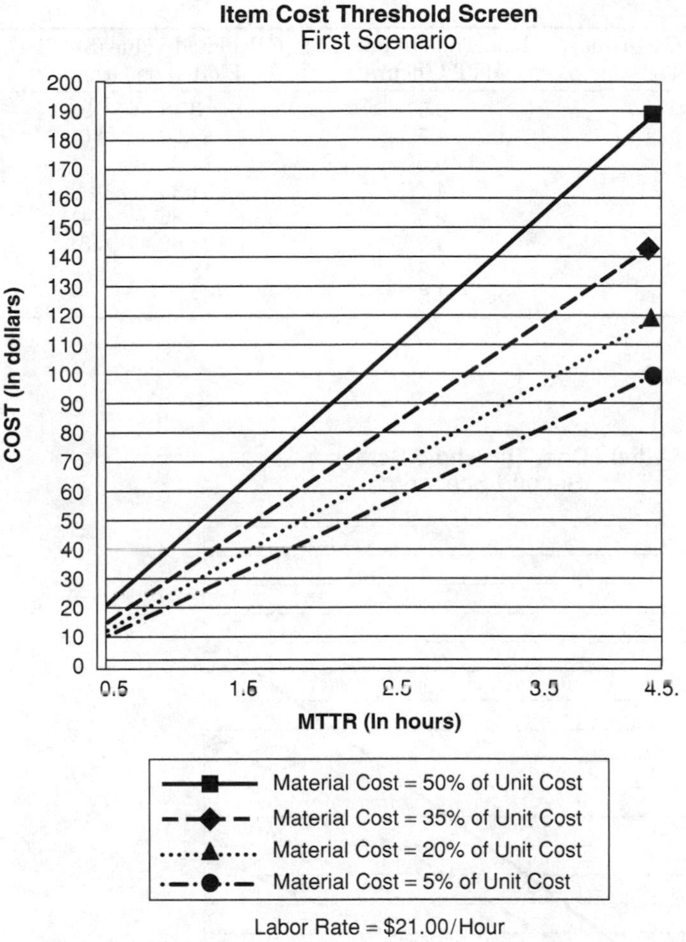

Item Cost Threshold Screen
First Scenario

Figure 23.1 legend:
- Material Cost = 50% of Unit Cost
- Material Cost = 35% of Unit Cost
- Material Cost = 20% of Unit Cost
- Material Cost = 5% of Unit Cost

Labor Rate = $21.00/Hour

Figure 23.1 Repair/discard screening graph based on Table 23.2.

Example Plots Showing MTTR, Labor Rate and Repair Materials/Item Cost Ratio

Second scenario. The typical summary in Table 23.3 and the example in Figure 23.2 illustrate a graph for the labor rate of $18.00 per hour. The defined MTTR plot parameters are from 0.5 (the minimum) to 4.5 (the maximum). The previously cited plot point formula applies.

Labor Rate: $18.00
MTTR Plots: 0.5 and 4.5

Plot scalar points for Repair Materials Cost/Unit Cost Ratio of 5, 20, 35, 50 percent.

Table 23.3

Repair Materials Cost/Unit Cost Ratio	Mean Time to Repair MTTR (hours)	Calculated Value ($) to be Plotted on Graph
0.05	0.5	9.47 (@ 9)
0.05	4.5	85.26 (@ 85)
0.20	0.5	11.25 (@ 11)
0.20	4.5	101.25 (@ 101)
0.35	0.5	13.86 (@ 14)
0.35	4.5	124.62 (@ 125)
0.50	0.5	18
0.50	4.5	162

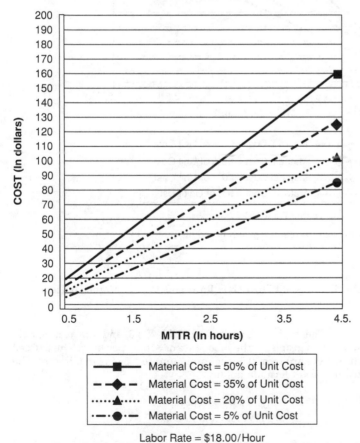

Labor Rate = $18.00/Hour

Figure 23.2 Repair/discard screening graph based on Table 23.3.

Example Illustration of Application of the Repair/Discard Graphs
Screening Item for Repair versus Discard—Case for First Scenario (Figure 23.1).

Labor: $21.00/Hour
MTTR: 3.5 Hours
Materials Cost: $38.50
Purchase Cost of Item: $110
Materials Cost/Item Cost Ratio = $38.50/$110 = 0.35 (35% of Item Cost)

Figure 23.3 shows that the plot point of the new item cost of $110 relative to
the MTTR of 3.5 hours is *below* the 35 percent of Unit Cost line on the graph
constructed by Figure 23.1. It is therefore considered cost-effective and economically feasible to purchase a new item and discard the reparable rather than
repair.

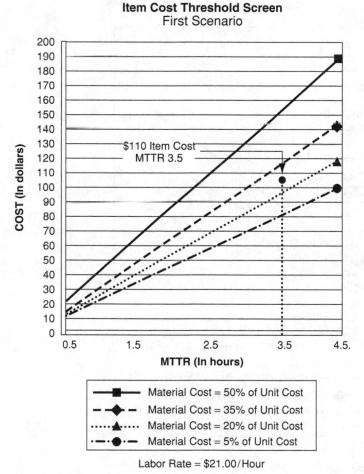

Figure 23.3 shows the plot points.

Figure 23.3 Repair/discard plot points based on the graph in Figure 23.1.

Screening Item for Repair versus Discard—Case for Second Scenario (Figure 23.2).

Labor: $18.00/Hour
MTTR: 2.5 Hours
Materials Cost: $60
Purchase Cost of Item: $120
Materials Cost/Item Cost Ratio = $60/$120 = 0.50 (50% of Item Cost)

Figure 23.4 shows that the plot point of the new item cost of $120 relative to the MTTR of 2.5 hours is *above* the 50 percent of Unit Cost line on the graph constructed by Figure 23.2. It is therefore considered cost-effective and economically feasible to repair the item rather than purchase a new replacement.

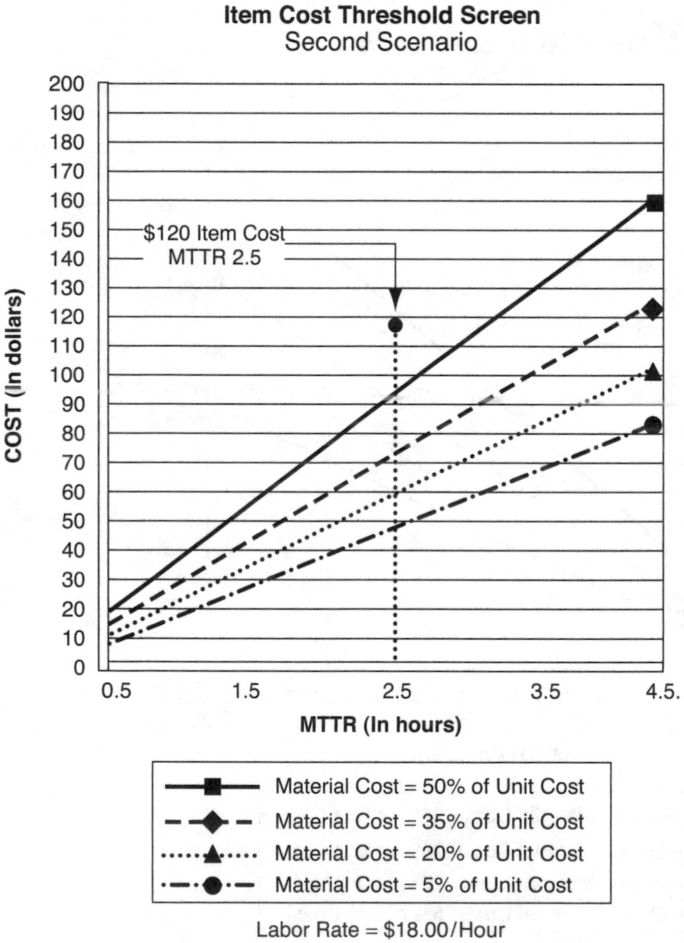

Figure 23.4   Repair/discard plot points based on the graph in Figure 23.2.

Conclusion

The methodology described herein has a wide range of applications. For example, the screening technique can be used to:

1. Make a repair-versus-discard decision at a designated facility assigned maintenance responsibility for an item.
2. Perform cost-effectiveness comparison analyses among a number of maintenance locations with varying maintenance hour (MTTR) requirements for specific item tasks and varying hourly labor rates preliminary to assigning maintenance responsibility for an item.

Factors favoring repair.

1. *Economic criteria*
   a. *Inventory.* The cost of maintaining inventory levels and warehousing storage capacity for replacements might be prohibitive (especially for environmentally sensitive items requiring climatically controlled storage).
   b. *Transportation.* Costs for shipment of replacement items could be disproportionate, relative to the unit value of the items.
2. *Noneconomic criteria*
   a. *Reliability database.* A repair capability can provide item reliability data.
   b. *Market responsiveness.* Local conditions may dictate an on-site repair capability to provide for emergency repairs and to assure responsiveness in a highly competitive market environment.
   c. *Weight.* Heavy items could necessitate local repair facilities, rather than incurring the need for special packaging, handling, and transport of the item.
   d. *Volume.* Limited storage could make it impossible to accommodate bulky items requiring unusually large cubic storage volume.

Factors favoring discard.

1. *Economic criteria*
   a. *Design.* An item with high reliability by virtue of its design would fail infrequently. It could be cost-effective to replace and discard such an item in the rare event of failure rather than maintain a repair facility.
   b. *Engineering.* A discard policy would afford savings in design and fabrication by eliminating the need to maintain accessibility to internal components of a throwaway module.

    c. Unit production cost. Increases in production quantities of replacement modules tend to reduce the unit costs of the items.

2. *Noneconomic criteria*

    a. Modification. Product improvement and upgrade is simplified by replacement of complete modular components

    b. Skills. Remove-and-replace procedures reduce the need for highly skilled technical personnel.

    c. Facilities. Discard policies reduce the need for expensive repair facility space, making the space available for more critical, high-priority commitments.

Maintenance Reporting

Every maintenance activity should utilize a basic management information system to monitor the types of maintenance actions, measure productivity, and project workload. There are in the logistics community numerous sophisticated maintenance data systems used to manage repairs and scheduled maintenance actions, as well as to track the causes of and factors involved in logistics delays. Information on these types of systems is readily available in current Department of Defense publications and in technical textbooks on the subject, some of which are referred to at the end of this book. This section will, therefore, focus on the basic elements involved in reporting corrective maintenance activity.

Corrective maintenance reporting elements

Figure 23.5 provides a sample format incorporating basic elements of maintenance information, which may be tailored to specific processes.

    It is advisable that the same data element structure and report format be consistently applied to all items within a defined system management group. This will facilitate categorization, aggregation, and analysis of the data to measure maintenance productivity and system performance trends.

Interactive Elements of the Corrective Maintenance System

The hierarchal maintenance system involves a process of interface and interaction between the organizational-, intermediate-, and depot-level activities. All levels are affected by demands generated within the systemic infrastructure. Individual activities are sensitive to fluctuations in productivity and usage rates at other levels of activity within the hierarchy. Figure 23.6 portrays this dynamic systemic interrelationship.

I. Maintenance event data[a]

Event	Description of action	Maintenance skill used	Event time[b]	
			Start	Stop
1. Failure detected				
2. Submitted to maintenance				
3. "How malfunction" diagnosed				
4. Replacement part(s) ordered[c]			(Time of order)	N/A
5. Replacement parts received[c]			N/A	(Time of receipt)
6. Skills, technical data, special tools, etc., requisitioned[d]			(Time of requisition)	N/A
7. Skills, technical data, special tools, etc., received[d]			N/A	(Time of receipt)
8. Repair completed				
9. Replacement reinstalled on system				
10. Verification and checkout				
11. Resubmitted to inventory			N/A	(Time resubmitted to inventory)[a]

FIGURE 23.5 Sample Corrective Maintenance Report Format.

II. Equipment descriptive data

Data item	Nomenclature	Drawing or part number	Manufacturer	Serial number	Hours operation on system
1. Modular component					
2. Next higher assembly (NHA)					

(All NHAs in the ascending system indenture levels between the reparable modular component and the system should be identified to provide traceability)

3. System
4. Replacement parts used

III. Support data
1. Description of support and test equipment used.
2. Description of technical instruction or maintenance procedure used.

[a]The interval between the failure detected element and the resubmitted to inventory element is the corrective maintenance cycle, excluding delay times in steps 4 to 7.
[b]To assure universal consistency, use of Greenwich Mean Time (GMT) standards is recommended.
[c]Order and shipping time is considered delay time, not maintenance time.
[d]Requisition of technical support time is considered delay time, not maintenance time.

FIGURE 23.5 (continued)

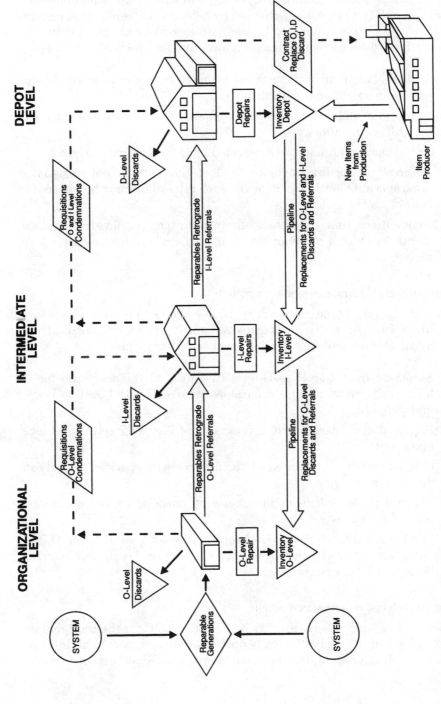

Figure 23.6 Interactive corrective maintenance system.

Organizational (O-level) maintenance activity

The corrective maintenance process starts at the organizational maintenance level. The process is activated by demands for repair and restoration of the system generated by the customer-users of the system. The essential O-level corrective maintenance functions basically include

1. Replacement of failed items in the primary system so as to restore the system to operational capability.
2. Repair of modular components designated as reparable for return to the O-level inventory.
3. Condemnation of items as prescribed by O-level discard criteria.
4. Referral to the intermediate level of those items not designated reparable at O level. (On occasion, such reparables may be referred to D level.)
5. Requisition from the intermediate maintenance level of replacements for O-level condemnations and referrals to intermediate-level maintenance.

Intermediate (I-level) maintenance activity

Intermediate-level maintenance reinforces the O-level support posture. A single I-level activity is typically assigned support responsibility for multiple O-level activities. Basic I-level activities include

1. Shipment from I-level to O-level inventory of replacements for O-level condemnations and reparables referred by O Level to I-level maintenance.
2. Repair of items designated as reparable at I level for return to I-level inventory.
3. Condemnation of failed items in accordance with specified I-level condemnation criteria.
4. Referral to depot-level maintenance of those items not designated reparable at I level.
5. Requisition from depot level of replacements for cumulative O-level and I-level condemnations plus O-level and I-level reparable referrals for higher-level maintenance.

Depot (D-level) maintenance activity

The D-level maintenance activity is the "safety net" for the total maintenance system. The depot typically provides direct support to all intermediate-level activities and "as required" support to O-level activities within the boundaries of the region or product line assigned to the depot. Essential D-level activities include

1. Shipment from D-level to I-level inventory of replacements for cumulative O-level and I-level condemnations plus O-level and I-level reparable referrals to higher maintenance levels.
2. Repair of failed items designated reparable at D level for return to D-level inventory.
3. Condemnation of items as prescribed by D-level discard criteria.
4. Procurement from the vendor or item producer of replacements for cumulative O-level, I-level, and D-level condemnations to replenish depot inventory.

Original item manufacturer

The source for the items ordered to replace those discarded pursuant to condemnation procedures might be a production facility collocated with the depot or a vendor-manufacturer under contract to provide the required modular components. The basic functions of a source producer in the maintenance system are

1. To manufacture those item quantities required to replenish the O-level, I-level, and D-level discards
2. To deliver the contractually specified items to the depot inventory

System overhaul and rebuilding

System overhaul and rebuilding is not addressed by Fig. 23.6. Although this may be a depot-level function, it is an aspect of the depot mission that involves restoration of the whole system rather than repair of modular components. System overhaul and rebuilding occur only after extended intervals of system operating time and, therefore, are not treated as part of the routine corrective maintenance process, which applies to periodic replacement and repair of failed modular components.

System overhaul and rebuilding are normally provided as a form of service to customer-users who determine that this maintenance option is a cost-effective alternative to purchase of a new system.

24

Design Interface of Logistics Elements

Overview

This chapter discusses an approach to analysis of logistics systems based on integration of the quantitative design characteristics of systems engineering with the functional logistics elements. Such a methodology is applied to ascertain the utility and potential value of concepts, practices, and procedures governing logistics supportability. To illustrate this integrative analysis approach, this chapter will describe the fundamental techniques of (1) system reliability prediction, (2) system maintainability prediction, and (3) log-normal maintenance analysis. In keeping with the tutorial purpose of this book, the technical description is oriented toward the entry-level professional, and these three major areas of analysis are addressed through non-complex case scenarios involving hypothetical systems which serve to illustrate the basic concepts and methodologies. The following will sequentially address the aforementioned major areas of analysis.

System-Level Reliability Analysis and Prediction Methodology

The basic element of the development of the system reliability is the Mean Time between Failure (MTBF). The MTBF factor is used to construct the "bottom up" reliability hierarchy described by the Assembly level, Subsystem level, and the System (top) level. The hierarchy reflected by the hypothetical system, XYZ, is portrayed in Figure 24.1.

Hierarchy of Assembly, Subsystem and System Level Attributes
for
Development of Reliability Prediction for System XYZ

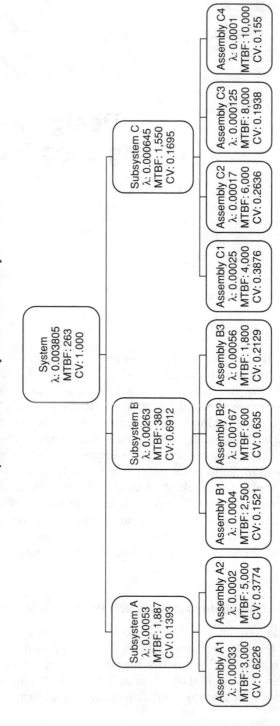

Figure 24.1 Hierarchy of system reliability attributes for development of system level reliability prediction.

The constituent data applicable to each component in the hierarchy include:

1. The MTBF value.
2. The Failure Rate (λ), which is the reciprocal of the MTBF.
3. The *criticality value* (cv), which is a ratio value based on failure rates, as explained later. This factor denotes the contribution of the failure rate of a subsidiary component to the failure rate of its parent module.

The hierarchal structure of system XYZ (shown in Figure 24.1) is defined in the following. Inasmuch as this involves a "bottom up" approach and for sake of clarity of methodology, the assemblies are assigned an MTBF value, based on available usage data or engineering estimates. The lower tier components, therefore, are the crucial determinants to the whole system-reliability prediction process.

System XYZ

    Subsystem A
        Assembly A1 (MTBF:3,000h)
        Assembly A2 (MTBF:5,000h)

    Subsystem B
        Assembly B1 (MTBF:2,500h)
        Assembly B2 (MTBF:600h)
        Assembly B3 (MTBF:1,800h)

    Subsystem C
        Assembly C1 (MTBF:4,000h)
        Assembly C2 (MTBF:6,000h)
        Assembly C3 (MTBF:8,000h)
        Assembly C4 (MTBF:10,000h)

Focusing on Subsystem B and its subsidiary assemblies for illustration, the data in the component blocks are developed as follows:

Assembly B1: λ = reciprocal of 2,500 = 0.0004
Assembly B2: λ = reciprocal of 600   = 0.00167
Assembly B3: λ = reciprocal of 1,800 = 0.00056
              Total   0.00263*

Note: The λ of the Assemblies' parent Subsystem B is, therefore, 0.00263.

Assembly B1 criticality value (cv) is 0.0004 ÷ 0.00263 = 0.1521
Assembly B2 criticality value (cv) is 0.00167 ÷ 0.00263 = 0.635
Assembly B3 criticality value (cv) is 0.00056 ÷ 0.00263 = 0.2129
                                                        1.00*

Note: The sum of the subsidiary group of critical values is always 1.00.

The Subsystem B MTBF is the reciprocal of its λ (0.00263) and = 380h.

The same methodologies applied for Subsystem C are applied to Subsystem A. Subsystem C and their respective subsidiary Assemblies, i.e.,

	Subsystem A	Subsystem C
Failure Rate (λ)	0.00053	0.000645
MTBF	1887h	1550h

The next step is to recapitulate and total the Failure Rates of Subsystems A, B, and C.

Subsystem A: Failure Rate (λ) = 0.00053
Subsystem B: Failure Rate (λ) = 0.00263
Subsystem C: Failure Rate (λ) = 0.000645
                                 Total   0.003805*

Note: The System XYZ Failure Rate (λ) is, therefore, 0.003805.
    The System XYZ MTBF is the reciprocal of its λ (0.003805) and = 263h.

Computation of the Subsystem/System XYZ Critical Values:

Subsystem A criticality value (cv) is 0.00053 ÷ 0.003805 = 0.1393
Subsystem B criticality value (cv) is 0.00263 ÷ 0.003805 = 0.6912
Subsystem C criticality value (cv) is 0.000645 ÷ 0.003805 = 0.1695
                                       Total   1.00* (System XYZ)

Note: The criticality value at system level (System XYZ) always = 1.00

The aforementioned computations are incorporated into the appropriate elements identified in the Hierarchy in Figure 24.1. It is significant that the highest criticality value at the Assembly level is Assembly B2, with a cv of 0.635, which strongly influences the Subsystem B cv of 0.692, the highest of all the subsystems.

The actual Reliability value (based on $R = e^{-\lambda t}$) cannot be determined until the operating cycle (t) is specified. One must also conclude from the Reliability prediction hierarchy that Assembly B2 is the "bad actor" in the context of the total system configuration and merits consideration for redesign or the substitution of a higher-quality item.

System-Level Maintainability Prediction Based on System Reliability Analysis Factors

For tutorial effectiveness and simplicity, this section will elaborate and utilize the factors and results from the Reliability prediction process depicted in Figure 24.1. The case example utilized to illustrate development for the system-level maintainability prediction will focus on Mean Corrective Maintenance Time. The constituent elements of this process are the following, based primarily on Figure 24.1.

- System XYZ Systems: Subsystem A, Subsystem B, and Subsystem C.
- Failure Rate (λ) for each subsystem, provided by computations of the factors in Figure 24.1.
- Criticality Value (CV) for each subsystem, provided by computations of the Factors in Figure 24.1.
- $\overline{M}_{ct}$/MTTR for each subsystem, which are generally based on available empirical data or engineering estimates.
  - Subsystem A MTTR: 4.5h
  - Subsystem B MTTR: 2.0h
  - Subsystem C MTTR: 3.5h

The matrix provided in Table 24.1 portrays the analytical structure reflecting the results of the process and methodologies using the aforementioned quantitative elements. With reference to Table 24.1, description of the specifically applicable mathematical procedures will follow.

The objective of this analysis is to determine the predicted System XYZ $\overline{M}_{ct}$ based on the calculated attributes derived from the Reliability prediction described in Figure 24.1.

- The column noted as "λi" lists the individual failure rates for the individual subsystems.
- The column noted as "Criticality Value (CVi)" restates the criticality values for the individual subsystems.
- The column noted as "$\overline{M}cti$" lists the above-cited $\overline{M}cti$/MTTR values for the individual subsystems.
- The column noted as ($\overline{M}cti$)(CVi) will provide the results of multiplying the Mean Corrective Time and critical value for each subsystem and determine the basis for determining the maintainability prediction, i.e.

$$\overline{M}cti \text{ multiplied by } CVi = (\overline{M}cti)(CVi)$$

The following summarizes the quantitative methodology as applied to the subsystems.

Subsystem A: $(0.1393)(4.5) = 0.62685h$
Subsystem B: $(0.6912)(2.0) = 1.3824h$
Subsystem C: $(0.1695)(3.5) = \underline{0.59325h}$
$\qquad\qquad\qquad\qquad 2.6205h$ (rounded to 2.6h)

Based on the above computations, the predicted System XYZ $\overline{M}ct$/MTTR is 2.6 hours. (Refer to Table 24.1 for a more detailed portrayal of the analytical results.)

Table 24.1　Development of System Maintainability Prediction Based on Reliability Data from Figure 24.1

Development of Prediction of System XYZ Mean Corrective Maintenance Time

	Subsystem	Mct/MTTR
	Subsystem A	4.5 h
	Subsystem B	2.0 h
	Subsystem C	3.5 h

i	Subsystem	λi	Criticality Value (CVi)	$\overline{\text{M}}\text{cti}$	$(\overline{\text{M}}\text{cti})(\text{CVi})$
i = 1	A	0.00053	0.1393	4.5	0.62685
i = 2	B	0.00263	0.6912	2.0	1.3824
i = 3	C	0.000645	0.1695	3.5	0.59325
		$\Sigma\lambda i = 0.003805$	$\Sigma\lambda\text{CVi} = 1.00$		$\Sigma\lambda(\text{Mcti})(\text{CVi}) = 2.6205$ (@ 2.6)

Predicted System $\overline{\text{M}}\text{mct} = 2.6\text{h}$

Log-Normal Analysis

The methodology of logarithmic analysis is applicable for stochastic predictions of maintenance action times and is based on the use of logarithmic relationships to measure and analyze the time values required to accomplish specified maintenance tasks within selected test sample groupings. This analytical technique is primarily utilized in determining two maintainability parameters that are frequently invoked in system testing plans and specifications: *median active corrective maintenance time* and *maximum active corrective maintenance time*.

Median active corrective maintenance time $(\widetilde{M_{ct}})$

Median active corrective maintenance time is that value which divides all the corrective maintenance time values in such a way that 50 percent of the task times are equal to or less than the calculated median and 50 percent are equal to or greater than the median. The calculated median indicates the most likely average location of the task time data values drawn from a sample. The equation for the median active corrective maintenance time is

$$\widetilde{M_{ct}} = \text{antilog} \frac{\sum_{i=1}^{i=n} \log M_{cti}}{n}$$

also

$$\widetilde{M_{ct}} = \text{antilog} \frac{\sum_{i=1}^{i=n} (\lambda i)(\log M_{cti})}{\sum_{i=1}^{i=n} \lambda i}$$

Note: The *antilogarithm* is the converse of the logarithm. It is the number resulting from applying the logarithm exponentially to the designated base number (e.g., $10^{\log}$ = antilog). Chapter 1 of this book provides an in-depth discussion of the principles of logarithms. Whereas the median for the *normal* distribution is the same as the population mean, the median for the *log-normal* distribution is denoted as the *geometric mean*.

Maximum active corrective maintenance time (M_{max})

M_{max} can be described as that value of corrective maintenance time below which a specified percentage of all corrective maintenance actions can be expected to fall. M_{max} is related primarily to the log-normal distribution, and the 90th or 95th percentile point is typically taken as the specified confidence point. Maximum active corrective maintenance time is expressed as

$$M_{max} = \text{antilog}\ [\overline{\log M_{ct}} + z(\sigma \log M_{cti})]$$

$$\text{where } \sigma \log M_{cti} = \sqrt{\frac{\sum_{i=1}^{i=n} (\log M_{cti})^2 - \sum_{i=1}^{i=n} (\log M_{cti})^2/n}{n-1}}$$

$\overline{\log M_{ct}}$ = the mean of the logarithms of M_{cti}

z = the *normal deviate* value corresponding to the designated percentage parameter

The normal deviate z is the expression of the number of standard deviation intervals above or below the mean within the normal distribution. The z point on the normal curve abscissa scale denotes the area under the curve demarcated by the z value relative to the total area under the normal curve. For example, if $z = +1.65$, this indicates that the point located at 1.65 standard deviation intervals above the mean defines a zone under the normal curve which includes 95 percent of the total circumscribed area. The analyst may use z values to infer from $z = 1.65$ that the percentile point is 0.95.

Example of computation of $\widetilde{M_{ct}}$ and M_{max}. A system test specification requires that the system achieve maintainability performance parameters of

$$\widetilde{M_{ct}} = 45\,\text{min}$$
$$M_{max} = 60\,\text{min}$$

A test sample of 25 maintenance actions is the basis of analysis.

The logistics manager must determine if the specified parameters were achieved, based on the following observed task time values from the test sample:

Sample Task Times
54, 48, 41, 46, 49, 51, 50, 53, 52
47, 45, 44, 43, 42, 40, 38, 39, 36
37, 34, 35, 33, 31, 32, 55

The following analytical sequence corresponds to the data displayed in Table 24.2.

1. Construct the test sample worksheet.

   a. In the first column, serially identify the 25 trials ($n = 25$) in the sample.

   b. In the second column, record the individual task times M_{cti} for the 25 trials.

   c. In the third column, record the logarithm log M_{cti} for each of the 25 trial tasks.

   d. In the fourth column, record the squares of the individual logarithmic values $(\log M_{cti})^2$.

2.

   a. Determine the sum of the log M_{cti} values:

   $$\log M_{cti} = 40.677$$

   b. Determine the sum of the $(\log M_{cti})^2$ values:

   $$\sum_{i=1}^{i=25} (\log M_{cti})^2 = 66.332$$

In steps 1 and 2, the calculation of the basic elements required for the $\widetilde{M_{ct}}$ and M_{max} equations has been completed.

3. Calculate $\widetilde{M_{ct}}$.

$$\widetilde{M_{ct}} = \text{antilog} \frac{\sum_{i=1}^{i=25} \log M_{cti}}{n}$$
$$= \text{antilog} \frac{40.677}{25}$$
$$= \text{antilog } 1.627$$
$$= 42.364$$

4. Calculate M_{max}.

$$M_{max} = \text{antilog } [\overline{\log M}_{ct} + z(\sigma \log M_{cti})]$$

$$\text{where } \sigma \log M_{cti} = \sqrt{\frac{\sum_{i=1}^{i=25}(\log M_{cti})^2 - \frac{\left(\sum_{i=1}^{i=25} \log M_{cti}\right)^2}{n}}{n-1}}$$

First,

$$\overline{\log M}_{ct} = \frac{40.677}{25} = 1.627$$

At the 95th percentile, $z = 1.65$. Now compute $\sigma \log M_{cti}$.

$$\sigma \log M_{cti} = \sqrt{\frac{66.332 - \dfrac{(40.677)^2}{25}}{24}}$$

$$= \sqrt{\frac{66.332 - 66.185}{24}}$$

$$= \sqrt{0.006125}$$

$$= 0.078$$

TABLE 24.2   Maintainbility—Logarithmic Analysis

i	M_{cti}	$\log M_{cti}$	$(\log M_{cti})^2$
1	31	1.491	2.224
2	32	1.505	2.265
3	33	1.519	2.306
4	34	1.531	2.345
5	35	1.544	2.384
6	36	1.556	2.422
7	37	1.568	2.459
8	38	1.580	2.496
9	39	1.591	2.531
10	40	1.602	2.567
11	41	1.613	2.601
12	42	1.623	2.635
13	43	1.633	2.668
14	44	1.643	2.701
15	45	1.653	2.733
16	46	1.663	2.765
17	47	1.672	2.796
18	48	1.681	2.827
19	49	1.690	2.857
20	50	1.699	2.886
21	51	1.708	2.916
22	52	1.716	2.945
23	53	1.724	2.973
24	54	1.732	3.001
25	55	1.740	3.029
		40.677	66.332

$n = 25$

Then compute M_{max}

$$M_{max} = \text{antilog } [1.627 + (1.65)(0.078)]$$
$$= \text{antilog } (1.627 + 0.1287)$$
$$= \text{antilog } 1.7557$$
$$= 56.977$$

Conclusion   The results of the sample test indicate that the system complies with the specified M_{ct} and M_{max} parameters of 45 mm and 60 mm, respectively.

Log-normal analysis applicability to preventive maintenance time M_{pt}

The procedures prescribed for logarithmic analysis of corrective maintenance M_{ct} apply equally to computations for preventive maintenance M_{pt}. The algorithmic methodologies are identical. The analyst evaluating M_{pt} using logarithmic analysis would substitute M_{pt} for M_{ct} wherever appropriate in the log-normal equations; the resulting M_{max} would describe maximum active *preventive* maintenance time.

25

Logistics Supportability Analysis

General

Supportability analysis is the currently established body of procedures and methodologies employed by the Department of Defense (and to some degree by nonmilitary federal agencies) to accomplish the planning, development, systems engineering, production, and management that are essential to ensure the logistical supportability of military systems and equipment delivered to the users. The paradigm of supportability analysis evolved from and replaced the Logistics Support Analysis (LSA) and LSA Record (LSAR) process, which was in effect until the mid 1990s. Like LSA, supportability analysis is integrated with the concomitant evolution of the military life-cycle concepts. Supportability analysis is not itself a disciplinary entity; it describes a process of orchestrating a multiplicity of logistics and systems engineering disciplines so as to achieve optimum logistics supportability for a system or equipment.

Purview of Supportability Analysis

While it is acknowledged that the supportability analysis process based on the economies of scale associated with typical military systems is not always feasible for normal commercial production programs, there are principles which could have beneficial applicability to nonmilitary systems. The commercial aspect is addressed in this chapter by a description of the application of the supportability analysis process to a typical commercial product scenario. Many of the military supportability analysis concepts that are transferable to the private sector are discussed in their corresponding contexts.

Supportability for Military Systems

Current Department of Defense supportability analysis policies and data development procedures are prescribed in MIL-HDBK-502, *Military Handbook, Acquisition Logistics,* which was promulgated in 1997 and MIL-PRF-49506, *Performance Specification, Logistics Management Information,* issued in 1996. The military handbook and military specification are companion documents. MIL-HDHK-502 replaced MIL-STD-1388-1, *Logistic Support Analysis;* MIL-PRF-49506 replaced MIL-STD 1388-2, *Logistic Support Analysis Record.* These documents set forth guidance for the executory details for integrating the various scientific and engineering analysis methods using the system engineering process depicted in Figure 25.1. The figure portrays the system engineering process, as tailored to achieve supportability in the system design and to formulate the logistical support infrastructure. The origin is the logistics requirement which, during the supportability analysis process, is defined in terms of functional requirements based on these stipulations. The requirements for support are synthesized and, through analytical iterations which include trade-off analyses, the supportability evolves with the design of the system.

The supportability analysis is, therefore, a planned series of tasks that examine all elements of a proposed system or equipment to determine the logistic support infrastructure required to keep that system or equipment usable for its intended purpose, and to influence the design so that both the system and its support can be provided at an affordable cost. The reader is referred to MIL-HDBK-502 and MIL-PRF-49506 for comprehensive address of military supportability.

Supportability Commercial Model

This section provides an illustration of how the use of a logistics analysis database in conjunction with the elementary application of factorial data

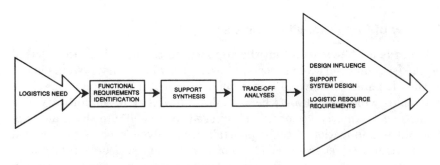

Figure 25.1 Supportability process.

elements can produce inventory management information. This case concerns assessment of the inventory levels required to sustain logistical support of organizational/retail-level (O-level), intermediate/service center-level (I-level) and depot/wholesale-level (D-level) maintenance activities. In keeping with analytical disciplines, the analyst follows a basic sequence:

- Design the analytical model.
- Define the scope of the analysis, in this case the item inventory posture of the O, I, and D levels of maintenance.
- Define the determinant attributes which empower the model to produce solutions.
- Develop the logistics database to incorporate the pertinent data elements.
- Formulate the computational algorithms.
- Accomplish the calculations necessary to support the analysis.
- Report the results.

The sample data sheet in Figure 25.2 provides a format for the pertinent attributes concerning the maintenance characteristics of a typical reparable item. The data sheet constitutes a matrix of data elements consisting of row element attributes, columnar element attributes, and data cells defined by the intersection of the row and columnar elements.

Definition of elemental attributes

Row elements. The row elements portray the individual maintenance levels.

Row A: Organizational (O-level) maintenance activity
Row B: Intermediate (I-level) maintenance activity
Row C: Depot (D-level) maintenance activity

Columnar elements. The columnar elements portray the maintenance characteristics of the maintenance level activities.

1. *Column 1, gross reparable generation rate.* The average number of reparable units of the item received by the maintenance activity during a defined period of calendar days. The computational factor is expressed in terms of units per day.
2. *Column 2, reparable referral rate.* The average rate at which reparable units of the item which are not designated for repair at the activity, but

Maintenance Level

COLUMNAR ELEMENTS (1,2,3,4,5,6,7,8,)

ROW ELEMENTS (A,B,C)	Maintenance Level	1 Gross Reparable Generation Rate	2 Reparable Referrals Rate	3 Local Reparable Generation Rate	4 Turnaround Time (Days)	5 Condemnation Rate	6 Pipeline (Days)	7 PALT (Days)	8 Safety Level (Days)
A	O - LEVEL	A1 *	A2 *	A3 *	A4 *	A5 *	A6 *	A7 N/A	A8 *
B	I - LEVEL	B1 *	B2 *	B3 *	B4 *	B5 *	B6 *	B7 N/A	B8 *
C	D - LEVEL	C1 *	C2 N/A	C3 *	C4 *	C5 *	C6 N/A	C7 *	C8 *

Figure 25.2 Sample supportability data sheet for maintenance-level inventory posture (data element/data cell matrix).

which are referred to a higher level for repair, are received. Referrals reflect demand on the supply system and are considered a stratum in the local inventory. The computational factor is expressed in terms of units per day.

3. *Column 3, local reparable generation rate*. The average rate at which reparable units which are designated for repair at the activity and become part of the local maintenance workload are generated. The local inventory must include allowance for drawdown of stockage of the item during the repair cycle. The computational factor is expressed in terms of units per day.

4. *Column 4, turn-around time (TAT)*. The repair cycle at the maintenance activity. The average time elapsed from removal of the failed component from the system, through the repair process, to return of the item to local inventory for reissue. The computational factor is expressed in terms of calendar days.

5. *Column 5, condemnation rate*. The rate at which reparable units are condemned for discard action by the activity in accordance with prescribed condemnation criteria. Condemnations represent demand on the central inventory and are considered a stratum in the local activity inventory posture. The computational factor is expressed in terms of units per day.

6. *Column 6, pipeline*. The average order and shipment (including in-transit) time between the requesting activity and the higher-level activity. This is essentially the time required for requisitioning and receipt of replacement items from the source of supply. The computational factor is expressed in terms of calendar days.

7. *Column 7, procurement and administrative lead time (PALT)*. This is also known as the depot ordering cycle. The time required to requisition, issue a contract for, and receive shipment of replacement units of an item from the item producer. Replenishment orders are typically for replacements of items discarded though condemnation procedures. Drawdown of depot assets during PALT must be compensated by a corresponding level in the depot inventory. The computational factor is expressed in terms of calendar days.

8. *Column 8, safety level*. Inventory cushion provided to absorb unanticipated surges in demand on the activity's inventory. The computational factor is expressed in terms of days, i.e., days of support based on prevailing normal demand rates.

Data cell factors.   Data cells are determined and identified by the matrix intersection of the row elements and columnar elements. Data cells for this model are designated as follows:

Data cell	Description
A1	Total reparable generation rate for units submitted by customer-users to O-level for repair (units/day)
A2	Reparable referral rate from O level to higher levels (I or D) of maintenance (units/day)
A3	Local reparable generation rate for O-level workload (units/day)
A4	Average TAT for O-level repair (days)
A5	Rate of O-level condemnations (units/day)
A6	Average pipeline time between O level and higher-level (I or D) source of supply (days)
A7	Not applicable to O level. PALT is applicable to D level only
A8	O-level safety stockage (days)
B1	Total reparable generation rate for units submitted by customer-users directly to I level for repair (units/day)
B2	Reparable referral rate from I level to D-level maintenance (units/day)
B3	Local reparable generation rate for I-level workload (units/day)
B4	Average TAT for I-level repair (days)
B5	Rate of I-level condemnations (units/day)
B6	Average pipeline time between I level and D-level source of supply (days)
B7	Not applicable to I level. PALT is a D-level parameter
B8	I-level safety stockage (days)
C1	Total reparable generation rate for units submitted by customer-users directly to D-level for repair (units/day).
C2	Not Applicable to D level. Depot is the highest maintenance level
C3	Local reparable generation rate for D-level workload (units/day)
C4	Average TAT for D-level repair (days)
C5	Rate of D-level condemnations (units/day)
C6	Not applicable at D level. Depot is the highest maintenance level and the location of wholesale inventory
C7	PALT, applicable to D level only (days)
C8	D-level safety stockage (days)

When appropriate, the above data cells may be tailored to reflect specific attributes peculiar to the analysis. For example, data factor A2 may be modified and subscripted as $A2_I$ to denote the total rate of reparable referrals received at I level from a multiplicity of O-level activities under the authority of the I-level activity.

Case study: Commercial item supportability   The maintenance characteristics of an item are reflected in the following empirical data collected for O-level, I-level, and D-level maintenance activities over a continuous 365-day working period. The objective is to establish an inventory posture at each level of maintenance. In the logistical network are 24 retail (O-level) activities, 6 service center (I-level) activities, and one depot (D-level) facility with authority over the total network. This analysis focuses on

1. The typical O-level posture, considering the average of the aggregate data from the 24 retail activities

2. The typical I-level posture, considering the average of the aggregate data from the 6 intermediate level service centers

3. The depot-level data based on the reported activity at the single, centralized hub of the logistical network

The case scenario is predicated on the following assumptions:

1. There are a total of 24 O-level maintenance activities, 6 I-level facilities, and one depot.

2. The typical I-level maintenance activity is responsible for support of 4 O-level sites.

3. The depot support responsibility includes the total network of 6 I-level and 24 O-level maintenance activities.

4. All reparable referals received at a maintenance level (I level or D level) are presumed reparable by the recipient activity, and are not reforwarded by another referral.

5. Reparables referred to a higher activity (i.e., by the O level and I level) for repair are drawn from those gross reparable generations submitted directly by customer-users and accepted by the activity. This accommodates the reality that customers may elect to submit failed items to any point in the maintenance hierarchy rather than follow the established maintenance level protocol.

The outputs of the following steps are entered into the data cells of the data element worksheet in Fig. 25.3.

1. Gross reparables generated by customer-users (365 days)
   - O-level maintenance: 4800 units
   - I-level maintenance: 900 units
   - D-level maintenance: 100 units

$$\text{Average O-level: } \frac{4800}{24} = 200$$

$$\text{Average I-level: } \frac{900}{6} = 150$$

COLUMNAR ELEMENTS

Row Elements		1 Gross Reparable Generation Rate	2 Reparable Referrals Rate	3 Local Reparable Generation Rate	4 Turnaround Time (Days)	5 Condemnation Rate	6 Pipeline (Days)	7 PALT (Days)	8 Safety Level (Days)
A	O - LEVEL	A1 .548	A2 .228	A3 .228	A4 7	A5 .091	A6 5	A7 N/A	A8 10
B	I - LEVEL	B1 .411	B2 .137	B3 1.05	B4 12	B5 .137	B6 7	B7 N/A	B8 10
C	D - LEVEL	C1 .274	C2 N/A	C3 .959	C4 21	C5 .137	C6 N/A	C7 60	C8 10

Figure 25.3 Supportability data element worksheet: maintenance-level supply posture

Data cell factors:

$$A1 = \frac{200}{365} = 0.548$$

$$B1 = \frac{150}{365} = 0.411$$

$$C1 = \frac{100}{365} = 0.274$$

2. Reparables not designated for local repair that are referred to higher maintenance levels (365 days)
   - From O to I level: 2000 units
   - From I to D level: 300 units

$$\text{Average O to I: } \frac{2000}{24} = 83.33 \text{ units}$$

$$\text{Average I to D: } \frac{300}{6} = 50 \text{ units}$$

Data cell factors:

$$A2 = \frac{83.33}{365} = 0.228$$

$$B2 = \frac{50}{365} = 0.137$$

3. Local reparable generations designated for repair at local activity (365 days)
   - O-level: 2000 units
   - I-level: 2300 units
   - D-level: 350 units

$$\text{Average O-level: } \frac{2000}{24} = 83.33 \text{ units}$$

$$\text{Average I-level: } \frac{2300}{6} = 383.33 \text{ units}$$

Data cell factors:

$$A3 = \frac{83.33}{365} = 0.228$$

$$B3 = \frac{383.33}{365} = 1.05$$

$$C3 = \frac{350}{365} = 0.959$$

4. Cumulative repair cycle (TAT) statistics
   - O-level: A total of 2000 units repaired over 14,000 cumulative repair cycle days
   - I-level: A total of 2300 units repaired over 27,600 cumulative repair cycle days
   - D-level: A total of 350 units repaired over 7350 cumulative repair cycle days

Data cell factors:

$$A4 = \frac{14,000}{2000} = 7 \text{ days (average O-level TAT)}$$

$$B4 = \frac{27,600}{2300} = 12 \text{ days (average I-level TAT)}$$

$$C4 = \frac{7350}{350} = 21 \text{ days (average D-level TAT)}$$

5. Total condemnations (365 days)
   - O-level: 800 condemnations
   - I-level: 300 condemnations
   - D-level: 50 condemnations

$$\text{Average O-level: } \frac{800}{24} = 33.33 \text{ units}$$

$$\text{Average I-level: } \frac{300}{6} = 50 \text{ units}$$

Data cell factors:

$$A5 = \frac{33.33}{365} = 0.091$$

$$B5 = \frac{50}{365} = 0.137$$

$$C5 = \frac{50}{365} = 0.137$$

6. Cumulative pipeline (O&ST) performance data
   - O to I level: 2800 requisitions requiring 14,000 total pipeline days
   - I to D level: 3400 requisitions requiring 23,800 pipeline days

Data cell factors:

$$A6 = \frac{14,000}{2800} = 5 \text{ days between O and I levels}$$

$$B6 = \frac{23,800}{3400} = 7 \text{ days between I and D levels}$$

7. PALT for Depot replenishment: 60 days
   Data Cell Factor: C7 = 60

8. Safety levels authorized
   - O-level: 10 days stockage cushion
   - I-level: 10 days stockage cushion
   - D-level: 10 days stockage cushion

Data cell factors:

$$A8 = 10$$
$$B8 = 10$$
$$C8 = 10$$

9. Tailored data factors

$A2_I$ = average rate of reparable referrals to typical I-level activity from subsidiary O-level sites

$$A2_I = \frac{2000}{6} \div 365 = 0.913 \text{ (per I-Level site)}$$

$A5_I$ = average rate of condemnations reported to typical I-level activity from subsidiary O-level activities

$$A5_I = \frac{800}{365} \div 365 = 0.365 \text{ (per I-level site)}$$

$A5_D$ = average rate of condemnations generated by and reported to D-level inventory by all O-level activities through I-level supply sources

$$A5_D = \frac{800}{365} = 2.192 \text{ (per D-level site)}$$

$B2_D$ = average rate of referrals to D level by subsidiary I-level activities

$$B2_D = \frac{300}{365} = 0.822 \text{ (per D-level site)}$$

$B2_D$ = average rate of condemnations generated and reported to D-level by subsidiary I-level activities

$$B5_D = \frac{300}{365} = 0.822 \text{ (per D-level site)}$$

On the basis of the completed data element worksheet, Fig. 25.3, the analyst constructs the supportability computational matrix, as illustrated by Table 25.1, to display the algorithmic application of the data cell factors. This will produce the quantitative data related to the stratification of the inventory postures at the organizational, intermediate, and depot maintenance levels. The computational worksheet in Table 25.2 captures the numeric values of the data cell factors provided by the computational matrix (Table 25.1).

The worksheet in Table 25.3 provides the results of the algorithmic calculations.

Results of the analysis   The results of the analysis, reflected in Table 25.3, provide the following:

TABLE 25.1 Supportability Computation Matrix with Algorithmic Application of Data Cell Factors for Maintenance-Level Supply Posture Structured by Constituent Stratification Levels

Inventory stratification based on indicated replenishment	Average O-level	Average I-level	D-level
Reparable referrals (for items designated for repair at higher levels of maintenance)	$(A2)(A6)$	$(A2_I)(A6)$ $(B2)(B6)$	$(B2_D)(B6)$
Local reparable generations (for items designated for repair at same maintenance level)	$(A3)(A4)$	$(B3)(B4)$	$(C3)(C4)$
Condemnations/discards	$(A5)(A6)$	$(A5_I)(A6)$ $(B5)(B6)$	$(A5_D)(C7)$ $(B5_D)(C7)$ $(C5)(C7)$
Safety level	$(A1)(A8)$	$B8(A2_I+A5_I+B2+B3+B5)$	$C8(A5_D+B2_D+B5_D+C3+C5)$

Average O-level inventory = $(A2)(A6) + (A3)(A4) + (A5)(A6) + (A1)(A8)$
Total O-level inventory = (Average O-level inventory)(number of O-level activities)
Average I-level inventory = $(A2_I)(A6) + (B2)(B6) + (A5_I)(A6) + (B5)(B6) + B8(A2_I+A5_I+B2+B3+B5)$
Total I-level inventory = (average I-level inventory)(number of I-level activities)
D-level inventory $(B2_D)(B6) + (C3)(C4) + (A5_D)(C7) + (B5_D)(C7) + (C5)(C7) + C8(A5_D+B2_D+B5_D+C3+C5)$

Note: Data cell factors B1 and C1 are not utilized, as they are not germane to the specific analysis. Based on the preliminary assumptions and stipulations, the I-level and D-level gross reparable generations are accounted for in the reparable referrals, local repairs, and condemnations addressed in the analytical approach.

TABLE 25.2 Supportability Computational Worksheet Based on Supportability Data Element Worksheet and Supportability Computational Matrix for Maintenance-Level Supply Posture Structured by Constituent Stratification Levels

Inventory stratification based on indicated replenishment	Average O-level	Average I-level	D-level
Reparable referrals (for items designated for repair at higher levels of maintenance)	(0.228)(5)	(0.913)(5) + (0.137)(7)	(0.822)(7)
Local reparable generations (for items designated for repair at same maintenance level)	(0.228)(7)	(1.05)(12)	(0.959)(21)
Condemnations/discards	(0.091)(5)	(0.365)(5) + (0.137)(7)	(2.192)(60) + (0.822)(60) + (0.137)(60)
Safety level	(0.548)(10)	10(0.913+0.365 +0.137+1.05+0.137)	10(2.192+0.822+0.822 +0.959+0.137)

Note: These computations incorporate the numerical values of the applicable data cell factors, as cited in the LSA supportability computational matrix.

TABLE 25.3 Supportability Computational Worksheet Results of Calculations for Maintenance-Level Supply Posture Structured by Constituent Stratification Levels

Inventory stratification based on indicated replenishment	Average O-level	Average I-level	D-level
Reparable referrals (for items designated for repair at higher levels of maintenance)	1.14	5.524	5.754
Local reparable generations (for items designated for repair at same maintenance level)	1.596	12.6	20.139
Condemnations/discards	0.455	2.784	189.06
Safety level	5.48	26.02	49.32
Total inventory posture	8.671 or 9 units	46.928 or 47 units	264.273 or 264 units

Summary:
Total O-level assets: (Average O-level posture)(number of O-level sites) = (9)(24) = 216
Total I-level assets: (Average I-level posture)(number of I-level sites) = (47)(6) = 282
Total depot assets: Total depot inventory posture = 264
                                                 Total logistics system assets   762

1. Inventory stratification of the postured inventory assets at the average O-level, average I-level, and D-level maintenance, defined as follows:
   - Stratum providing assets to accommodate reparable referrals
   - Stratum citing assets to support local reparables generated
   - Stratum identifying provisioning for condemnations/discards
   - Stratum noting asset requirements to fulfill safety levels

2. Total stratified inventory postures at the average O-level, average I-level, and depot maintenance activities

3. Total assets of all O-level activities

4. Total assets of all I-Level activities

5. Total D-level assets

6. Total logistics system assets, considering all activities at the three levels of maintenance

This case is designed to illustrate how empirical data for a maintenance-significant item can be captured, retained for record, and applied to aid in the logistics management decision process.

The opportunities to apply the military supportability concepts and methodologies embodied in MIL-HDBK-502 and MIL-PRF-49506 to commercial logistics are unlimited. The exploitation of the principles of logistics support analysis in any logistics scenario, whether it concerns military systems or commercial products, is constrained only by the imagination, creativity, and ingenuity of the product manager.

26

Configuration Management

General

Configuration management (CM) is the systems engineering management process that identifies the functional and physical characteristics of a product during its life cycle, controls changes in those characteristics, reports change processing and implementation status, and records the changes. CM is the means through which the integrity and continuity of the design, engineering, and cost trade-offs between technical performance, producibility, operability, and supportability are recorded, communicated, and controlled by program and functional managers. The purpose of CM is to provide for improved supportability, including updated technical manuals, identified critical spares, and identical or interchangeable equipment, and to validate product configuration baselines.

Configuration management provides the current baseline description of a developing hardware unit, software unit, system, etc., and provides traceability to previous baseline configurations of the same item. CM also provides complete information on the rationale for configuration changes, thus permitting analysis and correction of deficiencies when they arise. Configuration management involves four major functions:

- Configuration identification
- Configuration control
- Configuration status accounting
- Configuration audits

Product changes can be initiated as early as the conceptual and development phases of a system. Configuration changes occur throughout the life of the system as more knowledge of the system design, operation, and maintenance concepts is acquired; as operational requirements are redefined; or as nontechnical factors such as cost and schedule influence the design. These changes must be controlled to ensure, first, that they are cost-effective and, second, that they are properly documented so that customer-users are aware of the physical baselines of the currently marketed product. Within the private sector as well as the Department of Defense (DoD), CM procedures are governed by both military and commercial standards. The previously enforced MIL-STD-973, *Configuration Management,* was officially cancelled by DoD 30 September 2000 and replaced by ANSI/EIA-649: *National Consensus Standard for Configuration Management* (1998). DoD officially adopted the ANSI/EIA standard, based on the rationale that the commercial standard presents practices which make good business sense for both DoD and industry and are considered preferable to requirements that are imposed by an external customer.

The 2003 "white paper" developed by the Institute of Configuration Management observed that MIL-STD-973 and ANSI/EIA both emphasize the traditional CM functions of *configuration identification, change control, configuration status accounting, and configuration audits* as their framework. The study noted that the two documents have strong similarities, but with ANSI/EIA-649 lacking strength in certain areas that are emphasized by MIL-STD-973. Although MIL-STD-973 cannot be legally invoked for military contracts, it is still considered an excellent source of tutorial guidance on the details of practices, procedures, and documentation required for execution of system configuration management programs.

The more recent MIL-HDBK-61A, *Configuration Management Guidance,* effected 7 February 2001, remains active within DoD. This military handbook provides guidance and information to DoD acquisition managers, logistics managers, and other individuals assigned responsibility for configuration management. Its purpose is to assist them in planning for and implementing effective DoD configuration management activities and practices during all life cycle phases of defense systems and configuration items. It supports acquisition based on performance specifications and the use of industry standards and methods to the greatest practicable extent. MIL-HDBK-61A is a comprehensive, detailed document which encompasses the principles and methodologies incorporated in ANSI/EIA-649 and MIL-STD-973. In view of the aforementioned points of discussion, the material in this chapter is, accordingly, reflective of MIL-HDBK-61A and MIL-STD-973.

Because of the massive investment of financial and human resources, change control discipline is critical within the DoD and private industrial sectors to assure management control over the evolution of subsystems and components. It is important to ensure that there is not proliferation of nonstandard parts, there is control of the cost implications of modular and subsystem design, and all users of the equipment have timely and accurate technical documentation for system operation and maintenance. Accordingly, this chapter addresses the configuration management approach applicable to both military and commercial product technologies.

Configuration Identification

Configuration identification is based on the family of specifications and drawings that describes the system or configuration item (CI) during the design/development cycle, becoming more precise as the design progresses toward production. This identification serves throughout the CI's life cycle as a record and description of its required functional and physical characteristics.

Configuration identification documents technically describe the CIs and serve as the basis for change control, status accounting, and auditing. Configuration identification should be tailored to the requirements of the life-cycle process and contain sufficient detail to provide the data required by the product manager to maintain engineering integrity. Such detail permits logical preparation of the technical, administrative, and management documents (e.g., work breakdown structure, technical reports, parts lists) that enhance logistics support of the system.

Configuration items

An item selected for configuration management is referred to as a configuration item. The selection of a CI is based upon the need to control some or all of its functional or physical characteristics or to control its interfaces with other items, thus allowing the engineering design to be maintained and the appropriate integrated logistic support (ILS) products to be developed.

Once the CIs are identified, the specifications defining them can be produced. The CI identification function must ensure that

- All technical documentation describing the functional and physical characteristics of CM items is completely defined.
- Verified technical documents defining the baseline are current, approved, and available for use when needed.

Each configuration item should be identified by a drawing which identifies both hardware and software modular content, assigns a part number to the CI, and provides subsystem and modular interface identification. Additionally, each drawing should identify the functional and product engineering drawings, as well as other essential references such as interface control drawings, parts lists, etc., from which physical configuration audits (PCAs) and functional configuration audits (FCAs) are conducted. Each item shown on the face of the drawing should be linked to its corresponding technical documentation and selected logistic resources.

Functional configuration identification.   The functional configuration identification (FCI) is applicable to all items under development for which a product baseline has not yet been established. It is the technical documentation for a configuration item which prescribes

1. All essential system functional characteristics
2. Necessary interface characteristics
3. Special designation of the functional characteristics of key configuration items
4. All of the tests required to demonstrate achievement of each specified characteristic
5. Design constraints, such as envelope dimensions, component standardization, use of inventory items, and integrated logistics support documentation such as parts lists documentation

Product configuration identification.   The product configuration identification (PCI) prescribes the necessary "build-to" or "form, fit, and function" requirements for a CI. Each CI should have a PCI. The initial PCI is established upon successful completion of a physical configuration audit (PCA). The PCI includes, but is not limited to, detail design specifications, production drawings, and associated lists. To verify conformance, PCAs should be accomplished before the system product baseline is established. The product baseline consists of the specifications, drawings, computer programs, acceptance test criteria, and other documents of the complete design package. This baseline, with subsequently approved changes, defines the physical characteristics and interfaces controlled during the production, operation, maintenance, and logistics support phases of the system's life cycle for both hardware and software.

Establishing configuration baselines

One of the more important aspects of product configuration management is baseline management. Baseline management is required at the outset of a product program. The prevalent forms of baseline management involve the functional baseline and product baseline.

Functional baseline.   The functional baseline is ideally established at the end of the conceptual development phase. The system specification, or the development specification for smaller programs, defines the technical portion of the program requirements. The initial system specification provides the basis for controlling the system design during the early developmental phases. Once the system specification has been authenticated, formal configuration control is initiated. The authenticated system specification is the foundation for CM during the subsequent phases of the program.

Product baseline.   The product baseline is established by the detailed design documentation for each CI. It establishes the requirements for hardware fabrication and software coding (CI by CI) as the initial article of each item is reviewed and approved as satisfactorily meeting the specification requirements. This baseline will normally include product, process, and material specifications; engineering drawings; and other related data. The product baseline is established initially by approval of the product specifications. The CI product baseline is the basis of the production program and is verified by successful completion of the functional configuration audit (FCA) and PCA.

Configuration Control

The various forms of change control documentation provide the vehicles for initiating, evaluating, approving or disapproving, releasing, and implementing changes. They can also be used for reporting problems, requesting modifications, and submitting change proposals. They are a key source of information concerning the status of changes during change processing. Such governing change documentation generally pertains to formal engineering design changes, deviations in product specifications before production, waivers to specifications for in-production or completed products pending delivery, formal changes in product specifications dictated by approved product design changes, and notices of revision to product managers of other systems which incorporate the same components affected by approved changes (see Fig. 26.1). The characteristics and purposes of the various change control documents can be understood from a discussion of the forms of change vehicles employed for these purposes by the Department of Defense.

Engineering change proposals

Two forms of engineering change proposals (ECPs) are recognized, those for Class I (major) changes and those for Class II (minor) changes. Class I changes are changes which affect the following types of system attributes:

1. System specification
2. Authenticated performance specification, which includes

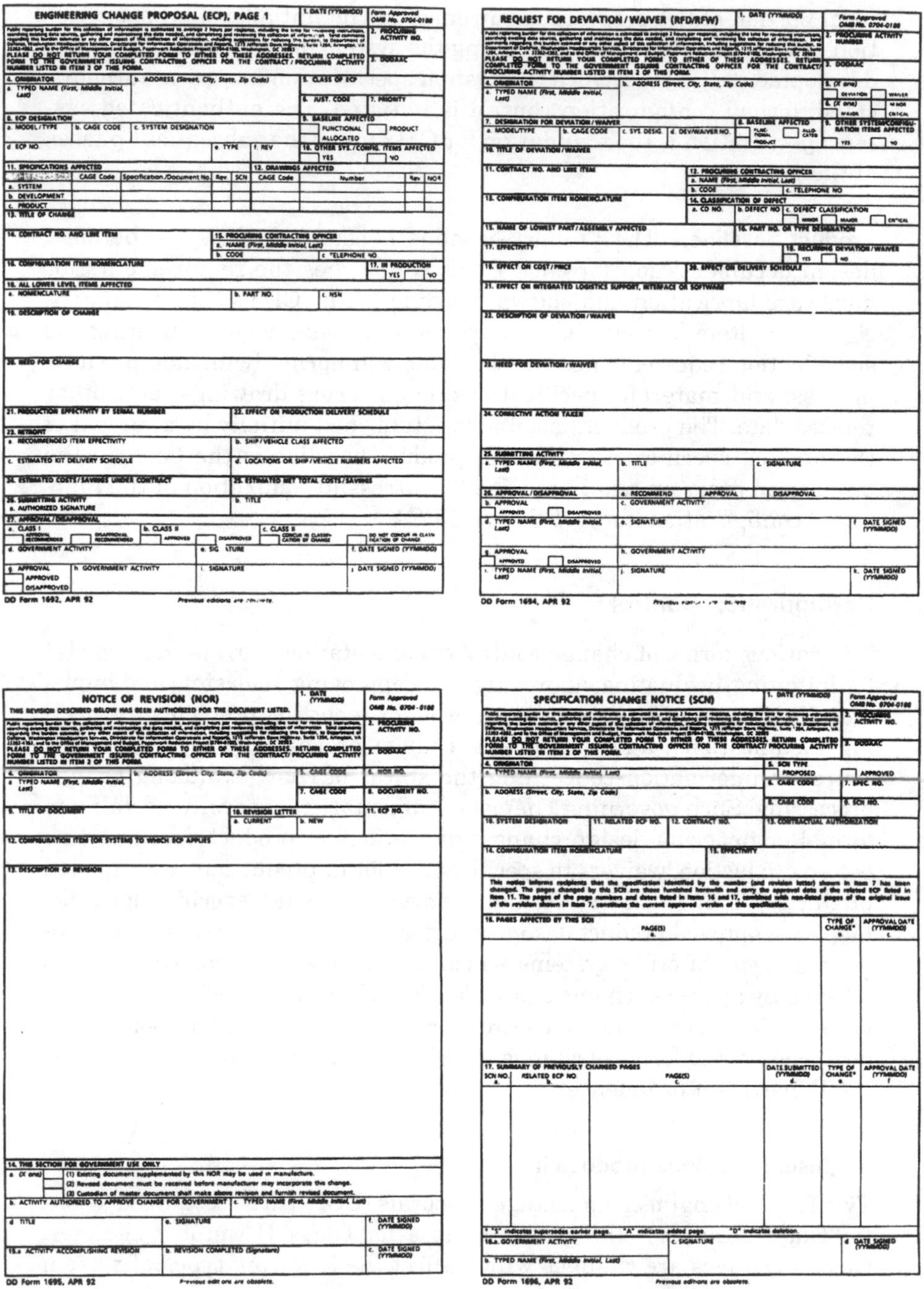

ENGINEERING CHANGE PROPOSAL (ECP), PAGE 1

1. DATE (YYMMDD)
Form Approved OMB No. 0704-0188
2. PROCURING ACTIVITY NO.
3. DODAAC
PLEASE DO NOT RETURN YOUR COMPLETED FORM TO EITHER OF THESE ADDRESSES. RETURN COMPLETED FORM TO THE GOVERNMENT ISSUING CONTRACTING OFFICER FOR THE CONTRACT/PROCURING ACTIVITY NUMBER LISTED IN ITEM 2 OF THIS FORM.
4. ORIGINATOR
a. TYPED NAME (First, Middle Initial, Last)
b. ADDRESS (Street, City, State, Zip Code)
5. CLASS OF ECP
6. JUST. CODE
7. PRIORITY
8. ECP DESIGNATION
a. MODEL/TYPE
b. CAGE CODE
c. SYSTEM DESIGNATION
d. ECP NO.
e. TYPE
f. REV
9. BASELINE AFFECTED
FUNCTIONAL
PRODUCT
ALLOCATED
10. OTHER SYS./CONFIG. ITEMS AFFECTED
YES
NO
11. SPECIFICATIONS AFFECTED
CAGE Code
Specification/Document No.
Rev
SCN
a. SYSTEM
b. DEVELOPMENT
c. PRODUCT
12. DRAWINGS AFFECTED
CAGE Code
Number
Rev
NOR
13. TITLE OF CHANGE
14. CONTRACT NO. AND LINE ITEM
15. PROCURING CONTRACTING OFFICER
a. NAME (First, Middle Initial, Last)
b. CODE
c. TELEPHONE NO
16. CONFIGURATION ITEM NOMENCLATURE
17. IN PRODUCTION
YES
NO
18. ALL LOWER LEVEL ITEMS AFFECTED
a. NOMENCLATURE
b. PART NO.
c. NSN
19. DESCRIPTION OF CHANGE
20. NEED FOR CHANGE
21. PRODUCTION EFFECTIVITY BY SERIAL NUMBER
22. EFFECT ON PRODUCTION DELIVERY SCHEDULE
23. RETROFIT
a. RECOMMENDED ITEM EFFECTIVITY
b. SHIP/VEHICLE CLASS AFFECTED
c. ESTIMATED KIT DELIVERY SCHEDULE
d. LOCATIONS OR SHIP/VEHICLE NUMBERS AFFECTED
24. ESTIMATED COSTS/SAVINGS UNDER CONTRACT
25. ESTIMATED NET TOTAL COSTS/SAVINGS
26. SUBMITTING ACTIVITY
a. AUTHORIZED SIGNATURE
b. TITLE
27. APPROVAL/DISAPPROVAL
a. CLASS I
b. CLASS II
c. CLASS II
d. GOVERNMENT ACTIVITY
e. SIGNATURE
f. DATE SIGNED (YYMMDD)
g. APPROVAL
APPROVED
DISAPPROVED
h. GOVERNMENT ACTIVITY
i. SIGNATURE
j. DATE SIGNED (YYMMDD)
DD Form 1692, APR 92
Previous editions are obsolete.

REQUEST FOR DEVIATION/WAIVER (RFD/RFW)

1. DATE (YYMMDD)
Form Approved OMB No. 0704-0188
2. PROCURING ACTIVITY NUMBER
3. DODAAC
PLEASE DO NOT RETURN YOUR COMPLETED FORM TO EITHER OF THESE ADDRESSES. RETURN COMPLETED FORM TO THE GOVERNMENT ISSUING CONTRACTING OFFICER FOR THE CONTRACT/PROCURING ACTIVITY NUMBER LISTED IN ITEM 2 OF THIS FORM.
4. ORIGINATOR
a. TYPED NAME (First, Middle Initial, Last)
b. ADDRESS (Street, City, State, Zip Code)
5. (X one)
DEVIATION
WAIVER
6. (X one)
MAJOR
MINOR
CRITICAL
7. DESIGNATION FOR DEVIATION/WAIVER
a. MODEL/TYPE
b. CAGE CODE
c. SYS. DESIG
d. DEV/WAIVER NO.
8. BASELINE AFFECTED
9. OTHER SYSTEM/CONFIGURATION ITEMS AFFECTED
YES
NO
10. TITLE OF DEVIATION/WAIVER
11. CONTRACT NO. AND LINE ITEM
12. PROCURING CONTRACTING OFFICER
a. NAME (First, Middle Initial, Last)
b. CODE
c. TELEPHONE NO
13. CONFIGURATION ITEM NOMENCLATURE
14. CLASSIFICATION OF DEFECT
a. CD NO.
b. DEFECT NO
c. DEFECT CLASSIFICATION
MINOR
MAJOR
CRITICAL
15. NAME OF LOWEST PART/ASSEMBLY AFFECTED
16. PART NO. OR TYPE DESIGNATION
17. EFFECTIVITY
18. RECURRING DEVIATION/WAIVER
YES
NO
19. EFFECT ON COST/PRICE
20. EFFECT ON DELIVERY SCHEDULE
21. EFFECT ON INTEGRATED LOGISTICS SUPPORT, INTERFACE OR SOFTWARE
22. DESCRIPTION OF DEVIATION/WAIVER
23. NEED FOR DEVIATION/WAIVER
24. CORRECTIVE ACTION TAKEN
25. SUBMITTING ACTIVITY
a. TYPED NAME (First, Middle Initial, Last)
b. TITLE
c. SIGNATURE
26. APPROVAL/DISAPPROVAL
a. RECOMMEND
APPROVAL
DISAPPROVAL
b. APPROVAL
APPROVED
DISAPPROVED
c. GOVERNMENT ACTIVITY
d. TYPED NAME (First, Middle Initial, Last)
e. SIGNATURE
f. DATE SIGNED (YYMMDD)
g. APPROVAL
APPROVED
DISAPPROVED
h. GOVERNMENT ACTIVITY
i. TYPED NAME (First, Middle Initial, Last)
j. SIGNATURE
k. DATE SIGNED (YYMMDD)
DD Form 1694, APR 92

NOTICE OF REVISION (NOR)

THIS REVISION DESCRIBED BELOW HAS BEEN AUTHORIZED FOR THE DOCUMENT LISTED.
1. DATE (YYMMDD)
Form Approved OMB No. 0704-0188
2. PROCURING ACTIVITY NO.
3. DODAAC
PLEASE DO NOT RETURN YOUR COMPLETED FORM TO EITHER OF THESE ADDRESSES. RETURN COMPLETED FORM TO THE GOVERNMENT ISSUING CONTRACTING OFFICER FOR THE CONTRACT/PROCURING ACTIVITY NUMBER LISTED IN ITEM 2 OF THIS FORM.
4. ORIGINATOR
a. TYPED NAME (First, Middle Initial, Last)
b. ADDRESS (Street, City, State, Zip Code)
5. CAGE CODE
6. NOR NO.
7. CAGE CODE
8. DOCUMENT NO.
9. TITLE OF DOCUMENT
10. REVISION LETTER
a. CURRENT
b. NEW
11. ECP NO.
12. CONFIGURATION ITEM (OR SYSTEM) TO WHICH ECP APPLIES
13. DESCRIPTION OF REVISION
14. THIS SECTION FOR GOVERNMENT USE ONLY
a. (X one)
(1) Existing document supplemented by this NOR may be used in manufacture.
(2) Revised document must be received before manufacturer may incorporate this change.
(3) Custodian of master document shall make above revision and furnish revised document.
b. ACTIVITY AUTHORIZED TO APPROVE CHANGE FOR GOVERNMENT
c. TYPED NAME (First, Middle Initial, Last)
d. TITLE
e. SIGNATURE
f. DATE SIGNED (YYMMDD)
15.a. ACTIVITY ACCOMPLISHING REVISION
b. REVISION COMPLETED (Signature)
c. DATE SIGNED (YYMMDD)
DD Form 1695, APR 92
Previous editions are obsolete.

SPECIFICATION CHANGE NOTICE (SCN)

1. DATE (YYMMDD)
Form Approved OMB No. 0704-0188
2. PROCURING ACTIVITY NO.
3. DODAAC
PLEASE DO NOT RETURN YOUR COMPLETED FORM TO EITHER OF THESE ADDRESSES. RETURN COMPLETED FORM TO THE GOVERNMENT ISSUING CONTRACTING OFFICER FOR THE CONTRACT/PROCURING ACTIVITY NUMBER LISTED IN ITEM 2 OF THIS FORM.
4. ORIGINATOR
a. TYPED NAME (First, Middle Initial, Last)
b. ADDRESS (Street, City, State, Zip Code)
5. SCN TYPE
PROPOSED
APPROVED
6. CAGE CODE
7. SPEC. NO.
8. CAGE CODE
9. SCN NO.
10. SYSTEM DESIGNATION
11. RELATED ECP NO.
12. CONTRACT NO.
13. CONTRACTUAL AUTHORIZATION
14. CONFIGURATION ITEM NOMENCLATURE
15. EFFECTIVITY
This notice informs recipients that the specification identified by the number (and revision letter) shown in Item 7 has been changed. The pages changed by this SCN are those furnished herewith and carry the approval date of the related ECP listed in Item 11. The pages of the page numbers and dates listed in Items 16 and 17, combined with non-listed pages of the original issue of the revision shown in Item 7, constitute the current approved version of this specification.
16. PAGES AFFECTED BY THIS SCN
PAGE(S) a.
TYPE OF CHANGE* b.
APPROVAL DATE (YYMMDD) c.
17. SUMMARY OF PREVIOUSLY CHANGED PAGES
SCN NO. a.
RELATED ECP NO. b.
PAGE(S) c.
DATE SUBMITTED (YYMMDD) d.
TYPE OF CHANGE* e.
APPROVAL DATE (YYMMDD) f.
* "S" indicates supersedes earlier page. "A" indicates added page. "D" indicates deletion.
18.a. GOVERNMENT ACTIVITY
b. TYPED NAME (First, Middle Initial, Last)
c. SIGNATURE
d. DATE SIGNED (YYMMDD)
DD Form 1696, APR 92
Previous editions are obsolete.

Figure 26.1 DOD change control forms.

a. Performance requirements outside previously stated tolerances

b. Reliability, maintainability, and survivability

c. Weight, balance, or moment of inertia

3. Type designator or drawing number (when operating limit/performance are extensively changed, new ones are required)

4. Major component interface characteristics

5. Contract or purchase price

6. Contract schedule or delivery rate

7. Contract guarantee

8. Government-furnished equipment (GFE) for military programs

9. Safety

10. Electromagnetic characteristics

11. Interface with other systems or developmental support systems

12. Support equipment (SE)

13. Capability of the CI or system with support equipment and trainers or training devices and equipment

14. Delivered operation and maintenance manuals for which change or revision funding is not provided under existing contracts

15. Life-cycle cost (development, production recurring, production nonrecurring, or operating and support cost)

16. Existing spare parts

17. Interchangeability, substitutibility, or replaceability of the item and all subassemblies and parts or repairable assemblies (excluding piece parts or nonreparable subassemblies)

18. Sources of assemblies or repairable parts at any level that are defined by source control drawings

19. Any system characteristic affected by a change which requires a retrofit (attrition or kit)

Changes that do not involve these attributes are designated Class II changes and are subject to less rigorous review, evaluation, and approval procedures than those for Class I changes. The executory vehicle for the ECP within the military is typically DD Form 1692. MIL-HDBK-61A and MIL-STD-973 provides guidance for documenting the pertinent rationale and descriptive details supporting the proposed change.

Deviations and waivers

Deviations and waivers are engineering change documents used to permit temporary departures from governing product design specifications when permanent changes are not acceptable.

Deviations. A deviation is a written authorization granted during engineering development, prior to production, that permits a departure from a particular performance or design requirement stipulated by the functional baseline documentation. The temporary departure from the established functional configuration identification must be adjusted to conform to the functional baseline before production commences.

Waivers. A waiver is a written authorization granted during production or when a system is pending delivery to the user. Like a deviation, a waiver is applicable when the item does not conform to particular requirements specified by the functional and product configuration identification. A waiver is granted in cases where the item is considered otherwise suitable for use or rework at a later point in time after delivery.

Execution of deviations and waivers. Within the military community, the typical vehicle of action for both deviations and waivers is DD Form 1694, procedural guidance for which is provided by MIL-HDBK-61A and MIL-STD-973. Whereas ECPs may be applicable to all units of a specified product, DD 1694 is limited in scope and effect. These limited-scope changes are regarded as temporary departures from the product design and specifications for a delimited series of system units and are, therefore, not applicable to the total product population. Such temporary departures are usually invoked for a limited period of time on defined lots or batches of units, normally identified and controlled by product unit serial numbers. The product baseline documentation will reflect those serialized items for which waivers to specifications are approved and effective. If it is determined the waivers provide enhanced product value or utility, they may be converted to formal engineering changes for installation on all in-production and delivered versions of the product.

Specification change notice

The specification change notice (SCN) is used to propose, transmit, and record changes in a product specification emanating from a formal engineering change proposal. The SCN is initially used to submit specification change pages in conjunction with a formal major (or military Class I) ECP for approval by the system manager. After the proposed specification changes are approved, the SCN is used to transmit the change pages to users of the specification. An effective vehicle for specification change notices is DD Form 1696, which is based on the procedures described by MIL-HDBK-61A and MIL-STD-973.

Notices of revision

The notice of revision (NOR) is used to activate revisions to those drawings that are part of the primary system configuration design, but that are not maintained by or under the control of the originator of the changes. The NOR enables the appropriate level of technical approval or coordinating authority within the corporate organization to direct or otherwise assure that necessary changes in technical documents maintained by other technical activities that are needed to accommodate the revisions required by the primary system are made. An effective vehicle for the notice of revision is DD Form 1695, which is processed based on MIL-HDBK-61A and MIL-STD-973.

Configuration Status Accounting

Configuration status accounting (CSA) is the engineering management information system that provides traceability of configuration baselines and changes thereto, and facilitates the effective implementation of changes.

CSA includes the listing of the approved configuration identification, the status of proposed changes to configuration items, and the implementation status of approved changes after the functional and product baselines have been established. It consists of reports and records documenting actions resulting from changes that affect the configuration item. Configuration status accounting provides traceability of configuration product baselines and subsequent changes to those baselines. CSA is a management tool used in accomplishing all tasks related to configuration changes.

The basic CSA documentation includes the configuration identification index, which describes the currently approved configuration, and the configuration status accounting report, which describes both the current configuration and the current status of proposed and pending approved changes. Configuration status accounts record the configuration of a system as it evolves, including proposed changes to the system as they are processed and approved alterations as they are implemented. The configuration of a system throughout its life cycle is thereby identified and recorded.

CSA elements

All CSA records and reports should support and be consistent with the prime system configuration identification documentation. These records are normally initiated at the establishment of the functional baseline and continue until the ultimate removal of the CIs from the operational inventory.

Configuration item drawings.   The basic building block of the configuration status accounting system is the configuration item (CI) drawing.

Configuration item drawing numbers provide a high-level summary of all equipment under the life-cycle cognizance of the system manager.

The CI drawing lists the location and configuration evolution, by part number, of each item of unit-level equipment within the CI. The current configuration is listed with the immediately preceding revision and the next scheduled revision. The configuration status information that CI drawings provide includes

- Unit number and name and/or nomenclature
- Part/drawing numbers for each unit
- Engineering changes (ECs) against each unit, if any
- Locations of all equipment

These drawings should be maintained by the cognizant design activity. Updates to these drawings are prepared as dictated by approved engineering changes.

CSA database.   CSA documentation should contain, at a minimum, the following elements of information, along with other data that may be deemed appropriate by the design activity.

1. Name, number, revision letter, and responsible organization for each specification that is part of the CI's baseline
2. Name, revision letter, and date of each assembly drawing that is part of the CI's baseline
3. Nomenclature and part-number history of each CI and of the next higher level CI of which it is a part
4. Change history
5. Change status (complete, incomplete, or partial for the CI)
6. CI/unit serial numbers
7. Applicable logistics resources, e.g., technical instructions, training material, parts lists, computer program specifications, etc.
8. History and status of changes to the system engineering support documentation
9. All other appropriate logistic support data

The CSA data may be maintained as a computerized database or in the form of a manual document.

Configuration identification manual

The configuration identification manual (CIM) is a fundamental element of the CSA and provides an effective vehicle for recording the configuration evolution of the configuration item. The CIM should be established upon confirmation of the product baseline immediately prior to commencement of the production phases. The original product baseline plus all subsequently approved engineering changes are documented in the manual.

The CIM should be organized the same way as the CSA database and should include at a minimum the CSA data listed above.

The CIM should be updated for every change in the configuration item as well as for other necessary corrections to the CIM database. The CSA process should be designed to consider and accommodate inputs from all users of the CIM.

Engineering release and correlation of manufactured products

The engineering release by the product manager of new or modified system technical documentation denotes that the documentation has been reviewed and approved, and is suitable for its intended use. The product engineering release system is established to provide an orderly process for issuing system design documentation to the participating functional activities (e.g., production, logistics, quality assurance, and contracting) and to authorize use of the system technical documentation associated with the currently approved configuration. Maintaining current and historical engineering release information for configuration documentation of all configuration items and their modular components is an important responsibility of the product engineering team. After release, each new or modified product document becomes part of the product configuration baseline.

Engineering release record.   The vehicle for management and control of the engineering release process is the engineering release record (ERR). Within the Department of Defense, DD Form 2617, as shown in Fig. 26.2, is normally used for release of system technical data for military systems, based on MIL-HDBK-61A and MIL-STD-973. Many commercial manufacturers utilize an engineering and design release record (EDRR) system to provide management control and visibility of product design information in a format similar to that of DD Form 2617.

Configuration Audits

An integral element of the system configuration management concept is the comparison of a CI's functional performance and physical characteristics with those specified in the product specification and current configuration identification. Such a comparison is described as a configuration audit. The audit is a formal examination to ascertain compliance with specifications and other technical requirements and to verify that a product and its support resources conform to the established configuration documentation. Two types of audits are performed: functional configuration audit and physical configuration audit. Both the functional and physical configurations of the items that make up individual systems and equipment are audited as part of the system configuration management process.

ENGINEERING RELEASE RECORD (ERR)

1. ERR NO.	2. DATE (YYMMDD)	3. PROCURING ACTIVITY NUMBER	4. DODAAC

5. BASELINE ESTABLISHED OR CHANGED (X one)
  ☐ FUNCTIONAL   ☐ ALLOCATED
  ☐ PRODUCT

6. TYPE OF RELEASE (X one)
  ☐ INITIAL
  ☐ CHANGE

7.a ECP NUMBER

b. EFFECTIVE DATE (YYMMDD)

8. FUNCTIONAL ASSEMBLY NOMENCLATURE

9. SYSTEM / CONFIGURATION ITEM

a. NOMENCLATURE	b. PART NUMBER

10. REMARKS / MISCELLANEOUS

11. DATA RELEASED OR REVISED

CAGE CODE a.	DOCUMENT				REVISION		RELEASE h		CHANGE i		OTHER j
	TYPE b.	NUMBER c.	PAGE d.	of PAGES e.	LETTER f.	DATE (YYMMDD) g.	IR	NAR	CH	CAN	

12. SUBMITTED BY (Signature)	13. APPROVED BY (Signature)

DD Form 2617, APR 92
232-100

Together with DD Form 2617, replaces DARCOM Form 1724-R, which is obsolete.

Page 1 of _____ Pages

Figure 26.2 Format of engineering release record.

Functional configuration audit

The functional configuration audit (FCA) is a means of verifying that development of a CI has been completed satisfactorily and that the item functions as required. It is a prerequisite to the physical configuration audit. The FCA is conducted on all CIs during the development or early phase of the production process to ensure that technical documentation accurately reflects the configuration item's functional characteristics, and that test and analysis data verify achievement of the performance specified in the functional configuration identifications. All functional references are reported and recorded as the CI is developed. The FCA is an evolutionary process over a prolonged time period rather than a specific event. This allows the product manager to monitor, in depth, the progress of product development and identify problem areas in a timely manner.

The FCA verifies that the CI has achieved the performance parameters specified in its functional configuration identification and that the technical documentation for the CI accurately reflects the CI's functional characteristics. Successful completion of certification testing of system or equipment performance serves as a functional audit of the system being tested. Each system or equipment change should be tested and certified prior to implementation of the change. The FCA is normally performed during the engineering development phase immediately prior to production, or, in some cases, concomitant with fabrication of the production prototype.

Physical configuration audit

The physical configuration audit (PCA) is the means of establishing the product baseline as reflected in the product configuration identification, and is used for the production and acceptance of units of a CI. The PCA may be accomplished during the latter part of the engineering development phase; however, it is usually delayed until the beginning of the production phase so that it may be accomplished on an early, representative production unit, such as a prototype or first production article. The objective of the PCA is to confirm that the "as-built" physical characteristics of a hardware or software CI match the approved physical configuration of the item. It verifies the accuracy of the technical descriptions and identifies documentation discrepancies upon audit completion. The PCA determines the actual configuration of the CI and identifies any differences between the technical documentation and the physical item. The results of the audit serve to verify the accuracy of the technical descriptions and identify baseline documentation discrepancies which must be corrected before commitment to full-scale production.

Configuration audit guidelines

Preliminary to conducting a configuration audit, the product design team should prepare an audit plan detailing the actions to be accomplished, the procedures to be followed, specific system attributes which need to be evaluated, and the individual expertise of the personnel assigned to the audit team. The audit checklist should, at a minimum, include the following tasks:

1. Review representative installation, arrangement, and assembly drawings. Use an appropriate sampling technique, down to the lowest replaceable subassembly level, to determine their accuracy and to ensure adequate description of the item.

2. Review the currently approved configuration item specification(s) and compare them with the CI to ensure that the specification(s) accurately describe(s) the configuration item.

3. Examine the CI to ensure that the current nomenclature, description, and part numbers agree with the specifications and drawings.

4. Review specifications, drawings, configuration item identification, and the engineering release record to ascertain whether all approved changes have been incorporated.

5. Review acceptance test procedures and documented test data to ensure that all specified tests were performed and were within specified parameters for acceptance.

6. Review sample organizational-, intermediate-, and depot-level operations and maintenance manuals for configuration baseline compatibility and completeness.

7. If software CIs are being audited,

   a. Review the computer program design, database design, interface design, and other equivalent design documents to ensure that all documentation is at the latest revision level.

   b. Examine the actual configuration items, data modules, tapes, etc., to ensure that labeling conforms to the master tape or data file.

   c. Perform tape or data file comparison, as required, with the appropriate master software control system.

The product manager should assume responsibility for audit planning and performance for the FCAs and PCAs. The audit of each CI should establish compatibility of the planning and manufacturing data with the released engineering and quality control records for the configuration item.

Engineering Change Order

The engineering change order is the vehicle for executing an approved engineering change proposal. Engineering changes are implemented through installation on the system using engineering change (EC) kits.

An engineering change kit may include a number of approved engineering change proposals, in essence incorporating a "block change."

Each engineering change kit consists of a publications package and all parts, materials, and special tools required to accomplish and support the change, including:

1. The engineering change order (ECO), containing at a minimum the following data elements:

   a. Sponsoring organization

   b. ECO identification number

   c. Estimated technician hours required

   d. Number of personnel required

   e. System elements affected

      - Equipment
      - Technical manuals and instructions
      - Computer programs
      - Spares and repair parts required
      - Preventive maintenance items
      - Preventive maintenance procedures
      - Other elements, as appropriate

2. All hardware/material required to accomplish the change

3. Technical manual changes

4. Preventive maintenance schedules and procedures

5. Support documentation (e.g., revisions to parts and components lists)

6. Spare/repair parts required

7. Components required for installation and checkout

8. All special tools and special test equipment required to accomplish the change

9. Revised diagnostic computer programs

10. Identification nameplates (if the change affects equipment identification)

11. Instructions for the proper disposition of materials

12. Test procedures, including accept/reject criteria, to ensure successful installation

13. Partially completed certification form for reporting the configuration change to CSA records

Any one or a combination of the above may be required by the engineering change order.

An evaluation engineering change kit is typically included as one of the first production kits. The evaluation kit is used for certification of the adequacy and completeness of the design, descriptive documentation, installation instructions, and physical composition of the change kits.

EC transition matrix

To clarify the narrative description of the change or changes being implemented by the engineering change order, a table showing the pre-change and after-change configurations is often helpful. Table 26.1, illustrates a typical engineering change transition matrix, using a military model as an example.

Putting CM into Practice

The product manager concerned with cost-effectiveness in development and production and long-term market viability is well advised to plan for configuration management of a product or system from the developmental phase through the market service life. The first symptoms of inadequate configuration management are usually manifest late in the developmental phase.

Failure to establish and maintain a strong configuration management system is a common source of risk in the transition from development to production. In a loosely implemented CM system, design changes can occur without proper maintenance of the configuration change documentation after the baseline is established. Lack of an effective CM system leads to many pitfalls, including an unknown design baseline, excessive production rework, poor spares support, stock purging rather than stock control, and inability to resolve problems of the customer-users during the market distribution period.

CM policy guidelines for the product manager

Poor CM is a leading cause of increased program costs and lengthened procurement schedules. To avoid these risks and the associated consequences to the corporation, the following guidelines should be adhered to:

1. An effective configuration management system should do the following:
   - It should provide an effective set of guidelines and standards to fit the attributes of the system, including hardware and logistics support elements.
   - Corporate or division policy should recognize the importance of proper configuration management in the development of a new program, and emphasize the need to generate an adequate plan for implementation.

TABLE 26.1 AN/BQS-15 Configuration Change Data for Engineering Change No. 004

The equipment configuration change data addressed by this table are the AN/BQS-15 product baseline plus engineering changes no. 001, 002, 003, and 004.

Ref. Des.	Configuration before change			Ref. des.	Configuration after change		
	Drawing/part no.	Rev. ltr.	Noun name		Drawing/part no.	Rev. ltr.	Noun name
P/O 1A2A1	B-60013053	A	Indicating panel	P/O 1A2A1	B-60034182	B	Indicating panel
2A5	B-6A013162	F	D/E motor controller amplifier	2A5	B-6A013162-1	G	D/E motor controller amplifier
2A5A3	B-6A013835	E	Decoder, command signal	2A5A3	B-6A034175	A	Decoder, command signal, D/E counter-D/E motor control

- The configuration management plan should be streamlined, yet adequately encompass the entire life cycle of the program, recognizing the requirements of each phase of the life cycle and the complexity of the system configuration.
- The configuration management plan should establish the mode of operation and interface relationships among vendors, subcontractors, and customers, as well as the prime item producer.

2. The configuration management organization should be properly staffed, with individuals having authority commensurate with their assigned responsibility.

3. The specification tree, engineering release, and drawing discipline should be governed by documentation requirements that have been established through the configuration management plan.

4. The staff should be trained in the operation of the configuration management system.

5. The configuration management program should be strictly enforced in order to organize and implement, in a systematic fashion, the process of documenting and controlling configuration.

6. Change control boards and status accounting systems should be responsive to the dynamics of product evolution and should be updated frequently by timely feedback from user activities.

7. Configuration control procedures should assure the establishment and maintenance of design integrity.

8. Configuration audits should be performed to establish the design baseline and to validate the drawing package before release for production.

9. Production engineering should interface with configuration control for work instruction planning.

10. The transition from development through production to market distribution should be made only when the product design is mature and there are sufficient logistics resources to support the product after delivery to the customer.

Formulating the CM process

Configuration management discipline should be invoked on a scale commensurate with the projected economic significance and technological complexity of the product. The CM organization for a major new system projected for global market distribution might require extensive input, participation, and coordination, integrating the expertise of all functional elements of the organization. The reduced scope of effort required for the redesign and modification of a subsystem or lower-indenture assembly to enhance the performance of the prime system might include the ad hoc commitment of a closely knit design team augmented by expertise contributed as required from other technical staff elements. The following describes those management tools and organizational techniques that are fundamental to execution of the change management process. This approach essentially tailors the pertinent elements of the military CM system to commercial applications.

It might be feasible for an individual CCB member to be responsible for more than one of the cadre of CCB disciplines; e.g., the logistics engineer could also represent the material control function or the product engineer could serve the function of manufacturing engineer. Where dictated by the scale of the program, the group could be expanded to include a panoply of subdisciplines and others with related corporate expertise as necessary to effectively accomplish the configuration management mission.

Configuration control board mission and organization. The hub of the configuration management system at any appointed corporate level and for any designated product is the configuration control board (CCB), in some organizations termed the change control board. The scope of the CCB's authority might be a specified product, a product line, a family of comparable systems, or some other grouping category deemed appropriate by corporate management. The CCB serves as a clearinghouse as well as having review and approval authority for all engineering changes generated within its assigned purview. The CCD is staffed by the minimum number of people that can provide the technical expertise required to effectively evaluate and render approval of proposed changes. The basic cadre of the configuration control board includes the following technical disciplines:

- Configuration manager (chairperson of the CCB)
- Product engineering
- Manufacturing (production) engineering
- Quality engineering
- Contracting
- Logistics engineering
- Material/inventory control

The CM process flowchart. The mission of the CCB is to exercise control over the product design evolution within its scope of authority. CCB procedures provide broadly for receipt of change proposals, logging and recording of proposal submissions, evaluation of proposed changes, collective determination of the acceptability and economic feasibility of the proposed changes, approval or disapproval of change proposals, and issuance of CCB directives promulgating engineering changes, typically in the form of engineering change kits. Figure 26.3 depicts the sequence of events in the CCB change control process.

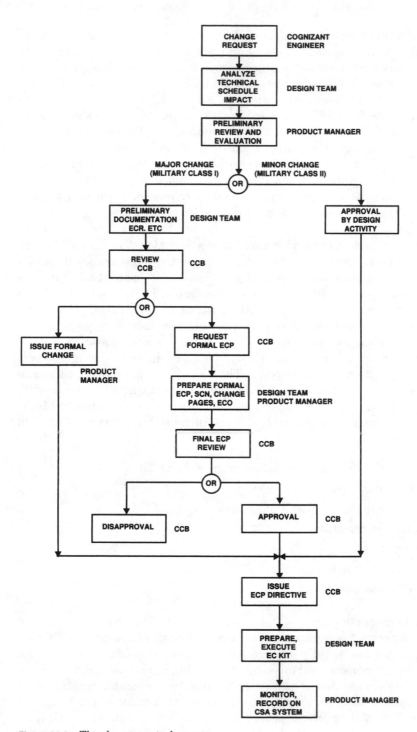

Figure 26.3 The change control process.

Formulation of master data list.   The master data list is published along with the engineering drawings. Figure 26.4, based on a military format, illustrates a typical data element structure. The data list is typically assigned a DL number corresponding to the identification number for the master item. The DL drawing typically consists of a cover page (Fig. 26.4*a*) and attachment pages (Fig. 26.4*b*) detailing the governing technical documents. The cover page includes the following major data elements:

1. Descriptive nomenclature of the item
2. Next higher assembly
3. What the item is used on, to denote systemic application of the item
4. Current revision
5. Revision history
6. Originator and approval blocks
7. DL document number

The data list attachment pages detail those documents that describe the master item. For each document, the data list attachment page provides

1. A title block describing the master item
2. The current revision designation of the document
3. Nomenclature of the module or component described by the document

In compliance with data discipline as practiced by the Department of Defense, which is equally applicable to commercial products, documents which detail specifications pertinent to a particular module or component are assigned the same numeric designation as the component. For example, the parts list (PL) and factory acceptance test (FAT) governing a particular assembly will have the same suffix number as the assembly drawing to facilitate correlation of the interrelated documents.

The configuration identification list.   The purpose of the configuration identification list is to identify the *currently approved* descriptive technical documentats which profile the functional or physical configuration of the system. The functional configuration identification includes the documents which detail the specified performance requirements and those developmental drawings (and, possibly, certain production drawings) and specifications which stipulate the operational features of the system. The functional configuration identification is effective prior to verification of the system design for production. The physical configuration identification describes the product form, fit, and function in its pro-

(a)

APPLICATION			REVISIONS				
NEXT ASSY	USED ON	ZONE	REV	DESCRIPTION		DATE	APPR
	SYS XYZ		A	REPLACES BASIC DWG PER ECP 10051		1/15/93	JWL
			B	REPLACES REVISION A PER ECP 10082		7/15/93	JWL

> THE SYSTEM DATA LIST CITES ALL CURRENT VERSIONS OF THE GOVERNING TECHNICAL DOCUMENTATION RELEVANT TO FABRICATION, TEST, AND LOGISTICAL SUPPORT OF THE SYSTEM.

CLASSIFICATION OF CHARACTERISTICS

AETNA DESIGN MANUAL 6/93

CRITICAL X

MAJOR

MINOR

Q A APPROVAL
MARY SCHULTZ
DIRECTOR QA

REV STATUS OF SHEETS	B	A	B	A	A	B	A	A	A	A												
	1	2	3	4	5	6	7	8	9	10												

UNLESS OTHERWISE SPECIFIED:
DIMENSIONS ARE IN INCHES TOLERANCES:
ANGLES
  1 PLACE DECIMALS:
  2 PLACE DECIMALS:
REMOVE BURRS
BREAK SHARP EDGES
MAX SURFACE ROUGHNESS 125

DESIGN LAB
AETNA CORP
LEXINGTON KY

ENGR	J. L. SMITH	4/30/92
ENGR	BJ JONES	5/15/92
CHECKED	R. K. BROWN	5/31/92
PREPARED		

DO NOT SCALE THIS DRAWING   JW LANGFORD 6/15/92

APPROVED	DATE	

SIZE	CASE	DWG NO.	REV
A	N/A	DL34567	B

SCALE	SCALE	SHEET 1 OF 10

(a)

(b)

DATA LIST	CONTRACT NO. N/A IN HOUSE	CAGE NO. N/A		DL 34567	REVISION LTR B DATE 7/15/93

TITLE XYZ ELECTRONIC SYSTEM	AUTHENTICATION. ———	REV AUTH NO. ———	SHEET 2 OF 10 SHEETS

CASE	DWG SIZE	DOCUMENT NUMBER	SHEET NUMBER	REV LTR	NOMENCLATURE OR DESCRIPTION
N/A	F	34567	1	B	SYSTEM XYZ
N/A	D	36567	4	A	TRANSFORMER ASSEMBLY
N/A	A	35657	2	A	PRIMARY SUBSYSTEM
N/A	D	37567	6	B	CABLE ASSEMBLY
N/A	A	PL36567		A	TRANS FORMER ASSY (PARTS LIST)

(b)

Figure 26.4   Data list. (a) Sample cover page. (b) Sample attachment page.

duction mode and represents the conformation of the completed production item ready for delivery to the customer-users. Figure 26.5 shows the essential data elements typically included in the physical configuration identification list.

The physical configuration identification list reflects the master data list. It captures the descriptive data of the product in its current production form and in market distribution. Approved engineering changes are translated into codified revisions of constituent component drawings, of which the current revision level is cited on the physical configuration identification list. The particular production units to which the approved changes apply are identified by serial number. The basic physical configuration identification list, as illustrated by Fig. 26.5, provides the data elements of the master item drawing as well as those of the constituent component drawings that support the master item. For each drawing, the following data should be incorporated:

1. Drawing number
2. Current revision level
3. Description of the item
4. Next higher assembly (NHA) of which the particular item is a constituent component
5. Serial number (S/N) effectivity, which identifies the serial number of the production units to which the specific drawing at the designated revision level is applicable

It is underscored that the physical configuration identification list reflects the *current, approved* product baseline, not those changes pending or in process of review by the CCB.

Design of the engineering design release record.   The engineering design release record (EDRR) is a companion to the physical configuration identification list and, likewise, a derivative of the master data list. The EDRR

SPECIFICATION

Dwg no.	Rev.	Description	NHA	S/N effectivity

ENGINEERING DRAWINGS AND ASSOCIATED LISTS

Dwg no.	Rev.	Description	NHA	S/N effectivity

Figure 26.5 Physical configuration identification list.

should provide a comprehensive summary of the evolution of the product design, including both historical and current activity. The typical EDRR includes the following data elements:

1 For an *engineering change request (ECR)*, the identification control number and date of submission
2. For an *engineering change proposal* (directed by the CCB based on concurrence with the ECR), the identification control number and date of submission
3. For an *engineering release*, the date of CCB release, pending assignment of a revision identifier or new drawing number (if appropriate)
4. For a *basic drawing number*, the date of issue and serial number effectivity
5. For *individual revisions to drawings*, the dates of issue and serial number effectivity of each

Figure 26.6 illustrates a typical EDRR format.

System: XYZ (top-level DWG 10500) Engineering Design Release Record (EDRR) Report date: 92-04-15			Page 1 of 10
Drawing/rev	**Change status (ECR/ECP/release)**	**Date**	**S/N effectivity**
10500	Release	92-03-15	All
10500A	Release	92-11-30	All
10500B	ECP 123	93-04-15	Unit 500
15615	Release	92-03-15	All
15615A	Release	92-10-15	Unit 500
15615B	ECR 112	93-02-15	N/A
15616	Release	92-02-15	All
15616A	ECP 101	93-02-20	Unit 500
DL10500	Release	92-03-15	All
DL10500A	Release	92-11-30	All
PL15615	Release	92-03-15	All
PL15615A	Release	92-10-15	Unit 500
PL15615B	ECR 112	93-02-15	N/A
PL15616	Release	92-03-15	All
PL15616A	ECR 102	93-02-30	Unit 500
FAT10500	Release	92-03-15	All
DL prefix refers to a *data list* PL prefix refers to a *parts list* FAT prefix refers to a factory acceptance test (FAT) Dates are year-month-day			

FIGURE 26.6   Typical EDRR format.

TABLE 26.2 System ABC Specifications and Drawings Family Tree

1	2	3	4	Nomenclature
	Indenture level			
6617042-1				ABC assembly
	6617043-1			Primary subassembly
		6617047		Molded front mass assembly 6616934 (Rubber compound) 6616936 (M-R adhesive) 6617074 (Front mass)
		6617053-1		Ceramic stark assembly
			6617077	Wire, electrical 6617096 (Electrical wire)
			6617056	Wire, electrical 6617096 (Electrical wire)
	6617048			Transformer assembly 6617071 (Flexible adhesive) 6617084 (Outer spacer) 6617085 (Transformer spec)
		6617044-1		Transformer assembly
			6617050	Potted coil assembly 6617062 (Epoxy resin) 6617091 (Pot core bobbin)
	6617045			Housing, molded 6616934 (Rubber compound) 6616936 (M-R adhesive) 6617093 (Housing)

Specifications and drawings family tree. The specifications and drawings family tree is the third element in the tripartite documentation structure that includes the physical configuration identification list and the EDRR. The family tree is a simple, structured pattern which portrays, in accordance with indenture levels established by work breakdown structure discipline, those items incorporated in the indenture structure. The typical family tree organizes the item drawing numbers into four columns, which correspond to the numbered indenture levels. Column 1 (indenture level 1) identifies the master item as the unique Level 1 item. The second column (indenture level 2) includes all items at the subsystem level that are constituents of the master item. Column 3 (indenture level 3) defines items at the assembly level that are constituents of the subsystems. Column 4 (indenture level 4) identifies subsidiary components of the Level 3 assemblies. Table 26.2 illustrates a typical specifications and drawings family tree which organizes the infrastructure of hypothetical system ABC according to the indenture hierarchy.

Selected References

1. American Society for Quality. *ANSI/ASQ Z1.4-2003 Standard Sampling Procedures and Tables for Inspection by Attributes.* Milwaukee, WI: ASQ Press, 2003.
2. Ballou, Ronald H. *Business Logistics/Supply Chain Management,* 5th ed. Upper Saddle River, NJ: Prentice-Hall, 2003.
3. Beizer, Boris. *Black-Box Testing: Techniques for Functional Testing of Software and Systems.* New York, NY: John Wiley and Sons, 1995.
4. Blanchard, Benjamin S. *Logistics Engineering and Management,* 6th ed. Upper Saddle River, NJ: Prentice-Hall, 2003.
5. Blanchard, Benjamin S. *System Engineering Management,* 3rd ed. New York, NY: John Wiley and Sons, 2003.
6. Blanchard, Benjamin S. and Walter J. Fabricky. *Systems Engineering and Analysis,* 4th ed. Upper Saddle River, NJ Prentice-Hall: 2006.
7. Bowersox, Donald, David Closs, and M. Bixby Cooper. *Supply Chain Logistics Management.* New York, NY: McGraw-Hill, 2002.
8. Chase, Richard B., F. Robert Jacobs, and Nicholas J. Aquilano. *Operations Management,* 11th ed., New York, NY: McGraw-Hill, 2005.
9. Coyle, John Joseph, Edward J. Bardi, and John Langley. *Management of Business Logistics: A Supply Chain Perspective,* 7th ed. Cincinnati, OH: South Western, 2002.
10. Department of Defense. *Defense Acquisition Guidebook.* Ft. Belvoir, VA: Defense Acquisition University, 2004.
11. Department of Defense. *Defense Federal Acquisition Regulations Supplement.* Washington, D.C.: Government Printing Office, 1998 (as augmented by subsequent DFARS change notices).
12. Department of Defense. *MIL-DTL-31000C: Technical Data Packages.* Washington, D.C.: U.S. Government Printing Office, 2004.
13. Department of Defense. *MIL-HDBK-61: Configuration Management Guidance.* Washington, D.C.: U.S. Government Printing Office, 2001.

14. Department of Defense. *MIL-HDBK-217F(2): Reliability Prediction of Electronic Equipment.* Washington, D.C.: U.S. Government Printing Office, 1995.
15. Department of Defense. *MIL-HDBK-470A: Department of Defense Handbook, Designing and Developing Maintainable Products and Systems* (Volume 1 and Volume 2), Washington, D.C.: U.S. Government Printing Office, 1997.
16. Department of Defense. *MIL-HDBK-472(1): Maintainability Prediction.* Washington, D.C: U.S. Government Printing Office, 1984.
17. Department of Defense. *MIL-HDBK-502: Department of Defense Handbook, Acquisition Logistics.* Washington, D.C.: U.S. Government Printing Office, 1997.
18. Department of Defense. *MIL-HDBK-881A: Work Breakdown Structures for Defense Materiel Items.* Washington, D.C.: U.S. Government Printing Office, 2005.
19. Department of Defense. *MIL-PRF-49506: Performance Specification-Logistics Management Information.* Washington, D.C.: U.S. Government Printing Office, 1996.
20. Department of Defense. *MIL-STD-105E: Sampling Procedures and Tables for Inspection by Attributes.* Washington D.C.: U. S. Government Printing Office, 1989.
21. Department of Defense. *MIL-STD-498: Software Development and Documentation.* Washington, D.C.: U.S. Government Printing Office, 1994.[1]
22. Department of Defense. *MIL-STD-882D: System Safety.* Washington, D.C.: U. S. Government Printing Office, 2000
23. Department of Defense. *MIL-STD-961E: Department of Defense Standard Practice, Defense and Program, Unique Specifications Format and Content.* Washington, D.C.: U.S. Government Printing Office, 2003.
24. Department of Defense. *MIL-STD-973: Configuration Management.* Washington, D.C.: U.S. Government Printing Office, 1992.[1]
25. Department of Defense. *MIL-STD-1916: DOD Preferred Methods for Acceptance of Product.* Washington D.C.: U. S. Government Printing Office, 1996.
26. Electronic Industries Alliance. *ANSI / EIA–649A: National Consensus Standard for Configuration Management.* Arlington, VA: EIA, 1998.
27. Freund, John E., and Benjamin M. Perles. *Modern Elementary Statistics*, 12th ed. Pearson Foundation, 2006.

[1]Military Standard has been officially cancelled by the Department of Defense and is no longer invoked in military contract programs. Notwithstanding official action, the document is yet considered a valuable source of guidance and procedures relevant to the principles and processes described in this book

28. Glaskowsky, Nicholas A., and Donald R. Hudson. *Business Logistics*, 3rd ed. Belmont, CA: Wadsworth, 1992.

29. Government Electronics and Information Technology Association. *GEIA-HB-649, Implementation Guide for Configuration Management.* Englewood, CO: GEIA/ IHS, 2005.

30. Institute of Electrical and Electronic Engineers. *IEEE/EIA Standard: Software Life Cycle Process.* Washington, D.C.: IEEE/ EIA, 1998.

31. Juran, J.M., and Frank M. Gryna. *Quality Planning and Analysis*, 3rd ed. New York, NY: McGraw-Hill, 1993.

32. Kumar, U. Dinesh, John Crocker, J. Knezevic, and M. El-Haram. *Reliability, Maintainability and Logistics Support: A Life Cycle Approach.* New York, NY: Springer-Vertag, 2004.

33. *Maintainability Toolkit.* Rome, NY: Reliability Analysis Center, 2000.

34. Murphy, Paul R., and Donald Wood. *Contemporary Logistics*, 8th ed. Upper Saddle River, NJ: Prentice-Hall, 2003.

35. O'Connor, Patrick D.T., David Newton and Richard Bromley. *Practical Reliability Engineering*, 4th ed. Chichester, England: John Wiley and Sons, 2002.

36. Office of Federal Procurement Policy, OMB. *Federal Acquisition Regulation.* Washington, D.C.: Government Printing Office, 2005 (as augmented by subsequent Federal Acquisition Circulars).

37. *Quality Toolkit.* Rome, NY: Reliability Analysis Center, 2002.

38. *Reliability Toolkit: Commercial Practices Edition.* Rome, NY: Reliability Analysis Center, 1996.

39. Sanders, Mark S., and Ernest J. McCormick. *Human Factors in Engineering and Design*, 7th ed. New York, NY: McGraw-Hill, 1993.

40. Stock, James R., and Douglas Lambert. *Strategic Logistics Management*, 4th ed. New York, NY: McGraw-Hill, 2001.

41. *Supportability Toolkit*, Rome, NY: Reliability Analysis Center, 2005.

Internet Websites, Logistics Associations, and Sources of Information

1. The American Society   *http://www.astl.org*
   of Transportation and Logistics
2. APICS—The Educational Society   *http://www.apics.org*
   for Resource Management
3. American Trucking Association   *http://www.truckline.com*
4. Institute for Supply Management   *http://www.napm.org*
5. International Warehouse   *http://www.iwla.com*
   Logistics Association
6. National Industrial   *http://www.nitl.org*
   Transportation League   *http://www.nitl.org*
7. SOLE—The International   *http://www.sole.org*
   Society of Logistics
8. Supply-Chain Council   *http://www.supply-chain.org*
9. Warehouse Education   *http://www.werc.org*
   and Research Council
10. Council of Supply Chain   *http://www.cscm.org*
    Management Professionals
11. International Council   *http://www.incose.org*
    on System Engineering
12. Society of Reliability Engineers   *http://www.sre.org*
13. Reliability, Maintainability,   *http://www.rmspartnership.org*
    Supportability Partnership

14. Defense Acquisition University (DAU)   *http://www.dau.mil*

15. U.S. Navy Human Performance Center (Spider) (Credible source for current information and data on Human Factors Engineering)   *http://www.spider.hpc.navy.mil*

16. U.S. Army Publishing Directorate—USAPA (Credible source for current status and availability of Army logistics documents)   *http://www.usapa.army.mil*

17. Logistics and Material Readiness (Office of Secretary Defense— current status on logistics policy and publications)   *http://www.acq.osd.mil/log*

18. DOD 5000 Series Directives Resource Center (DAU)   *http://akss.dau.mil//dapc*

19. FAR (All Parts, Subparts, Sections, and Appendices of the Federal Acquisition Regulation)   *http://farsite.hill.af.mil/ VFFAR1.HTM*

20. ASSIST: QUICK SEARCH (Provides status, availability and access to active, superseded and cancelled Military Standards, Military Handbooks, Military Specifications and related documentation.)   *http://assist.daps.dla.mil/ quicksearch*

21. Dictionary—Measures, Units, and Conversions (Provides data on interrelationships between SI (metric) and U.S. measurement systems)   *http://www.ex.ac.uk/ cimt/dictunit/dictunit.htm*

Note: In addition to the above-cited Internet sources, the search engine Google is always available as a possible back-up to assist the reader in researching logistics data and documentation.

Crosswalk Guide for Conversion of U.S. Measurement Units to Système International (SI) Measurement Units

The following describes the interrelationships and conversion methodologies employed and incorporated in this book to correlate, as applicable, the U.S. measurement units with their equivalent SI (metric) units.

Expression in U.S. Measurement Unit(s)	*Expression in Corresponding SI Measurement Unit(s)*
inch	0.0254 metres (2.54 centimetres)
foot (12 inches)	0.3048 metres
yard (3 feet)	0.9144 metres
square foot	0.0929 square metres
cubic foot	0.2832 cubic metres
mile	1.60934 kilometres
gallon	3.7584 litres
square mile	2.589988 square kilometres
pound	0.453592 kilograms
ton (2,000 pounds)	0.907185 tonnes
	(Note: tonne = 1,000 kilograms)
ton-mile	1.4663 tonne-kilometres
miles per gallon (mpg)	mpg multiplied by 0.4252 equals kilometres per litre
x miles per gallon	Divide 235.2 by x to determine litres per 100 kilometers
temperature: (degrees Celsius) to (degrees Fahrenheit): $(9/5)C + 32$	temperature: (degrees Fahrenheit) to (degrees Celsius): $(F-32) \times (5/9)$

The following describes the standard, basic SI units and their corresponding acronymic citations/abbreviations utilized and incorporated, as applicable, herein.

SI Basic Unit	*Pertinent Notes*
mass (or weight): kilogram (kg).	1 kilogram = 2.2046 pounds (US)
length: metre (m)	*a.* Kilometre (km) (1,000 metres) is a derivative application of the metre. *b.* 1 metre = 1.0936 yards (US) *c.* 1 kilometre = 0.6214 miles (US).
area: square metre (m^2)	*a.* Square kilometre (1,000,000 square metres) is a derivative application of square metre *b.* square kilometre = 0.386 square miles (US)
volume: cubic metre (m^3)	*a.* Litre (l or L) is a derivative of the cubic metre; 1 cubic metre equals 1,000 litres *b.* 1 litre = 0.2661 gallons (U.S.)
Fuel consumption: litres per 100 kilometres (Ls/100 kms)	It is noted, for illustration, that 10 litres per100 kilometres equates to 2.661 gallons per 62.14 miles
Temperature: Celsius (C)	*a.* Degrees *Celsius* is frequently cited as degrees *Centrigade*. *b.* The *Kelvin* (K) scale, when occasionally used, starts at absolute zero, which is –273.15 degrees on the *Celsius* scale and –459.67 degrees on the *Fahrenheit* scale.

Index